GW01218190

MASTER HANDBOOK OF
HAM RADIO CIRCUITS

MASTER HANDBOOK OF HAM RADIO CIRCUITS

EDITORS OF 73 MAGAZINE

FIRST EDITION

FIRST PRINTING—DECEMBER 1977
SECOND PRINTING—JANUARY 1979

Copyright © 1978 by TAB BOOKS

Printed in the United States of America

Reproduction or publication of the content in any manner, without express permission of the publisher, is prohibited. No liability is assumed with respect to the use of the information herein.

Library of Congress Cataloging in Publication Data

Main entry under title:

Master handbook of ham radio circuits.

Includes index.
1. Radio—Amateurs' manuals. I. 73 magazine for radio amateurs.
TK9956.M344 621.3841'66 77-20872
ISBN 0-8306-7801-8
ISBN 0-8306-6801-2 pbk.

Preface

We amateur radio operators are an inventive lot. When a certain circuit is needed for a specialized job, we look far and wide for a diagram or even an idea which might help in our efforts to combine that maze of transistors, resistors, and a few homebrew components to end up with a functioning unit which performs all of the tasks we require of it. If the circuit we need hasn't been invented yet, we attempt to conceive, nurture and finally, to give birth to such a device.

We have attempted to make this entire process a bit easier by providing hams everywhere with out *Master Handbook of Ham Radio Circuits.* This edition contains many circuits which, until a few years ago, were only glimmers in the eyes of their ham inventors. Some of the circuits are new twists in the eyes of their ham inventors. Some of the circuits are new twists on projects that have withstood the years of new innovations, only to emerge victorious as a proven and always-worthwhile addition to any shack, and other circuits are so old that they are nearly forgotten.

Whether your interests range from ragchewing to EME; from CW to slow-scan televison; from DX to county nets, you'll find this handbook a welcomed addition to your amateur library. All circuits were made for hams by hams. This publication will be found useful not only for direct references and building, but also for obtaining new information, new ideas for even finer projects designed to open the amateur radio hobby to even greater fields of endeavor.

Contents

Chapter 1

CW Circuits

SIMPLE QRP TRANSMITTER

This two transistor transmitter is not difficult to build, and can supply up to one watt RF output for the QRP or portable sportsmen. The transistors are not excessively priced and are very reliable even though operated beyond their ratings. The use of an unusual type of keying provides a chirpless note with a negligible backwave. The transmitter operates from a 12 volt source making a car or lantern battery ideal for portable use. With this configuration, tuning is not critical and the transmitter is completely stable. The output tank circuit is relatively inefficient but provides good selectivity and is simple to tune and construct.

The Circuit

The schematic is shown in Fig. 1-1. This is just about as simple as a transmitter can be made to provide this type of performance. Q1 is the oscillator transistor. This oscillator circuit was designed by trial and error for simplicity and reliability. Keying, however, is very poor. If the final, Q2, is only keyed by turning the B+ on and off, the signal from the oscillator will feed through the transistor. This is an inevitable result of the basic characteristics of the transistor, and the RF path must be broken. The obvious way to do this is to key the signal from L1 to Q2. This completely eliminates the backwave and chirp problem.

The final stage is operated as a straight through amplifier. A resistor and bypass capacitor might be added to the emitter to bias it farther into class C, but this would add extra components and reduce the simplicity.

The tank coil, L3, is connected to the antenna and Q2 by fixed link coupling. The small number of turns on L2 and L4 allow L3 to have a higher impedance for better selectivity. The antenna and transistor have a lower

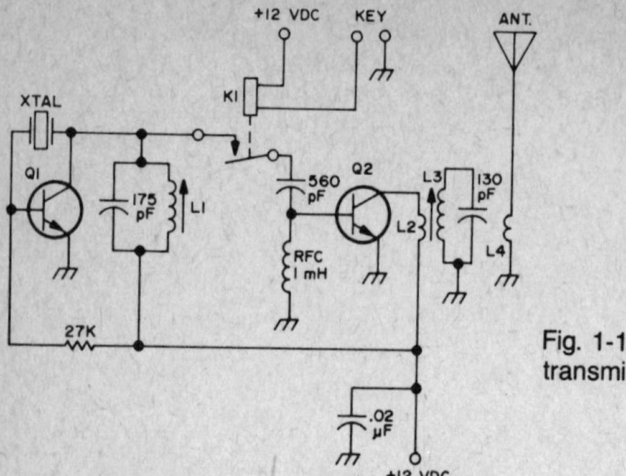

Fig. 1-1. QRP transmitter schematic.

XTAL	7 MHz fundamental
Q1	40080
Q2	40081
L1	20 turns No. 28 on ¼" dia. slug tuned form
L2, L4	5 turns No. 24 on L3
L3	28 turns No. 28 on ¼" dia. slug tuned form
K1	sensitive spst relay for 12V (see text)

impedance which is the source of many of the matching and tuning problems in transmitters of this design.

Construction

The transmitter was built on 1/8" thick double clad printed circuit board. Figure 1-2 shows the layout of the bottom of the board. The top of the board is left with the copper clad sheet on except in places where leads come through. This allows the copper on the top disc to be used as a common ground connection that is easily soldered to anywhere on the top of the board. This double clad board is also very mechanically rigid and leaving the copper on provides a means of structural connections by soldering.

The only unusual component is the relay, K1, which is a sensitive reed type that is ideal for this purpose. The current of the coil is only a few percent of the total current that the transmitter draws. Other types are available in the catalogs, and the only major requirements are compatible size, low coil current, and fast switching. Good quality contacts are also necessary to minimize clicks.

The board can be processed by the methods outlined in any good printed circuit board kit, or the builder may wish to use his own methods. An easy method is to use paint as the resist and obtain ferric chloride from a chemical supply house for the etch. The paint can be applied with a small brush where the conductor is to remain. The parts to be used should be kept handy to check for size and position. Clean the copper before applying the resist by rubbing with steel wool. Keep the steel wool away from any electrical equipment. During etching, the etch and board are placed in a shallow pan or tray. The solution must be kept warm to hasten the chemical reaction. This can be done with a heat lamp or a stove. A beginner should start with a good kit and develop his own methods after a little experience.

The capacitors are soldered directly on to the coils and are placed wherever the shortest leads will result. Components are soldered in with little trouble. Placement and layout are not critical.

The transistors must have heat sinks because of the power dissipated. Since there is little room for the commercial varieties with radial fins, homebrew heat sinks are made. They can be simply bent from copper that is stiff enough to retain a good grip on the transistor case. Silicon grease could be used to provide a good thermal contact, but this has not been necessary. If the transmitter is to be used in dark surroundings (inside a cabinet), the heat sinks can be painted black to increase heat radiation.

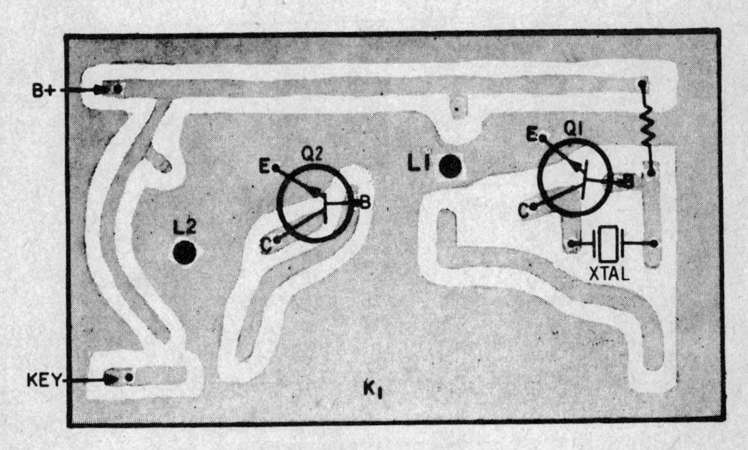

Fig. 1-2. Printed circuit board layout.

Results

The tests conducted on the final version showed tuneup to be non-critical on the oscillator. Oscillation will result with the slug of L1 in almost any position, but there is a point that will produce maximum stability. Adjust the slug of L1 while keying the final to check the loading effects on oscillator stability. When the oscillator is operating, current will be considerably smaller than the non-oscillating condition. The oscillator can be tuned for a dip in collector current, but it will be necessary to detune from the dip to provide optimum stability. There is plenty of output from the oscillator to drive the final. This amount of drive can be estimated by the amount of final collector current. Tune for maximum final collector current with good stability. The tuning of the final tank coil should produce a sharp peak when moved through resonance. The link coupling was designed experimentally for optimum power transfer to a 50Ω load. Other final tank circuits provided more output but this circuit was easiest to tune and provided good selectivity against harmonics.

While tuning, avoid prolonged key down periods, because the transistors can become overheated in a few minutes. In a test, the input to the final was about three watts with output at about one watt. This means that the transistor must dissipate the remaining two watts into the air, which is the maximum power dissipation rating of this transistor with a good heat sink. The use of CW allows the transistor to cool during key up periods, but it will still run quite warm.

Keying is quite sharp and free from chirp. A look at the scope pattern confirms this, but the keying tends to be "clicky." This has not been a serious problem, however, and the simplicity of design and freedom from chirp makes it an acceptable compromise.

This transmitter will make an amusing weekend project for the builder with average experience, and QRP is an increasingly popular sport. Contacts of thousands of miles have been established with much less than a watt, and one million miles per watt can be achieved with microwatt transmitters over short distances. Anyone can buy a kilowatt and work the world, but doing it with milliwatts is a personal achievement that can provide real satisfaction.

SIXTY WATTS, FOUR BANDS, AND ONLY TWO TUBES

It is a relatively simple matter to build a transmitter with twenty to sixty watts of input power on 7, 14, 21, or 28 MHz. The circuit described (see Fig. 1-3) is so well proven that very few difficulties should arise to the home builder. Few other circuits can give so much with so little complication.

The 5763 is a crystal oscillator which provides RF output on the crystal frequency or on a multiple thereof. A 7 MHz crystal in this circuit will also oscillate at 14, 21, or even 28 MHz.

L1 is the plate coil which is tuned by C1. The output from this portion of the circuit is fed to the grid of the 6146 through C5, a 100 pF coupling capacitor. This 6146 final amplifier tube may be easily damaged when it is operated in an off-resonance condition or without proper bias, so grid current is measured at all times on a small meter. C1 is tuned while observing this meter for a reading of 2 mA which indicates a bias of 54 to 65 volts.

After the parasitic suppressor (L2/R6), RF reaches the pi tank L3 with power amplifier tuning capacitor C2 and output capacitor C3-4. This will match into a useful range of impedances.

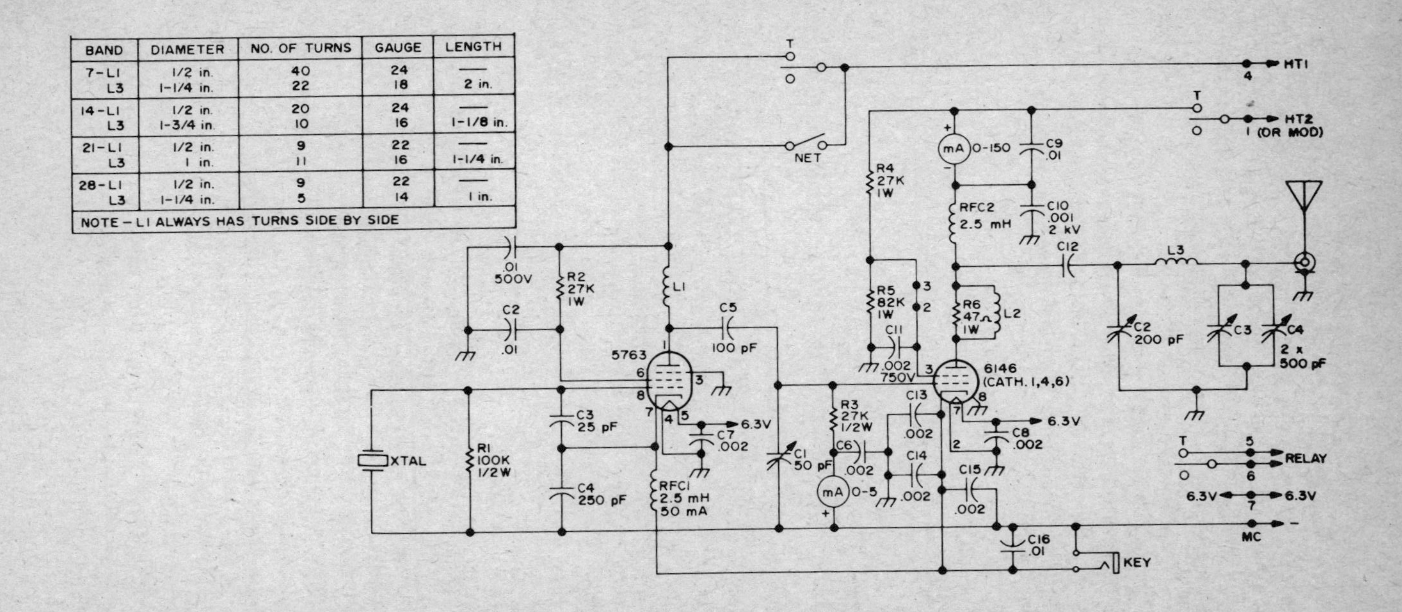

BAND	DIAMETER	NO. OF TURNS	GAUGE	LENGTH
7-L1	1/2 in.	40	24	—
L3	1-1/4 in.	22	18	2 in.
14-L1	1/2 in.	20	24	—
L3	1-3/4 in.	10	16	1-1/8 in.
21-L1	1/2 in.	9	22	—
L3	1 in.	11	16	1-1/4 in.
28-L1	1/2 in.	9	22	—
L3	1-1/4 in.	5	14	1 in.
NOTE – L1 ALWAYS HAS TURNS SIDE BY SIDE				

Fig. 1-3. 4-band transmitter.

Keying is via both cathodes, giving complete cutoff with the key up. R5 and R7 are to keep screen-grid and cathode voltages within limits with the key up. With the key unplugged, the key jack closes for tuneup of grid current or amplitude modulation working. Closing the NET switch puts B+ on the 5763 only. This gives an indication of the crystal frequency with the receiver and will also allow tuning for grid current.

When the transmit/receive switch is at transmit, B+ reaches both stages. Turning this switch to receive takes B+ off. Spare contacts can operate the antenna changeover relay. This is useful for phone contacts because this switch can control modulator B+, transmitter, antenna, and receiver muting circuits through a relay.

Construction

A chassis 10 × 6 × 3 in. is suitable, with pi tank and other output components on top (Fig. 1-4). To obtain a short RF return, a stout lead joins rotor tabs of C2 and C3-4 together, and to a tab at the 6146 holder.

The meters, C1, NET switch, T/R switch, and key jack fit on the front runner.

Leads not intended to carry RF (those to heaters, B+ circuits, key jack and meters) run against the chassis. Bypass capacitors go directly from the holder terminals to the chassis lugs. In particular, have very short leads from the 6146 to C13, C14, C15, and C11, and to chassis (Fig. 1-5).

The B+ circuits must be well insulated, especially for the higher voltage. Should you practice high-level modulation of the 6146 with a 600V supply, the peak B+ swing will be 1.2 kV; hence, the need for insulation for this voltage, and the use of high voltage reliable mica capacitors for C10 and C12.

Power Requirements

The heaters need 2A at 6.3V. The 5763 crystal oscillator will work with about 200–300V. It draws 30 mA at 300V. A separate receiver type power supply is good for this stage, though it is possible to work with a single supply for both stages.

The 6146 supply may depend on what is available. Good results at a lower input rating are obtained with 300–400V, though 500–600V will naturally give more input and output power.

Though any usual supply is practical, the best type will have a choke input from the rectifier, with bleeder current consumed by a resistor across the B+ line. Then the voltage will not soar during intervals when the transmitter is not drawing current.

A capacitor-input power pack gives more voltage from a particular transformer secondary, but unless a heavy bleeder current can be spared, the voltage rises badly "off load."

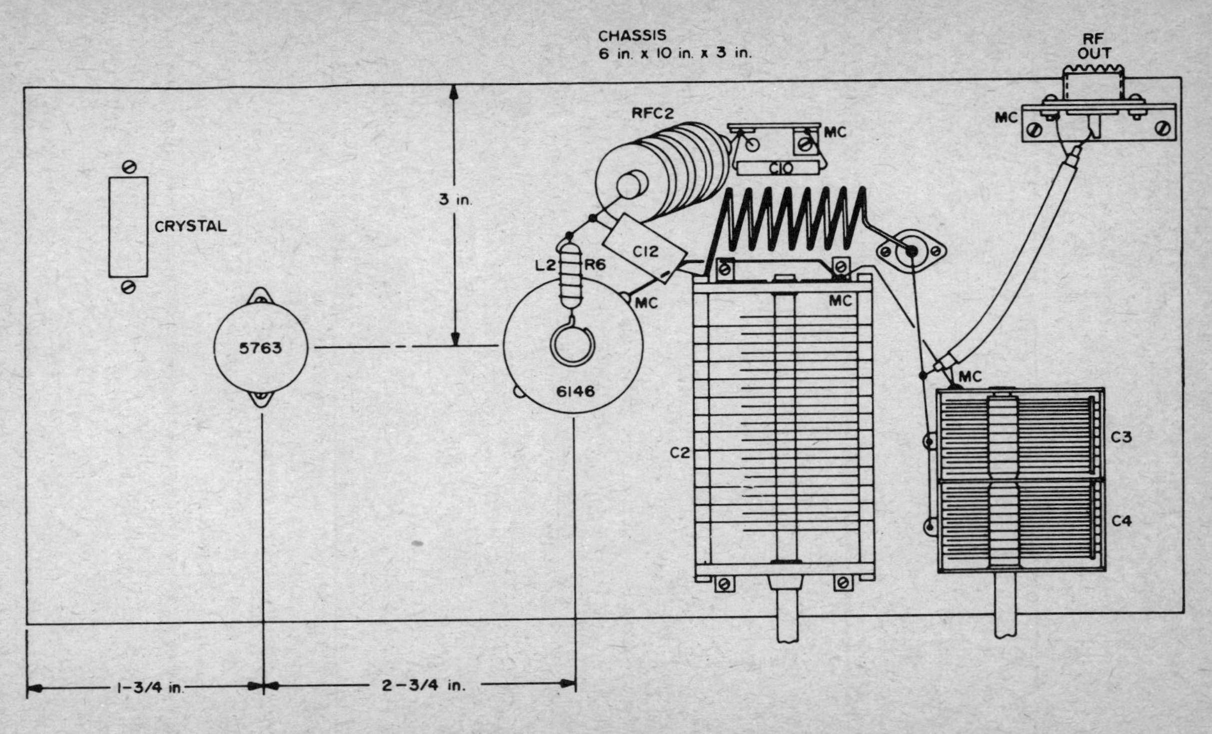

Fig. 1-4. On top of the chassis.

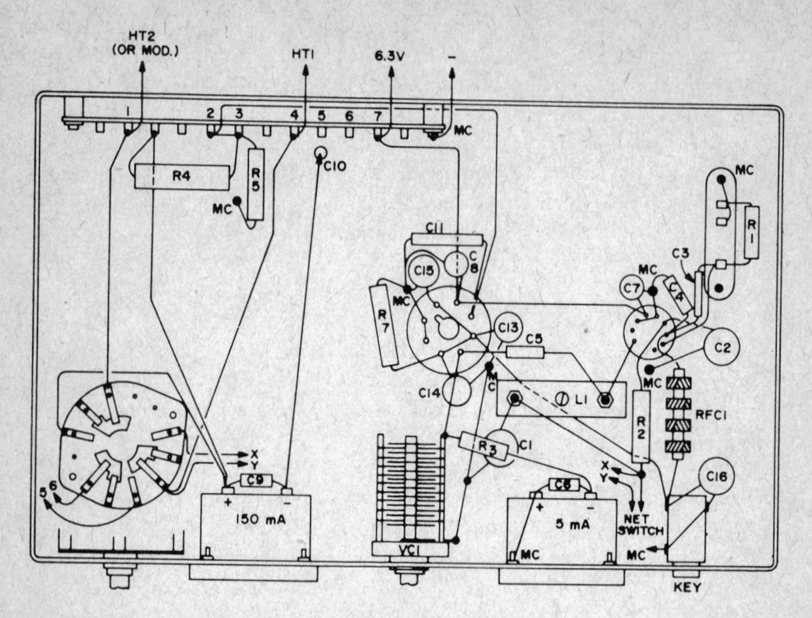

Fig. 1-5. Under the chassis.

Inductors

L2 is 5 turns of 18-gauge, 3/8 in. outside diameter and 5/8 in. long, with R6 inside. L1 and L3 are fitted in pairs. Coils for L1 can be close-wound on insulated tubing, or self-supporting for 28 and 21 MHz. If turns are as in Fig. 1-3, wrong harmonics are not likely to be tuned. L3 tank coils are self-supporting for the HF bands.

Operating Procedure

Those crystals which multiply up into the wanted part of a band can be used where they provide enough grid current; some low-band crystals will allow working on one or more HF bands as well. As an example, a 3.55 MHz crystal will do for this frequency, and also 7.1 MHz. Similarly, a 7.05 MHz rock would do for 7.05, 14.1 and 21.15 MHz. A 3.5 MHz crystal gave better than 4 mA grid current on 14 MHz, and a 7 MHz crystal about 4 mA on 21 MHz. Grid current is greater at the fundamental, or with only 2X multiplication.

Always tune for at least 2 mA grid current, with the NET switch putting B+ on the oscillator only. If enough grid current is not obtained, the crystal is unsuitable. A 3.5 MHz crystal will not give enough output on 21 or 28 MHz.

A 60W domestic lamp as load will check RF output from the transmitter, but not for keying, due to changes in cold and hot resistance. For this, a 75Ω dummy load can be connected.

Close C3-4. Tune for grid current. Open the NET switch. Switch to transmit and immediately tune C2 for minimum current as shown by the meter. Increase input by opening C3-4 while readjusting C2 for the HT dip. As this proceeds, plate current will rise, and the lamp will light with good brilliance.

Check grid current as necessary, readjusting C1. The 6146 is not neutralized, and tuning its plate circuit causes some change in grid current, especially on the higher frequencies. This has no great effect on operation, as grid current is always seen and easily adjusted.

If adjustment of C3-4 does not allow satisfactory loading, then the antenna impedance is outside the range of the transmitter. The best solution is to add a match box.

The 6146 should not be left operating in a condition of high DC input, but little RF output, because the power is then dissipated *inside* the tube. This is avoided by always having at least 2 mA grid current and dipping C2 for minimum current.

To use a VFO, remove the crystal and C3. C4 can be 0.002 μF. Take the VFO output through a screened coax lead to the 5763 grid. Plugs in the crystal sockets will do this and ground the coax shield at the transmitter. It is usually best to apply drive at a frequency which lets the 5763 multiply. For example, at 7 MHz for 14 and 21. A two-stage VFO with fundamental operation on 1.75 and 7 MHz was found to be adequate for all bands, 3.5 to 28 MHz.

MAGIKEY FOR AUTOMATIC DIDAHS

An electronic key, the Magikey, is a wonder for making self-completing dots and dashes with controlled spacing. Only three plastic transistors are used to generate the keying waveform, and the lot can be bought for about two dollars. A keying relay and driver are available for about another three bucks if you need them, and the single pole key can be anything from a "Vibro-key" to a piece of hacksaw blade.

The Magikey uses the Programmable Unijuction Transistor (PUT) to develop the timing function and keying waveform. The PUT is a peculiar device introduced by GE. It exhibits a negative resistance between anode and cathode whenever the anode-gate junction is forward biased and a critical anode current, Ip, is permitted to flow. That is, after the anode is made more positive than the gate, the regeneration inherent in the PNPN structure causes the anode-cathode and gate-cathode voltages to drop to less than a volt above cathode voltage. The device remains "on" until the anode current is reduced below Iv. After the anode current is reduced below Iv, the device turns "off," and the anode cathode terminals become essentially an open circuit.

The ratio of Ip to Iv can be from 5 to 50, depending on the gate voltage source resistance. The highest ratios seem to occur when the source resistance is in the range from 10K to 100K. The lowest ratios occur when the source is in the range of 1K.

The Magikey, shown schematically in Fig. 1-6, can be divided into three parts: the timing section, Q1; the keying waveform generator, Q2 and Q3; and the keyer, Q4.

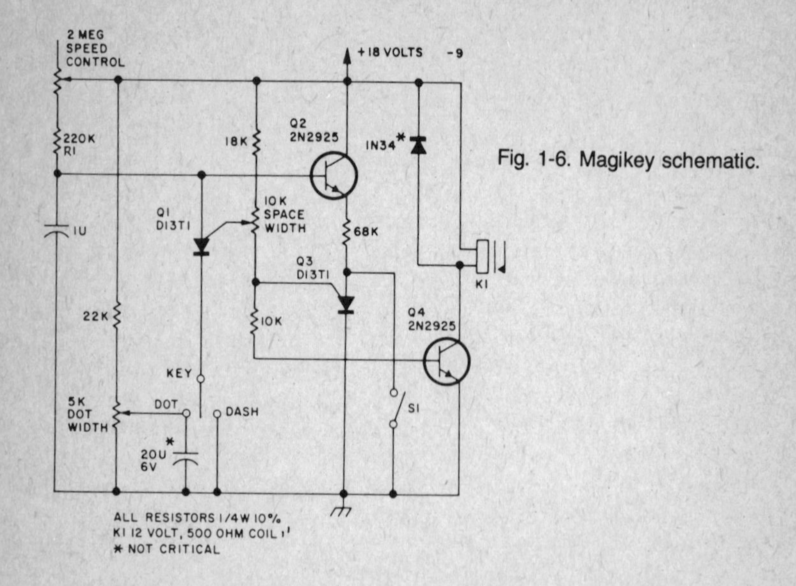

Fig. 1-6. Magikey schematic.

The timing circuit basically is a UJT relaxation oscillator. When the cathode is grounded, it operates as follows: The timing capacitor, C1, charges through the speed control R1, and the PUT anode voltage rises at a rate determined by C1, R1, and the supply voltage. R1 is chosen so that its current is greater than the Ip of the PUT but less than Iv for all conditions. When C1 has charged to the gate voltage, Vg, the PUT turns on and discharges C1 to about cathode voltage. After C1 has completely discharged, the only current available to the anode is through R1. Since this is less than Iv, the PUT turns off. The capacitor then begins to recharge to repeat the cycle.

The anode voltage of the PUT, then, is a sawtooth with the maximum positive voltage approximately Vg, and the least positive voltage approximately equal to cathode voltage. When the cathode circuit is open, the anode voltage rises to Vg and is limited there by conduction of the anode-gate junction, and oscillations are interrupted. As soon as the cathode circuit is closed, C1 discharges to cathode voltage at a rate determined by C1 and the impedence in the cathode circuit. In the circuit given, any closure longer than 1 ms is enough to allow C1 to completely discharge.

To generate dash timing, the cathode is closed to ground and C1 discharges to about 1 volt. The time required for C1 to recharge to Vg is equal to the time of a dash and a space. To generate dot timing, C1 is not closed to ground but to a positive voltage, so that the time required to recharge to Vg is reduced to that of a dot and a space. The sawtooth pattern for the word "an" is shown in Fig. 1-7.

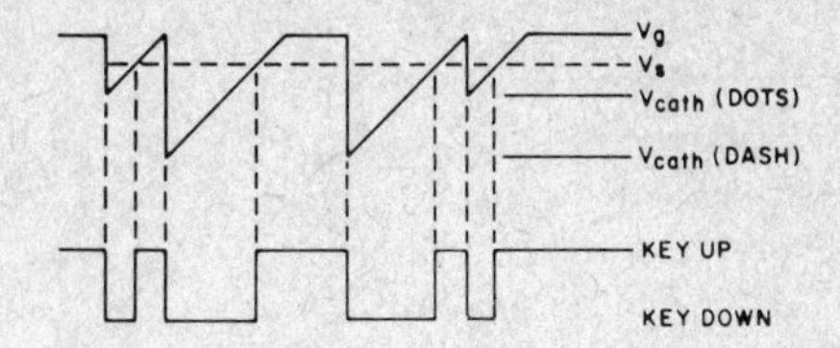

Fig. 1-7. Keying waveform for word "an."

The keying waveform is generated by sensing whether the sawtooth voltage is above or below a particular voltage Vs. When the sawtooth is more positive than Vs, a "key-up" voltage is produced. When the sawtooth is less positive than Vs, a "key down" voltage is produced. The time required for the sawtooth to rise from Vs to Vg is the time of a space, or the minimum "key up" time that can be generated.

In the keying waveform generator section, the PUT, Q3, does the voltage level sensing, while Q2 isolates the timing capacitor from the loading effects of Q3. When the voltage on the emitter of Q2 is higher than Vg of Q3 and causes Ip to flow through R4, the PUT turns on, and the anode and gate voltages fall to about 1 volt. This voltage level represents "key up."When the voltage on the emitter of Q2 is reduced, (the timing capacitor is discharged), the anode current available to Q3 is less than Iv, and the PUT turns off. The voltage at the gate of Q3 switches from about 4 volts "key down" to 1 volt "key up." This is more than enough to drive a keying transistor or even a low-power audio side-tone generator.

In the keyer section, Q4 is the keying transistor. The "key down" current varies with power supply voltage, from .25 mA with a 12 volt supply, to .5 mA with a 24 volt supply. This current is sufficient to drive a low current transistor. In Fig. 1-6, the keying transistor is shown driving a keying relay, but the most practical and economical approach depends on the keying arrangement used in your transmitter. A high voltage transistor costs about half as much as the relay, but has limited current carrying capability.

When driving a relay, the power dissipation in the transistor can be very low. The dissipation is less than 75 mW when driving the 500 ohm coil. The problem is not dissipation; it's the inductive kick when the transistor turns off. The diode across the relay eliminates that problem, so that a low voltage transistor can be used to drive the relay. The diode does extend the relay release time, but it should be significant only if you're batting along at 30 wpm or so. You won't be working me, though, at those speeds, so I'm not too picky. If you are, use a high voltage transistor that can stand the kick.

The circuit given has a keying speed range from about 5 wpm to 50 wpm. If this is a greater range than you need, you can rejuggle the R and C combination in the timing circuit. Just keep the total resistance between 200 K and 2 megohms, and pick a value for C1 that gives the speed you want. C will depend on the supply voltage, among other things, but generally it will be in this range: C (in μF) = words per minute/3 × R (in megohms).

The switch S1 across Q3 provides a continuous "key down" condition for tuning the transmitter, if you need it. I tune up with a string of dashes, and mentally correct the meter readings to account for the fact that the key is "up" one-fourth the time.

Adjusting the Magikey for proper character formation is a snap. You can use a scope for a voltmeter at the gate of Q3, or an ohmmeter across the relay contacts. Either method is better than trying to go by ear. The adjustments are straightforward when made in this order:

1. Connect the ohmmeter across the relay contacts, or a voltmeter to the gate of Q3. Offset the meter's zero to read zero with "key up." Note the meter reading with "key down," S1 closed.
2. Set the keying speed control for near maximum speed so that the meter can not follow key closures.
3. Close the key to the dash position and adjust R3, Q1 gate voltage, to cause the meter to deflect to exactly 75% of the "key down" reading. On the ohmmeter this is 75% of full scale, since the contacts are closed three-fourths of the time when making properly formed dashes.
4. Close the key to the dot position and adjust R2, Q1 cathode voltage, to cause the meter to deflect to exactly 50% of the "key down" reading. On the ohmmeter this is 50% of full scale, since the contacts are closed exactly half the time when making properly formed dots.

That's all there is to it. The dot/dash/space ratio holds for all keying speed settings.

The power demands of the Magikey are minimal, a nominal 18 volts at 1.3 mA. A pair of small 9 volt batteries like Eveready No. 216 will give about 300 hours of operating time if the keying transistor is powered from the transmitter. If you steal the power for the keyer from the transmitter, it can be anything from 12 volts to 24 volts. Regulation isn't extremely important, but character weight varies with supply voltage. So a zener diode regulator for Q1 and Q3 is a good idea, if you steal the power from an uncertain source.

Construction is not critical, and if you don't use the keying relay, the parts easily can fit on a 2" × 3" PC board. The small size makes it a natural for

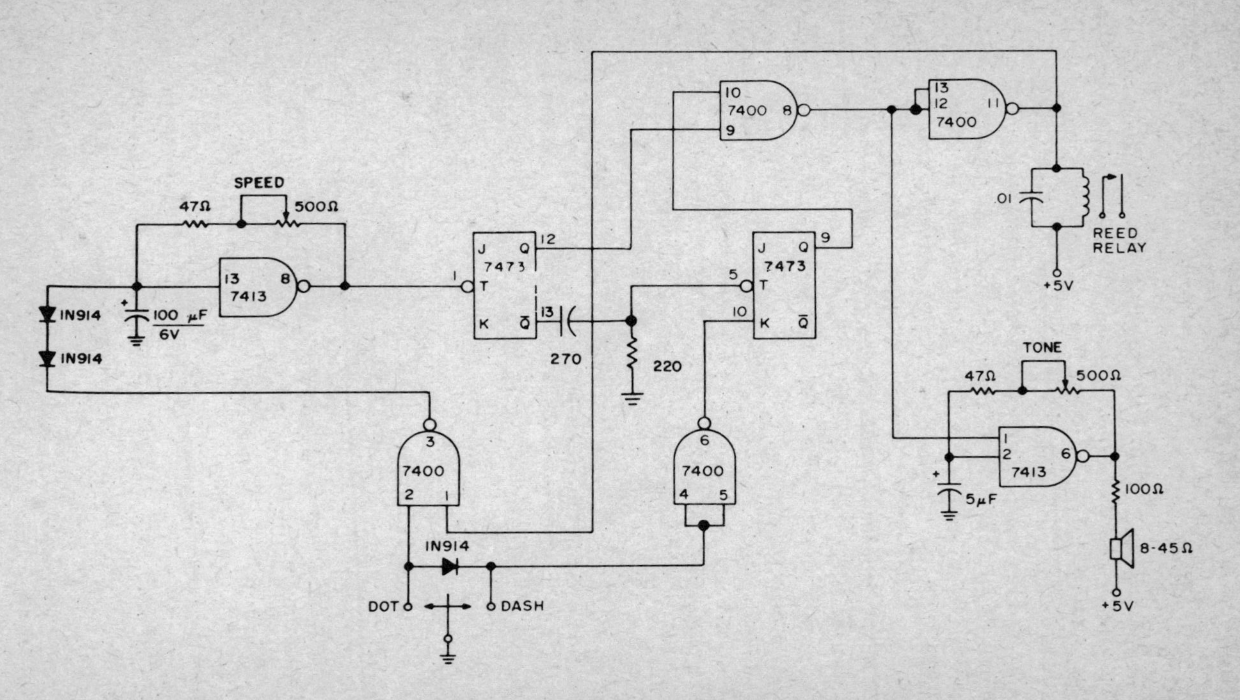

Fig. 1-8. Schematic diagram of the IC keyer.

adding to the base of your key, or even building into a vest pocket CW transmitter.

AN IC KEYER

Figure 1-8 is the schematic diagram for an electronic keyer that features self-completing, built-in monitor and instant start. The bulk of the circuit is straightforward, using gating and flip-flop functions to generate dots and dashes. The really unique aspect of the keyer is the gated clock.

The gating action of the clock must operate in the following sequence. The clock is enabled by either the paddle or the output which is the method of self completing. When the output goes low after the completion of a dot or dash and the paddle is not depressed, the clock is disabled and must remain disabled for at least one dot time duration, after which it should be able to start on demand. This provides a minimum spacing between dots and dashes regardless of the motion of the paddle.

The oscillator uses a SN7413 Schmitt trigger as a relaxation oscillator. Typical waveforms appear in Fig. 1-9.

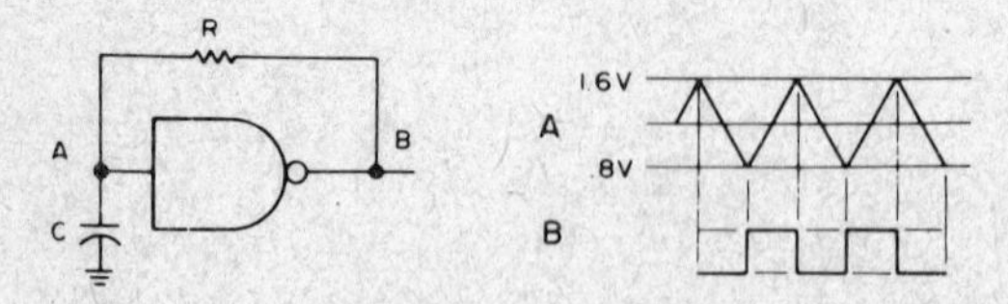

Fig. 1-9. Relaxation oscillator and associated waveforms.

These criteria require a control signal to disable the clock, prevent any negative transition for one dot duration and then be ready on demand. The circuit in Fig. 1-10 accomplishes this function with the minimum components.

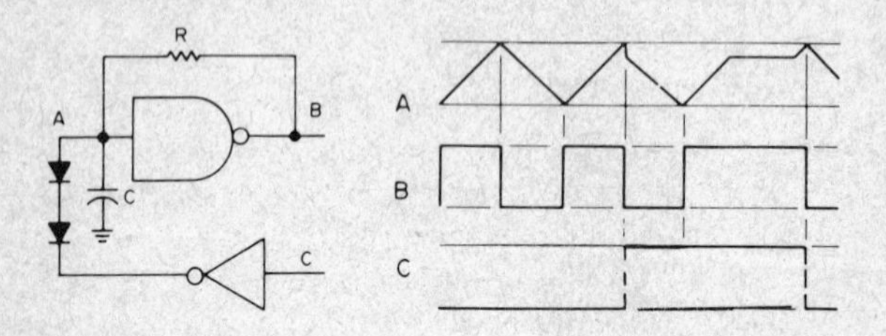

Fig. 1-10. Control circuit for the clock.

When the control line is low the circuit oscillates normally. When the control line goes high, which in the keyer coincides with the negative transition of the clock, the diodes conduct and the capacitor voltage quickly discharges from 1.6 to about 1.5 V, which is the sum of the "0" output voltage of the gate and the two .7V diode drops. The capacitor continues to discharge at the usual rate until it reaches the SN7413's threshold of .8V, when the gate changes state and the capacitor voltage heads up gain. The capacitor charges up to 1.5V and becomes clamped by the diodes. The oscillator is now ready for operation. When the control line goes low the capacitor can now charge up 100 mV to the threshold and begin oscillation.

The oscillation is not "instant" start, but "fast" start. Approximately 10% of a dot duration is required for start-up.

The unit is easy to build and careful shopping for surplus components can place the cost below $10.00 for everything, including cabinet and power supply.

DIGITAL CODE SPEED DISPLAY

The first acquaintance with semiconductors for many hams came through the building of an electronic keyer. This gave the ham a chance to experiment and learn with a fairly simple and non-critical circuit. With the advent of integrated circuits (ICs), the same thing happened all over again. The natural circuit for application of these ICs for the ham was again the keyer. The first ICs were digital, either on or off, and were easily adapted to the on/off characteristic of code.

Now that many hams have a keyer, it may be time to add an extra goodie to it. This section describes a circuit made up of a few ICs that display your sending speed in words per minute. Halfway between a speed marked 5 and 15 on your keyer never did mean 10 wpm. Now you can really know your sending speed and impress the neighbors with this new gadget in your ham shack.

Block Diagram

The ARRL Handbook gives a formula for calculating code speed.

$$\text{Speed (wpm)} = \frac{\text{dots/min.}}{25}$$

$$\text{or Speed (wpm)} = \text{dots/2.4 sec.}$$

Consequently, if we can develop a pulse precisely 2.4 seconds long, and use this pulse to let dots from the keyer count-up a counter, then display the contents of the counter, we will have accomplished our goal.

A block diagram of the code speed display unit is shown in Fig. 1-11. The source of the timing is the 60 Hz from the AC line. A filament transformer steps down the voltage and the shaper prevents the noise that may be present on the AC line from producing false trigger pulses. The 60 Hz is then divided by 144 which yields a pulse each 2.4 seconds. This pulse goes to a flip-flop which develops a positive pulse for 2.4 seconds. The positive pulse enables a gate which lets dots to the counter. The timing circuit also resets the counter and enables the circuit to operate without manual reset. The counter is made up of two decode counters which enable it to count to 99. The decode counter outputs go to a binary coded decimal (BCD) to seven segment decoder which in turn drives the display.

This description will give those with a knowledge of digital circuit techniques sufficient ideas to get going and build the circuit with their own variation of components. However, for those without too much experience, the design will be reviewed.

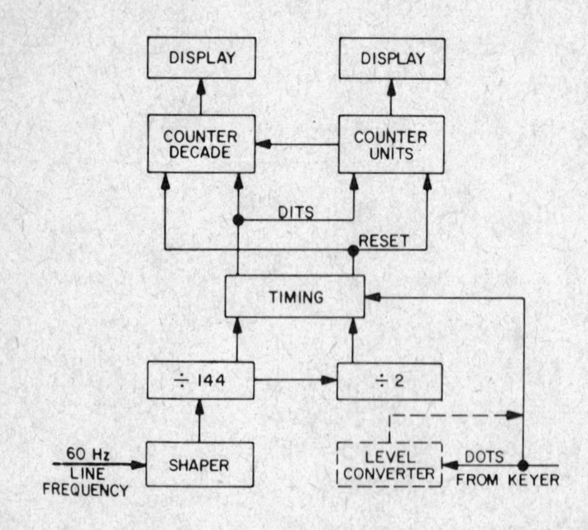

Fig. 1-11. Block diagram of the code speed display.

Circuit Description

Figure 1-12 is the schematic of the code display unit. The shaper uses a Fairchild 9602 single shot. When pin 4 receives a voltage greater than 1.2 but less than 5V, a pulse appears at pin 6. The width of the pulse at pin 6 is determined by the RC combination. With pin 7 fed back to pin 5, the single shot is inhibited from putting out another pulse even with noise on the line, until the pulse at pin 7 has gone away.

The period of the incoming line frequency is 16.6 ms. The RC time constant was chosen to produce a pulse at pin 6 equal to 12 ms, this will insure that the pulse has gone away by the time the next AC line cycle appears. The shaper, therefore, puts out a 12 ms pulse for each cycle of the AC line. When making a voltage divider for the filament transformer, use peak voltage for the calculation to insure that voltage peaks at the shaper input never exceeds 5V. The pulses on pin 7 will be the same as those on pin 6 only inverted in phase.

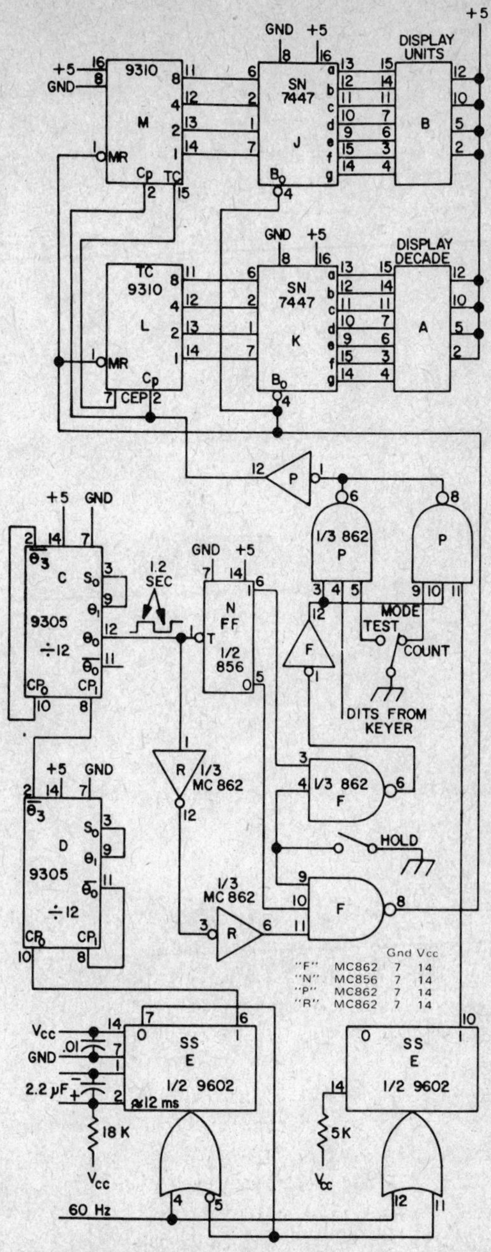

Fig. 1-12. Schematic of the display.

Divider

The Fairchild 9305 IC can be wired to divide by various numbers; the ICs used in this application are wired to divide by 12. Hence two such circuits divide the incoming line frequency by 144, which results in a pulse at pin 12 of IC "C" each 2.4 seconds. This pulse train goes to pin 1 of flip- flop "N" to yield a pulse at pin 6 of "N" which is positive for 2.4 seconds and negative for 2.4 seconds. The same inverted sequence appears at pin 5 of "N." With the HOLD switch open as shown, the 2.4 second pulse appears at the output of inverted "F" pin 12. We now have the 2.4 second gating pulse at "P" pin 3 and 10. With the switch in the COUNT position, dots from the keyer pass through the gate during the 2.4 seconds and cause the units and decode counters to count the dots. The number contained in the counters is then decoded by the J and K ICs to drive the seven segmented display. After displaying the wpm for 1.2 seconds, the counters are reset and the cycle begins over again. Once the count has been made and displayed, it may be held by closing the HOLD switch. With the HOLD switch closed, the gating pulse and reset pulse are blocked and the display remains constant.

Test

This circuit contains a built-in test feature which is an asset for those who lack the necessary test equipment. With the HOLD switch open and the MODE switch in the TEST position, the line frequency pulses from single shot "E" pin 10 are steered to the counters for a period of 2.4 seconds. These pulses substitute for the dots. There are 144 line frequency pulses per 2.4 seconds. Therefore, when the gate is enabled for 2.4 seconds, 144 pulses will enter the counters. The counters are only capable of counting up to 99, then they go to zero and continue to count incoming pulses. In this case, the counters will overflow and display 44. Consequently, a count of 44 in the TEST MODE indicates a good working unit.

Logic Levels

This unit is made up of TTL and DTL logic families. These logic families require positive (0 to 5V) input levels with a threshold of about 1.2V when the

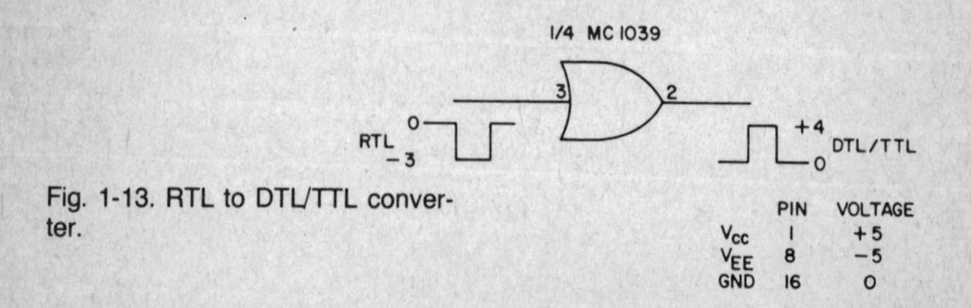

Fig. 1-13. RTL to DTL/TTL converter.

logic switches. Many keyers have been made using the RTL logic family. If your keyer is of this family and develops pulses which go from 0 to around −3V the above circuit will not work. Don't despair; with one more IC the 0 to −3V pulse can be converted to a respectable 0 to 5V pulse. The circuit for this level conversion is shown in Fig. 1-13.

Timing

Figure 1-14 gives all the timing for the code speed display unit. The reset pulse comes just before the count begins. Evidence of the reset pulse can be seen since it blanks the displays during the reset period.

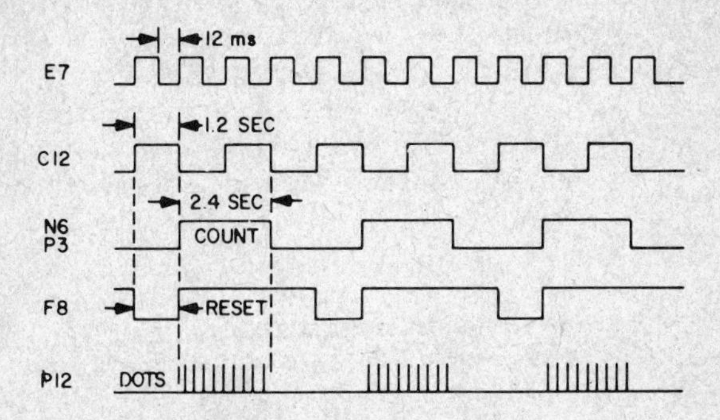

Fig. 1-14. Code speed timing.

Power Supply

A word of caution to those who would rob power from their keyer to power the code speed display unit. This circuit requires 400 mA at 5V, so make sure your existing supply can handle this new load. It it can't, you can build the one shown in Fig. 1-15.

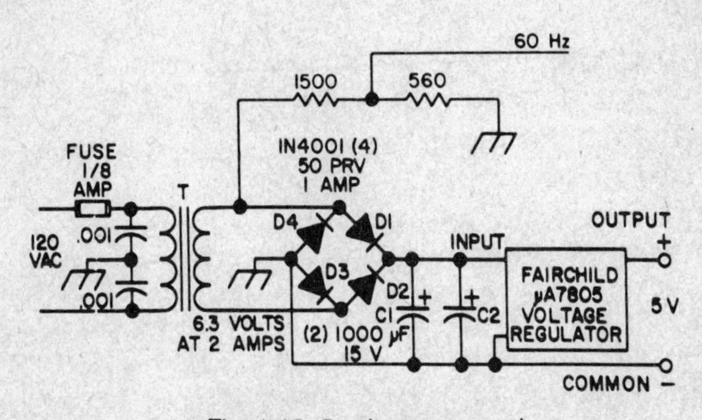

Fig. 1-15. 5-volt power supply.

Operation

The best way to set your desired code speed is to adjust the speed with the transmitter off (no RF) then switch to HOLD and continuously display the wpm until you want to make another code speed change.

One thing to remember when using digital counters, whether it be this one or a frequency counter, is that the least significant bit may be incorrect by one unit. Since the opening and closing of the count gate is rarely synchronized with the dots passing through the gate, there always exists a ±1 count gating error. In Fig. 1-16, line (c) represents the opening of the signal gate. If the phase relationship of the incoming dots with respect to the gate is as shown in (a), six pulses will be counted. If, however, the phase relation is as shown in (b) only five pulses will get through the open count gate. If the dots from your keyer were derived from the line frequency, this would not occur.

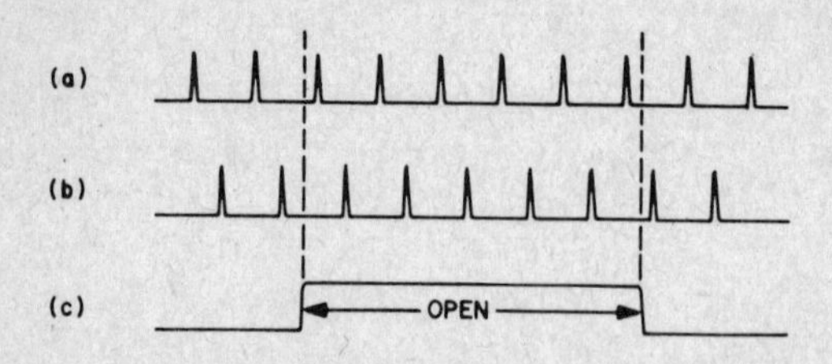

Fig. 1-16. Gating error.

It is not important to use the same ICs as presented here to arrive at the same results. Any IC may be used that can fulfill the requirements of the block diagram.

No specific information is given regarding the lighted display units, because each type has its own connection outline. It is only required that "a" through "g" of the SN7447A go to the corresponding segments of your chosen display unit.

AN IC AUDIO NOTCH FILTER

A bridged-T audio notch filter combined with an IC amplifier produces a highly versatile, wide-range audio rejection filter with both variable frequency and variable "Q" controls.

Audio filters are certainly nothing new. They have long been used to improve the selectivity of a receiver or transceiver when it was not desired to "dig" into the IF circuitry and improve the selectivity at the RF level. The disadvantage of such a method of selectivity improvement is that the selectivity takes place late in the receiver processing chain. Therefore, when one is listening to a weak station, a strong station near in frequency can control the AVC or overload the receiver stages.

Nonetheless, audio selectivity is easy to apply and can take the form of either an audio frequency peaking or notching type function. Audio peaking can easily be provided by a number of fixed frequency filter designs and numerous inexpensive units are available from surplus outlets. The disadvantage of the peaking approach is that most filters which are of any real use produce a "ringing" effect. The sound is unnatural and definitely very tiring if the filter is constantly left in the circuit without any provision for disabling it. Stations can also be lost when scanning a band unless tuning is done very slowly when using the filter on CW.

The notching type filter, on the other hand, is usable on both CW and phone. It does not cause any ringing effect and does not mask any signals when quickly scanning a band. A single notch filter can only eliminate the one frequency to which it is set, but when a receiver already has a good phone filter—such as the multiple crystal or mechanical types found in SSB transceivers—one notch frequency possibility seems to suffice in most QRM situations. Audio notch filters built around passive components only are fairly old, but their use has disappointed many operators because to achieve reasonable narrowness and high attenuation at the notch frequency, expensive capacitors were necessary and the filter could only be used in very high impedance circuits. The use of an integrated circuit amplifier with a notch filter in a feedback arrangement, however, produces a notch filter of very high Q using inexpensive components. It is neither critical as to circuit impedance nor does it introduce any overall circuit insertion loss.

Basic Circuit

The circuit of the IC notch filter is shown in Fig. 1-17. The actual notch filter consists of the bridged-T network—the ganged 50 K potentiometer and

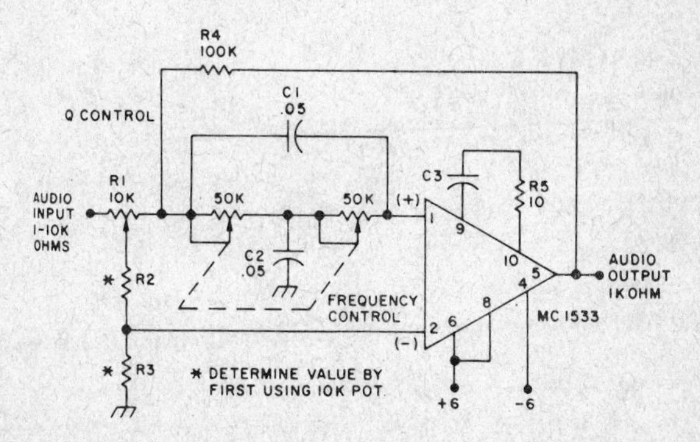

Fig. 1-17. Schematic diagram of the variable frequency and variable bridged-T notch filter. Various other IC units may be used besides the unit shown.

the two .05 mF capacitors. The circuit presents a very high attenuation at one frequency which is related to the time constants of the circuit. The frequency of maximum attenuation can be changed by either changing the value of the resistive or capacitive legs. As was mentioned, however, unless special precautions are taken, the bridged-T network alone will tend to produce a very broad rejection notch which is particularly unsuited to CW work. The integrated circuit, however, corrects this situation in the following manner: the input signal passes through the bridged-T network to one input of a differential operation amplifier (a Motorola MC1533 in this case). Feedback through R4 is coupled from the amplifier output back to the signal input point. The other amplifier input (–) then receives a combination of the original input and feedback signals. Since the nature of a differential amplifier is such that when the (+) and (–) inputs receive equal level signals and the output is zero, R2 and R3 are chosen such that this condition exists and infinite attenuation takes place at the notch frequency.

There may be situations where a somewhat broader rejection notch is desired with correspondingly less maximum attenuation at the notch frequency. This adjustment is provided by making R1 variable, as shown in Fig. 1-17. As the wiper arm on R1 moves from right to left, the feedback voltage around the operations amplifier decreases and the effective Q of the bridged-T network is reduced. Thus, if desired, R1 can be made variable and functions as a "Q" control. Otherwise, R1 can be a fixed value resistor and R2 is connected to the junction of R1 and R4 to produce a single frequency, deep notch audio filter. In this case, the only variable control would be the dual 50K ohm potentiometers which vary the time constant of the bridged-T network and hence the notch frequency. The dual potentiometer is capable of varying the notch frequency over about a 10:1 frequency range—300 to 3,000 cycles approximately. This range should certainly suffice for most applications but, if desired, the range can be changed by using different (but equal) values of capacitance for C1 and C2.

The differential operational amplifier used may seem an unusual component to many readers. Although the basics of integrated circuits cannot be explored in this section it should be realized that the integrated circuit used is only a multi-stage transistor amplifier packaged into a housing the size of the usual single transistor. The main feature that separates the differential amplifier from a conventional amplifier is its input circuitry. The "differential" input has a non-inverting (+) and inverting (–) input. A positive-going voltage applied to the non-inverting input produces a positive-going output voltage. The same voltage applied to the inverting output produces a negative-going output. Thus, the same value and polarity voltage applied to both inputs will produce no output.

Figure 1-17 shows the use of a Motorola MC1533 operation amplifier, but almost any similar unit will suffice. A number of inexpensive "surplus" IC operational amplifiers are available from such suppliers as Poly Paks, Lynnfield, Massachusetts 01940. Other units may differ in their voltage requirements and rolloff compensation needs (the RC network between pins 9 and 10 on the MC1533), but this information is usually supplied with the unit. It should be noted that a simple integrated circuit audio amplifier cannot be used; such units do not have differential input circuits.

Construction

There are very few precautions to be observed in constructing the unit because of the nature of its operation. One possible method of construction which the author explored is shown in Fig. 1-18. All of the circuit components are mounted on a piece of Vectorboard which in turn is mounted on the rear potentiometer of the dual 50K ohm potentiometer. A piece of foam plastic material is glued between the underside of the vectorboard and the rear potentiometer to achieve the mounting.

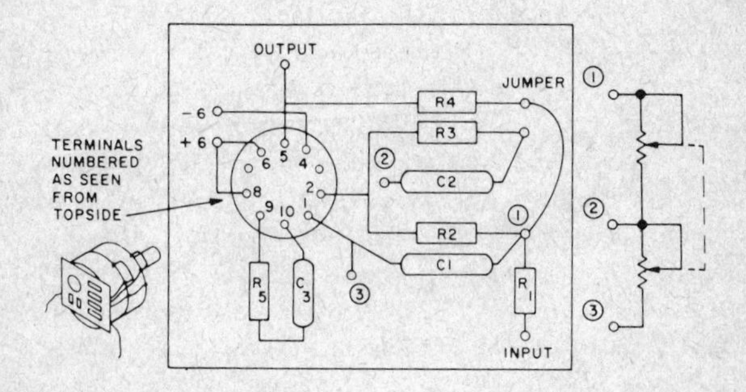

Fig. 1-18. No particular circuit layout is necessary but the components can be grouped so compactly onVectorboard that the entire circuit mounts on the back of the frequency control potentiometer.

The parts layout shown for the vectorboard in Fig. 1-18 need not be followed exactly, although it was the simplest which the author could devise. If an integrated circuit packaged in a dual-inline containter is used (rectangular with 5-7 connections on each long side), the IC can be mounted on the left side of the Vectorboard and R5 and C3 both placed under C1. No "Q" control is provided for in the parts layout shown, and R1 is a fixed value resistor. R1 need only be made variable, as shown in Fig. 1-17, if this feature is desired. The values of R2 and R3 can only be properly chosen by first using a 10K ohm potentiometer in their place (with the wiper arm goingto terminal 2 on the IC) due to component value variation. The procedure is fairly simple. The wiper arm on R1 (if a variable unit is used) is first set towards the junction of R1 and R4. Then, using an input signal containing a frequency which the notch filter can reject, the dual 50K ohm potentiometer is adjusted for maximum rejection. Leaving this control set, the temporary potentiometer used in place of R2 and R3 is adjusted for complete signal rejection. At this point, the arms of the 10K ohm potentiometer are measured and replaced by equivalent value fixed resistors.

The resistors used in the notch filter can either be 1/4 or 1/2 watt sizes. The capacitors need be rated no higher than the maximum value of the supply voltage used for a particular IC. The capacitors used in the bridged-T network should be of good quality to achieve the sharpest notch selectivity. Disc ceramic types are acceptable, although, if possible, low-loss types such as the Aerovox P123ZN series are preferred. The current demand from the power supply is in the order of a few milliamperes, but the operating voltages should be obtained from a well-filtered source to avoid any possible hum problems.

The dual 50K ohm potentiometer used in the bridged-T circuit is a standard linear taper type. Although one can purchase such potentiometers from various supply houses, particular attention should be paid to being cost-conscious about this item.

Mounting and Operation

The mounting or placement of the notch filter in a receiver or transceiver is fairly flexible. The unit can be inserted between almost any two audio stages. DC blocking capacitors must, of course, be used to prevent other than audio frequencies passing through the filter. Alternatively, the unit can be used completely external to a receiver or transceiver. The audio output from the receiver or transceiver can be taken from a headphone jack or from the loudspeaker terminals through a transformer (4-10 ohms to 1K ohm or more). The gain of the operations amplifier, in the latter installation, is more than sufficient to drive any pair of medium to high-impedance headphones.

Operation of the notch filter is extremely simple. When not in use, the filter is adjusted for maximum low-frequency attenuation. As QRM develops, the filter is used (in conjunction with tuning of the receiver bandpass) to eliminate the most severe interfering beat (on SSB or CW). The result is an almost complete attenuation of the interfering signal while still retaining the full fidelity of the desired signal. The difference between this method of QRM elimination and that which depends upon a severe reduction in bandpass to accomodate only the desired signal is quite startling in terms of fidelity and ease of tuning.

PC BOARD IMPROVEMENT

Before starting to wire your next printed circuit board, kit, or homebrew, take time to save yourself trouble later. Drill out all holes where wires connect to the proper size and install Vector terminals. This eliminates future problems involved in trying to reconnect wires to holes which are out of sight beneath a pool of solder. The possibility of lifting the foil during repairs is also minimized.

Another place where this technique bears fruit is in a situation where a component must be removed in order to make circuit adjustments. Some transceiver front ends are a prime example. In order to neutralize the RF stage it is necessary to remove plate voltage. This is done by removing a resistor from the PC board at a point which is at best impossible to reach. Reinstallation is even worse, calling for at least three hands, long nose pliers, and a truck driver vocabulary. After the Vector terminals are installed on the *bottom* of the board, removal and installation is a distinct pleasure. SB-110A owners please note.

A CW FILTER BOX

As most hams know only too well, CW operators, especially those using the Novice band segments, are a long suffering lot. There is QRN and foreign broadcast interference to contend with, to say nothing of adjacent-station QRM due to crowded bands. Rarely is a contact completed without at least one station issuing a strident CQ only a few cycles from the frequency in use. Many a good QSO has ended in frustration, fading away into this constant background din.

There is a simple and inexpensive solution to problems brought on as a result of too many stations occupying too little band space. An audio filter, selective enough to single out a discrete audio note, can make a nearly impossible snarl sound like code practice.

The Circuit

The audio filter is based upon three major components: a pair of toroidal coils which combine small size and high inductance, and a solid-state audio amplifier module.

Receiver audio, consisting of CW signals of many different audio frequencies, enters the filter through coupling capacitor C1 (Fig. 1-19). Capacitor C2 and coil L1 comprise a tank circuit resonant at 758 Hz. This tank will shunt to ground signals that are not at the resonant frequency. A 758 Hz

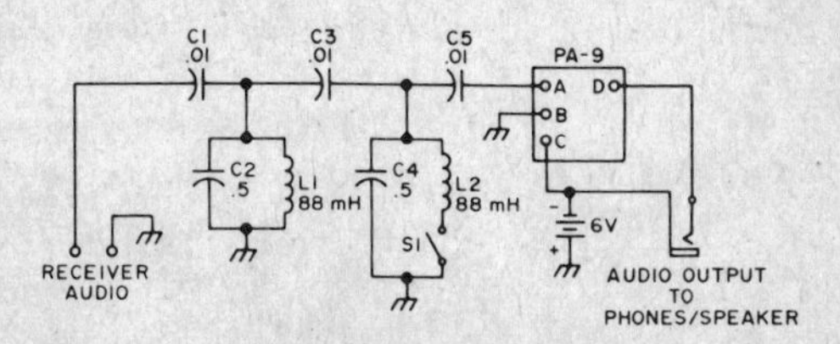

Fig. 1-19. CW audio filter.

signal will cause signal voltage to be developed across the tank. This signal is coupled through C3 to the top of tank circuit C4–L2. If switch S1 is closed, off-resonance signals are further attenuated. With S1 open, the second filter circuit is disabled and the signal passes directly to C5. The function of S1 is to provide two levels of selectivity for ease of tuning.

From C5, the filtered signal enters module PA-9 where it is amplified to drive headphones or speaker. Volume of the input audio from the receiver determines output volume of the amplifier.

Filter output frequency may be changed to suit the ear of the individual. Some may desire a lower tone, some a higher pitch note. By changing the value of C or L, the resonant frequency of the filter may be altered.

Components

Voltage ratings of the components are not important in this circuit. Receiver audio voltage never will approach usual component ratings. Capacitors may be disk ceramic or any other type found in the junkbox or surplus electronics store. If size is no object, an economical substitute for the two 0.5 μF capacitors is a surplus double oil-filled type. Most have flanges for chassis mounting.

Coils are toroids, available through many sources. You may come across them advertised for sale in the classified section of electronics magazines. Toroids often are used in Teletype equipment as filters to differentiate between two audio tones that carry RTTY information.

PA-9 is a "public address amplifier module" available through many sources. Higher powered, higher priced modules are readily available from many sources where more audio output power is desired.

The battery may be of any type, considerations being size versus life.

Your preference is the rule here.

Construction

A Minibox, a few terminal lugs or strips, a soldering iron, drill, and a screwdriver will get you through this project in fine form. The device, however, need not be built in a Minibox. You can breadboard it or even build it right into your receiver.

Toroids are sold centertapped with these leads open. You must solder them together before installation. The coils may be mounted one above the other with a bolt running through the centers. You can, however, mount them in any way that will best fit the space available.

Mounting the module should present no problem even though it has absolutely no provisions for mounting. A few rubber grommets glued to the bottom of the module and then glued to the Minibox will hold nicely. Leads should be covered with spaghetti tubing and run through a rubber grommet placed in a hole in the box under the module.

CW TRANSMITTER FOR 160 METERS

The transmitter consists of a VFO that tunes from 3.25 MHz to 3.45 MHz, the output of which is heterodyned with a 5.25 MHz crystal oscillator to produce a 1.8 MHz to 2.0 MHz output. Both oscillators are continually running. Keying is accomplished by blocking the grid of the mixer and the final amplifier.

The VFO used was a surplus Collins 70-K1 which was found in a local surplus store for less than 10 dollars. Since these may not be readily available elsewhere, the circuit of an alternate VFO (which was tested) is included.

The final amplifier is a pair of 1625s operating in parallel into a pi section network. 1625s were chosen because of their abundance at about two bits each, plus the fact they are quite rugged.

The power supply uses a surplus receiver power transformer with a high voltage secondary of 340-0-340V. This is bridge rectified to produce about 940V DC at no load. A choke input filter from the center tap provides + 300V for the other stages. A small 6.3V filament transformer is connected backwards across the 6.3V winding of the power transformer to provide bias for the mixer and final amplifier.

While the bias voltage could be keyed directly, surplus reed relays were used for keying, receiver muting and side tone keying. Antenna switching is done by a simple TR switch.

The VFO mixer sub assembly was built on a 4 × 6 × 2 inch chassis. The alternate VFO circuit is shown in Fig. 1-20. A surplus link variable inductor from a Navy ATD transmitter tuning unit (3 – 9MHz) was modified by removing two turns from the rotatable link. The remaining 2 1/2 turns were then mechanically adjusted for maximum linearity. The final tuning range was 500 kHz.

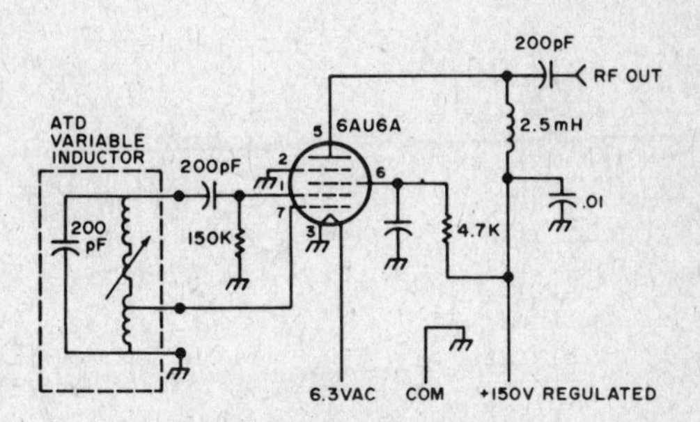

Fig. 1-20. Alternate VFO

Transmitter construction was done in three stages: first the power supply stage (Fig. 1-21) followed by the VFO-mixer (Fig. 1-22) and then the driver—final amplifier (Fig. 1-23). Most components were mounted on simple

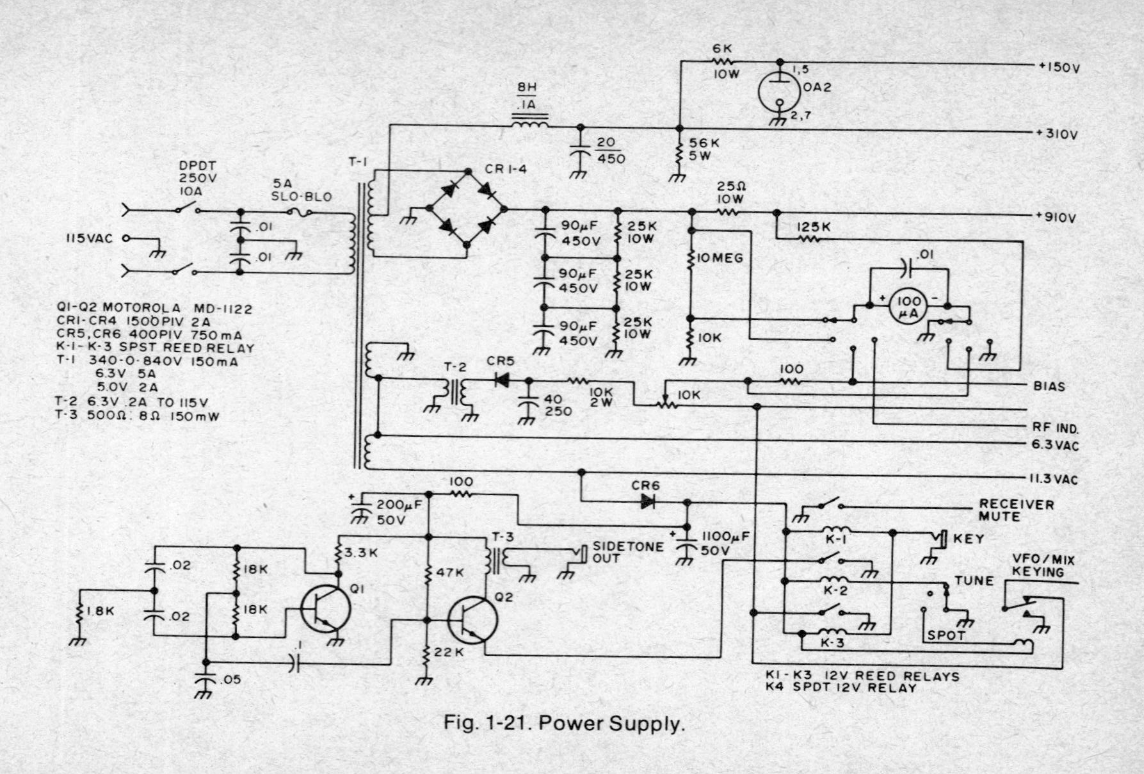

Fig. 1-21. Power Supply.

tag strips. The sidetone oscillator, keying relays and meter multipliers are mounted on small pieces of fiberglass Vectorboard which are rigidly supported on standoff bushings.

Using the component values shown, the meter reads 1000V full scale (fs) in position 1, 500mA fs in position 2, 15mA fs in position 3 and relative amplitude in position 4. The 10M resistor was made up of three 3.3M 2W resistors.

A word of caution on selecting a meter switch! Avoid the low cost phenolic imports as they won't handle the voltage.

Filter capacitors are mounted on a heavy bracket under the chassis. A separate shield assembly is provided for the TR switch. A bottom plate and screen cage around the final tank assembly completes the shielding.

Alignment and check-out is simple. The VFO-mixer assembly is best aligned and calibrated using an oscilloscope and counter. However, satisfactory work can be done with a grid dip oscillator and either an LM or BC-221 frequency meter and a receiver. Adjust the slug in the mixer plate coil for a maximum output at 1.9 MHz, checking to insure that the output is really in the 160 meter band! The VFO dial can now be calibrated.

Next, the driver and final amplifier are aligned. Set the drive level control at minimum. Place the spot—tune switch in the tune position and adjust the drive level control (checking to see that the driver tuning control is adjusted for maximum grid drive) for plate current of 200 mA with the power amplifier resonated into a 50 ohm dummy load. Incidentally, power input can be reduced using the drive control to meet FCC regulations for the various geographical areas.

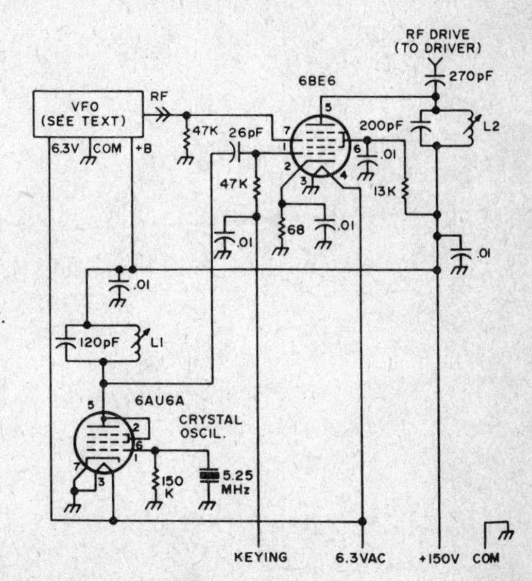

Fig. 1-22. VFO Mixer.

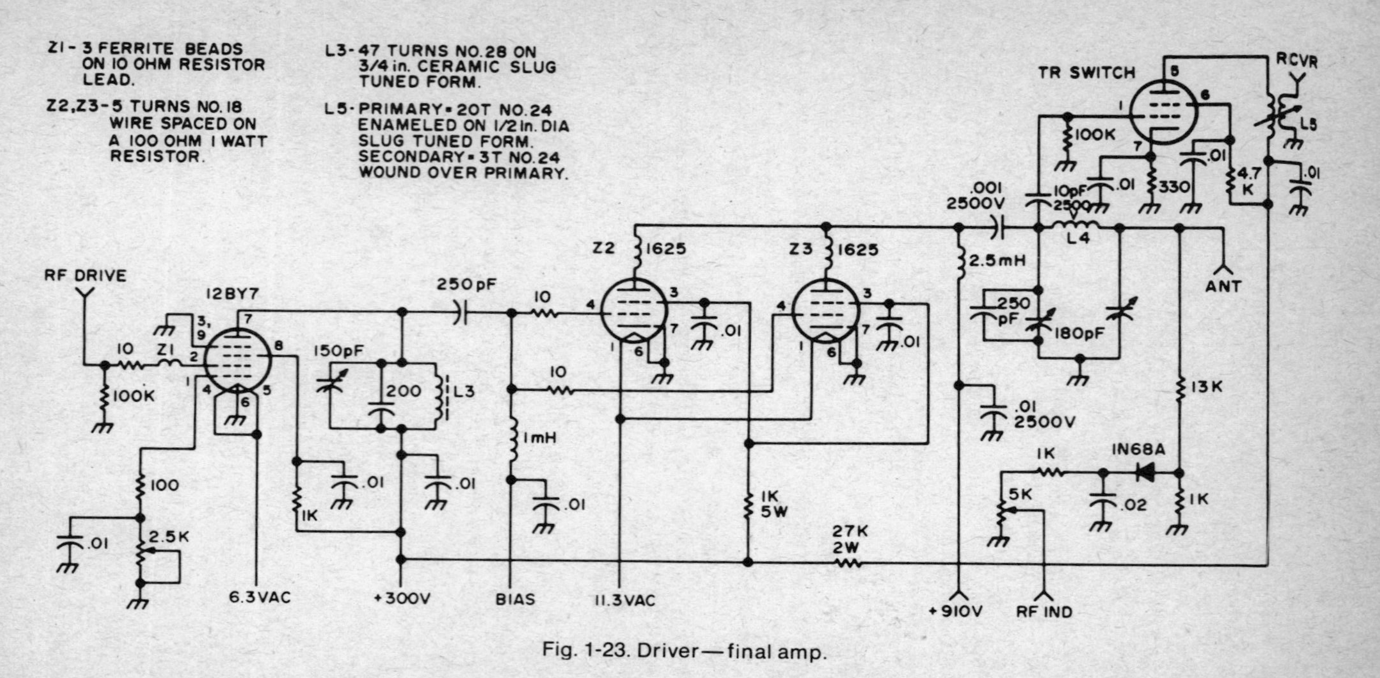

Fig. 1-23. Driver—final amp.

A note on the VFO dial construction is in order. While any vernier drive assembly can be used, try this inexpensive method. The illuminated dial mechanism is made using a National Velvet Vernier planetary drive from a scrap CB-375 tuning unit.

The dial itself is made from a piece of aluminum sheet cut out with a fly cutter. After spraying white, calibrations are applied using rub-on transfers. After the calibration is checked, the dial plate is given several coats of Krylon. The cursor is made by scribing a line on a small piece of Lucite which is cemented into a panel cutout with model airplane cement.

Place a black baffle behind the dial plate. A 5/8 inch (standard) hole in the baffle accepts a standard push-in lamp socket. Complete the assembly with a small skirted drive knob. The combination of the main dial calibration with skirt dial markings allows 1 kHz readout.

Unless you are fortunate enough to have an antenna system that presents a 50 to 75 ohm load at these frequencies, an antenna tuner will have to be used. A simple rotary inductor from a BC-375 transmitter works well. Tuning is accomplished by tuning up the transmitter into the 50 ohm dummy load then switching to the antenna. The variable inductor is then adjusted for maximum output power.

DIGITAL "HI" GENERATOR

Probably one of the most used greetings/expressions in amateur radio, be it a car with call letter plates passing another, or the passing of an OSCAR satellite overhead, is the simple world, "Hi." I do not know how it all started but I know that if I want to greet another ham on the road, I want to honk "Hi."

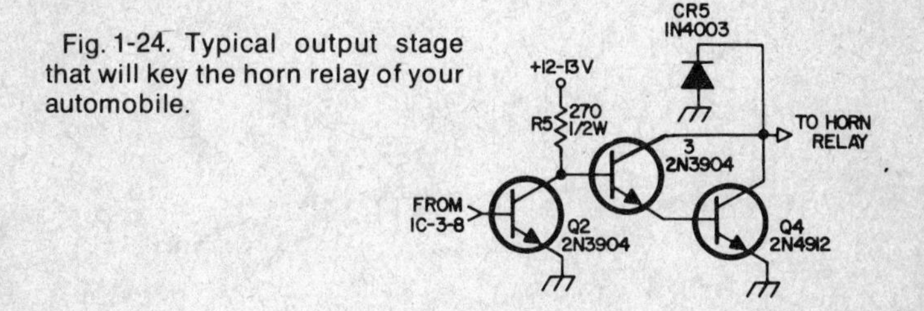

Fig. 1-24. Typical output stage that will key the horn relay of your automobile.

Therein lies the problem. My car is one with the horn incorporated in the steering wheel rim, and it is next to impossible to honk any kind of code, much less the rapid succession of four dots and then two dots. This prompted me to design a circuit which will digitally produce the word "Hi." The scope of this section will be to describe the basic digital "Hi" generator, which can be used in any application desired by the addition of the appropriate output keying state.

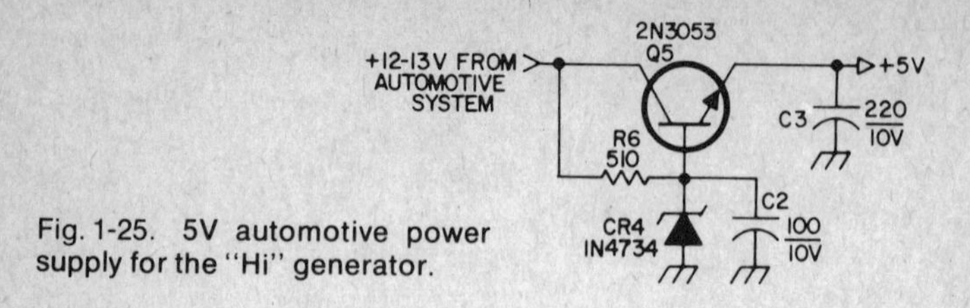

Fig. 1-25. 5V automotive power supply for the "Hi" generator.

Figure 1-24 shows a typical output stage which can be used to key the horn relay of a car. In this application, the driver would be simply required to touch a momentary contact push button type of switch and the "Hi" generator will start and self-complete the word "Hi." This circuit uses a total of four integrated circuits and eight discrete components, and when built will easily fit in the palm of your hand. Figure 1-25 is a suggested 5V regulator.

Referring to Figs. 1-26 and 1-27, a unijunction transistor oscillator located on plug-in-board CB-1 generates a timing pulse which is fed into a 7404 inverter (IC-4-1) and is shaped and inverted to form a narrow, negative-going pulse. This pulse is fed into a JK flip-flop (IC-1) which, when allowed to run by the setting of the "run" flip-flop, acts as a divide-by-two counter whose output is a symmetrical square wave. This square wave, which will be referred to as the clock term, is wired to two locations. The first location is the clock input of the

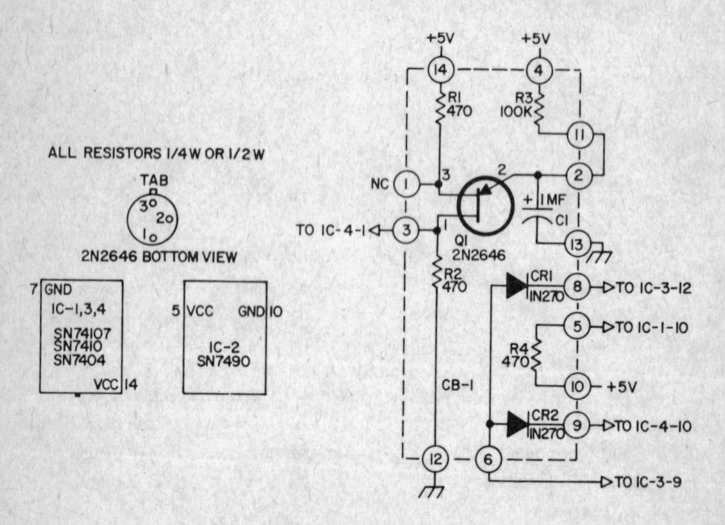

Fig. 1-26. "Hi" generator.

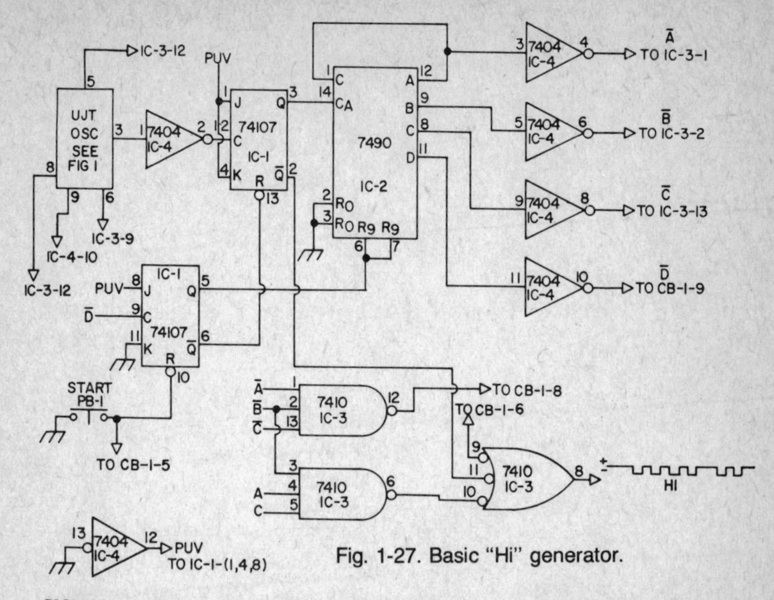

Fig. 1-27. Basic "Hi" generator.

SN7490 decade counter that "counts" the clock pulses, and the second location, the output gauge, (IC-3), which uses the oppose phase or clock. This term is simply the Q side of the clock flip-flop. The purpose of the clock term driving one input of the output gate is to chop anything passing through this gate into a series of dots. Since this gate, as described thus far, will produce an endless string of dots, all we have to do is inhibit the unwanted dots and we will have the word "Hi" remaining. Referring to the timing diagram (Fig. 1-28) it will be noted that to form the word "Hi" we must get rid of count zero, count five, and counts eight and nine. To do this we do not need to decode all bits; all we need is enough information to electrically describe the period we are interested in. You will see that count zero has a unique condition when terms A, B, and C are all low. These three terms are inverted by IC-4 so as to provide the high level needed for the 7410 to decode an AND condition of A B C. In the same manner, the count of five is decoded using the terms A, $\overline{B}$, and C.

Since we do not want counts eight or nine, we use the term "D," which occurs only during these counts, to directly inhibit them. A secondary function of this term is to reset the "run" flip-flop back to the waiting state. So we can now say that we will have an output key unless clock is low *or* D is low *or* five decode is present. Add this all together and all that is remaining is..."Hi."

A SIMPLE INTEGRATED CIRCUIT Q MULTIPLIER

Almost any integrated circuit operational amplifier can be used to build this Q-multiplier. Its advantages are extreme circuit simplicity and a useful

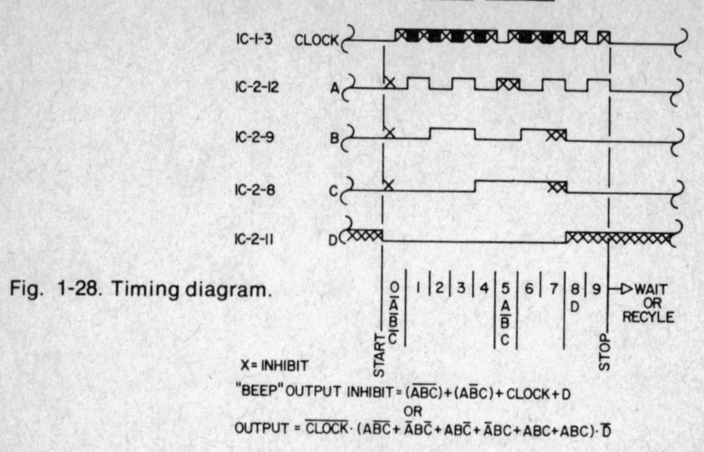

Fig. 1-28. Timing diagram.

frequency range that extends from audio frequencies to almost all IF frequencies. Both the peaking frequency and Q can be made variable.

Both vacuum tube and transistor Q-multiplier circuits find wide application in improving the selectivity of transceivers and receivers, particularly on CW when the Q-multiplier is used to peak the IF response following a steep-skirted crystal or mechanical SSB filter. The Q-multiplier provides a very narrow bandpass but if used alone does not provide steep skirt selectivity. When used in conjunction with an SSB filter, however, the latter provides the necessary skirt selectivity (see Fig. 1-29).

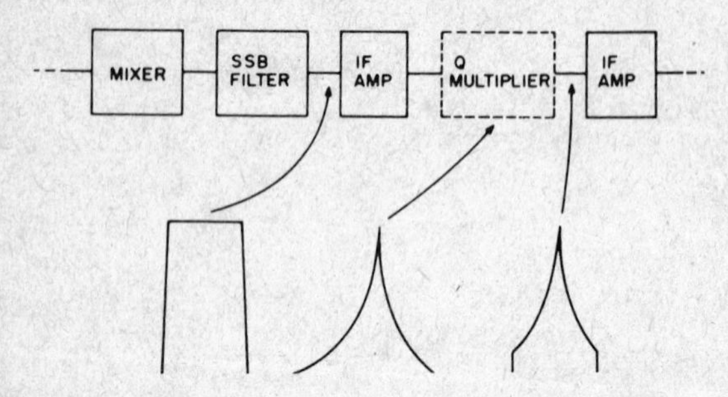

Fig. 1-29. Placement of Q-multiplier after SSB filter provides effective narrowband response for CW reception.

The Q-multiplier described in this section is meant to be inserted between any IF stage in a transceiver or receiver following an SSB filter. It can be easily switched for broadband operation so it does not affect normal SSB operation. Since it is adjusted for unity gain, it does not upset any gain relationship in the IF stages. As compared to vacuum tube and transistor-type Q-multipliers, the circuitry of the unit is extremely simple due to the use of an integrated circuit operational amplifier. The unit can be successfully used on frequencies far higher than those normally used with vacuum tube or transistor Q-multipliers up to 5 MHz or more, depending upon the integrated circuit used. The circuit can also serve at audio frequencies, if desired. One can also build an audio selectivity unit for outboard use, avoiding internal modifications to a transceiver or receiver.

Circuit Description

The Q-multiplier is constructed around an integrated circuit operational amplifier. Many such amplifiers are available on the market at prices starting at a few dollars. The main requirements for choosing a suitable unit are that it have a differential input (inverting and noninverting inputs), and a single-ended output and a bandwidth sufficient for the frequency of operation. For example, Fairchild 709T amplifiers are available for about $3 and are usable up to at least 500 kHz. A Fairchild 741 is usable up to 1–2 MHz. Other amplifers such as a Motorola MC1530 can be used up to 10 MHz.

Figure 1-30 shows the schematic of the type of operational amplifier which is used and the formula for the output voltage of an ideal amplifier. The Fairchild 741 amplifier requires no external frequency compensation components. Other amplifiers may require a few external components for this purpose as specified on their data sheet. The frequency rolloff components should be chosen such that the amplifier gain starts to decrease just above the frequency where it is used as a Q-multiplier. There is no advantage to having the gain "rolloff" at any higher frequency and would just make the amplifier more susceptible to oscillation due to a stray feedback path via external components. As noted from the gain formula, the gain of the amplifier depends upon the ratio of R2 and R1. If R2 is made equal to R1, the gain is unity. The Q-multiplier effect is based upon replacing R2 with a parallel resonant circuit which will present a very high impedance at one frequency

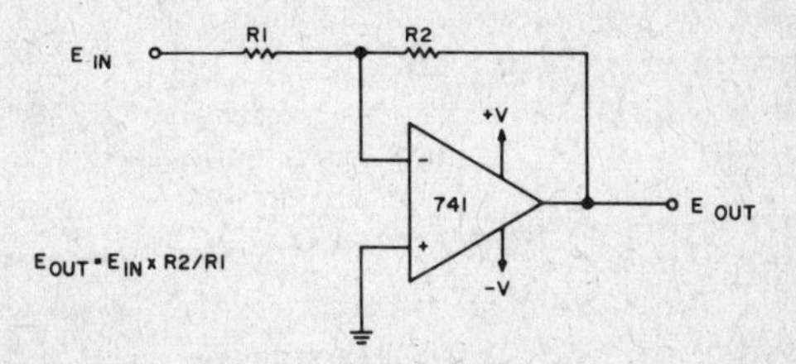

Fig. 1-30. Basic type of integrated circuit operational amplifier with a differential input and single-ended output that is needed for the Q-multiplier circuit.

and, therefore, maximize the overall gain at that frequency. Positive feedback is also used to enhance the Q-multiplying effect of the circuit.

Figure 1-31 shows the schematic of a practical Q-multiplier circuit with the LC circuit resonant at 455 kHz for use in a 455 kHz IF chain. Positive feedback is supplied via a 1 kΩ potentiometer. As with other Q-multiplier circuits, the Q can be increased by regulating the feedback until a point is reached when the unit will break into oscillation. The Q-multiplying effect is most effective when components are used for the resonant circuit which in themselves have a good Q. The inductor used for the 455 kHz Q-multiplier is a molded type which provides a Q of about 55 at 455 kHz by itself. The circuit can multiply this value by about 50 times or more. Another suitable inductor can be obtained by using only one or two sections from a regular 1 mH RF choke. The trimmer capacitor is used to set the frequency of the unit. The input resistor, although shown with a nominal value of 1 kΩ, should be chosen so that the gain of the overall circuit is approximately unity.

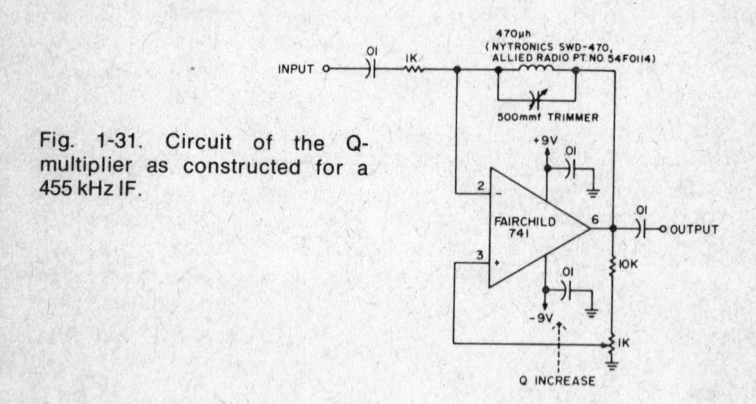

Fig. 1-31. Circuit of the Q-multiplier as constructed for a 455 kHz IF.

Construction and Adjustment

There is nothing critical about the construction of a unit utilizing the circuit described as long as the various lead lengths are kept short. For instance, the various components can be directly wired together on a small piece of Vectorboard. The board itself can be directly mounted near the IF chain in a transceiver, receiver, or together with the potentiometer used for feedback control if the latter is panel mounted.

If the Q-multiplier is not made tunable, the initial adjustment consists of peaking up the LC circuit. This can be done with the unit connected in an IF strip and using any test signal centered in the IF bandpass. The adjustment is best done with the feedback control set for minimum Q and with a barely audible CW test signal. The input resistor value should then be chosen for approximate unity gain. The adjustment is not difficult and need not be made exactly. The output level produced by a test signal without the Q-multiplier connected is noted. Then when the Q-multiplier is used, the input resistor is chosen so that the output level remains approximately the same. With most transistor IF stages, the value of the resistor needed will be about 1–5Ω.

Bypassing of the Q-multiplier can be done in a number of ways in order to allow for normal SSB reception. If the Q-multiplier is made frequency tunable, it can simply be tuned outside the IF bandpass. In order not to have the IF gain decrease too far when doing this, the input resistor must be chosen for unity gain to take place when the Q-multiplier is just tuned outside the IF bandpass. This will result in some increase in gain when the Q-multiplier is tuned to the center of the IF bandpass but normally the result will not be objectionable. Another way to bypass the unit is to replace the LC circuit with a simple resistor equal in value to the input resistor (as in Fig. 1-31). The switching action can be accomplished by using a 1 kΩ potentiometer for the feedback control which also incorporates an SPDT switch. The switch must be wired such that the resistor replaces the tuned circuit when the wiper arm of the potentiometer is at ground potential.

Summary

The simple Q-multiplier circuit described can be used for a variety of purposes besides that of improving IF selectivity. It is useful for improving the Q and selectivity of a variety of tuned circuits as they might be used in FSK converters, audio filters for distortion test, etc. Stagger-tuned circuits used in series can be formulated to provide a variety of bandpass shapes, often replacing more expensive components where bandpass shape factor is not important.

Chapter 2
Amateur Receivers and Converters

THE 2Q COMMUNICATIONS RECEIVER

A strong desire to duplicate the popular Drake 2-B receiver in transistor form prompted the building of the receiver shown. It is the result of over two years of experimental design, building, rebuilding, testing and listening. The block diagram in Fig. 2-1 closely resembles that of the Drake, and for that reason it's named the 2Q.

The completed receiver is a triple conversion superheterodyne, covering all amateur bands 10 through 80 meters. It has excellent sensitivity, selectivity and stability. Cross modulation has been reduced to a minimum by the use of FET transistors in both the RF and first mixer stages. Such features as bandpass tuning, FET detector, S-meter, AGC and a 100 kHz crystal calibrator are included.

The circuit shown in Fig. 2-2 is actually the result of two that were built. The first design, following the usual transistor circuit theory, matching impedances, etc., resulted in a receiver that lacked the necessary sensitivity and selectivity. Cross modulation was also a problem because bipolor transistors were used in the front end. The second design is the result of a concentrated effort toward obtaining maximum selectivity by the use of small capacity coupling where possible, high Q tuned circuits, and tapping collectors down on the coils to perserve their Q. Cross modulation was reduced to a minimum by using FET transistors in both the RF and first mixer stages and by using a separate RF gain control.

The Circuit

Capacitive coupling is used throughout the front end (preselector). It uses high-Q toroid coils and slug-tuned coils. The simple switching provides the necessary selectivity and ease of adjustment desirable when compact construction is used. Ami-Tron toroids were not used for the 15- and 10-meter

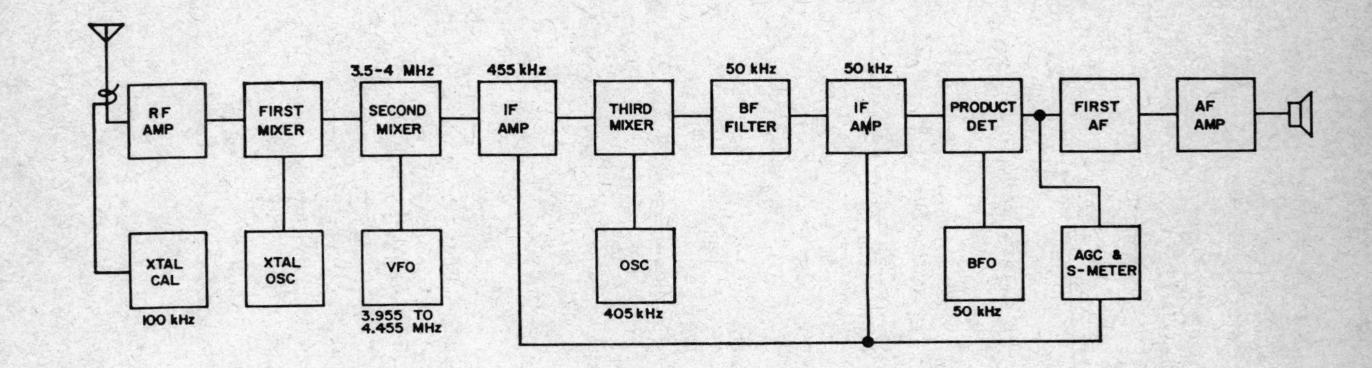

Fig. 2-1. Block diagram of the 2Q, a completely transistorized communications receiver of modern design using FETs in the front end.

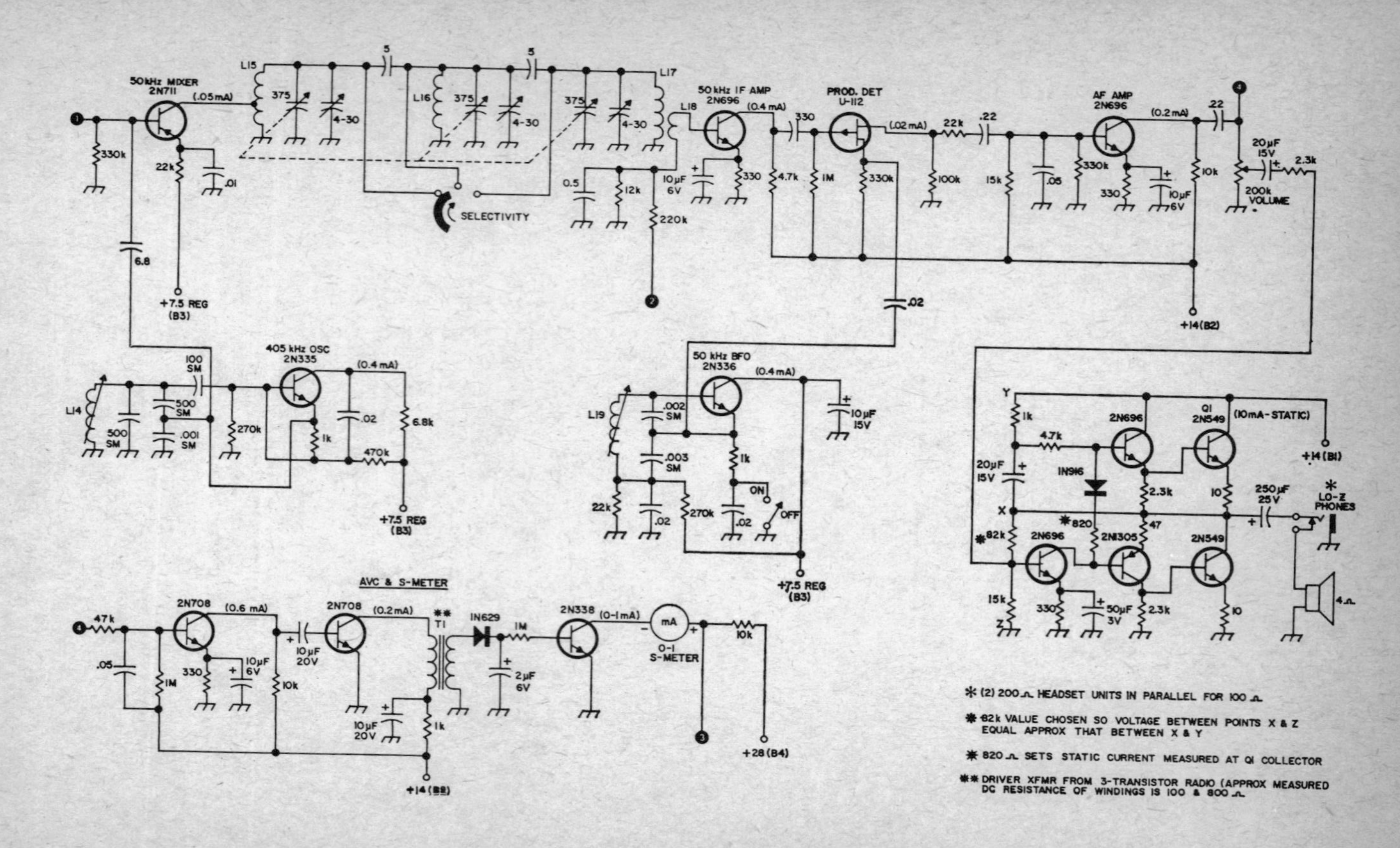

50kHz MIXER 2N711
(.05mA)
405 kHz OSC 2N335
(0.4 mA)
SELECTIVITY
50kHz IF AMP 2N696
(0.4 mA)
PROD. DET U-112
(.02 mA)
AF AMP 2N696
(0.2 mA)
200k VOLUME
+7.5 REG (B3)
+14 (B2)
50 kHz BFO 2N336
ON
OFF
AVC & S-METER
2N708
(0.6 mA)
(0.2mA)
IN629
2N338
(0-1mA)
0-1 S-METER
+28 (B4)
+14 (B8)
Q1 2N549
(10mA-STATIC)
2N696
2N1305
2N549
IN916
LO-Z PHONES
+14 (B1)
* (2) 200 Ω HEADSET UNITS IN PARALLEL FOR 100 Ω
* 82k VALUE CHOSEN SO VOLTAGE BETWEEN POINTS X & Z EQUAL APPROX THAT BETWEEN X & Y
* 820 Ω SETS STATIC CURRENT MEASURED AT Q1 COLLECTOR
** DRIVER XFMR FROM 3-TRANSISTOR RADIO (APPROX MEASURED DC RESISTANCE OF WINDINGS IS 100 & 800 Ω

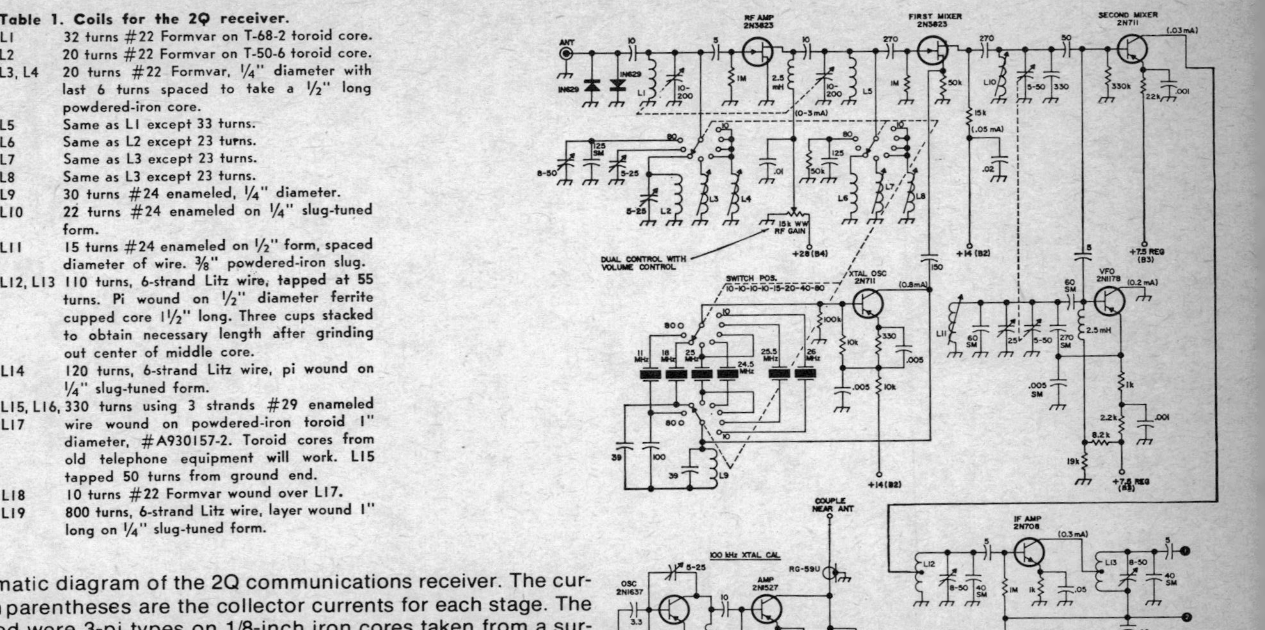

Table 1. Coils for the 2Q receiver.

L1	32 turns #22 Formvar on T-68-2 toroid core.
L2	20 turns #22 Formvar on T-50-6 toroid core.
L3, L4	20 turns #22 Formvar, 1/4" diameter with last 6 turns spaced to take a 1/2" long powdered-iron core.
L5	Same as L1 except 33 turns.
L6	Same as L2 except 23 turns.
L7	Same as L3 except 23 turns.
L8	Same as L3 except 23 turns.
L9	30 turns #24 enameled, 1/4" diameter.
L10	22 turns #24 enameled on 1/4" slug-tuned form.
L11	15 turns #24 enameled on 1/2" form, spaced diameter of wire. 3/8" powdered-iron slug.
L12, L13	110 turns, 6-strand Litz wire, tapped at 55 turns. Pi wound on 1/2" diameter ferrite cupped core 1 1/2" long. Three cups stacked to obtain necessary length after grinding out center of middle core.
L14	120 turns, 6-strand Litz wire, pi wound on 1/4" slug-tuned form.
L15, L16, L17	330 turns using 3 strands #29 enameled wire wound on powdered-iron toroid 1" diameter, #A930157-2. Toroid cores from old telephone equipment will work. L15 tapped 50 turns from ground end.
L18	10 turns #22 Formvar wound over L17.
L19	800 turns, 6-strand Litz wire, layer wound 1" long on 1/4" slug-tuned form.

Fig. 2-2. Schematic diagram of the 2Q communications receiver. The currents shown in parentheses are the collector currents for each stage. The RF chokes used were 3-pi types on 1/8-inch iron cores taken from a surplus computer board although miniature 2.5 mH units should work ok. Coil values are given in Table 2-1. Later experimentation indicates that the bias network used with the 2N708 455 kHz IF amplifier is not too tolerant of different transistors. Removing the IM base-bias resistor and replacing it with a 330K resistor and a 27K resistor from base to ground should help.

bands due to the lack of space for the necessary trimmer capacitors, but their use is definitely recommended for all bands. The selectivity and stuffing ratio gained by their use is very necessary. The tuning capacitor, a two-gang TRF unit, was reduced to 200 pF per section. Space for the RF choke was solved by placing it in the crystal oscillator compartment.

The FET mixer, using source injection, is capacitively coupled to the second mixer. This circuit was preferred to a gate injection circuit. The only FETs available were N-channel 2N3823s, but possibly some of the cheaper ones will work as well. Alignment of the front end is simply a matter of adjusting turns, spacing, and trimmer capacitors, until the amateur bands are staggered across the preselector dial.

The 3.5 MHz-4.0 MHz variable IF mixer and oscillator section, consists of a high-C Colpitts oscillator, and a base injected mixer, with an output at 455 kHz. Only the highest quality components should be used here, since it is a major frequency determining circuit.

Only one stage of amplification was found necessary for the 455 kHz IF section. The mixer is capacitively coupled with base injection at 405 kHz from a high-C Colpitts oscillator giving a 50 kHz output. Here again the oscillator is a major frequency determining circuit and care should be used in its construction. The 455 kHz IF coils can be any high-Q center tapped units, preferably using toroids or cup cores. This is a good spot for a mechanical filter.

The bandpass tuner was constructed using coils wound on 1″ diameter powdered iron toroids from an old telephone company audio filter. The ones used were blue and numbered A9301572. The tuning is done with a three-gang TRF type broadcast tuning capacitor, with a stop added to limit its travel to about 20 degrees, starting from maximum. The switch uses a hollow 1/4 inch shaft, with the TC shaft being operated through it.

The FET detector using a P-channel U112 or 2N2497 has plenty of BFO injection and works very well on SSB.

Good S-meter action and a certain amount of gain control is provided by the circuit shown by simply reducing the amount of voltage applied to the IF transistors.

From this point, the rest of the receiver is simply audio, six transistors in all, with a transformerless audio circuit taken mostly from a GE transistor manual. The power supply, one left over from another project, is no doubt over-filtered. Any well-filtered DC source of 14V and 28V will do (see Fig. 2-3). The receiver draws 20-125 mA depending on volume. The dial lamps use an additional 40 mA each. The receiver will work well on only 12-14V, but the S-meter and AGC will be out of the picture.

Choice of Transistors

The transistors used are by no means the only ones which will work in the receiver. Choice was made largely from tests with the ones which were

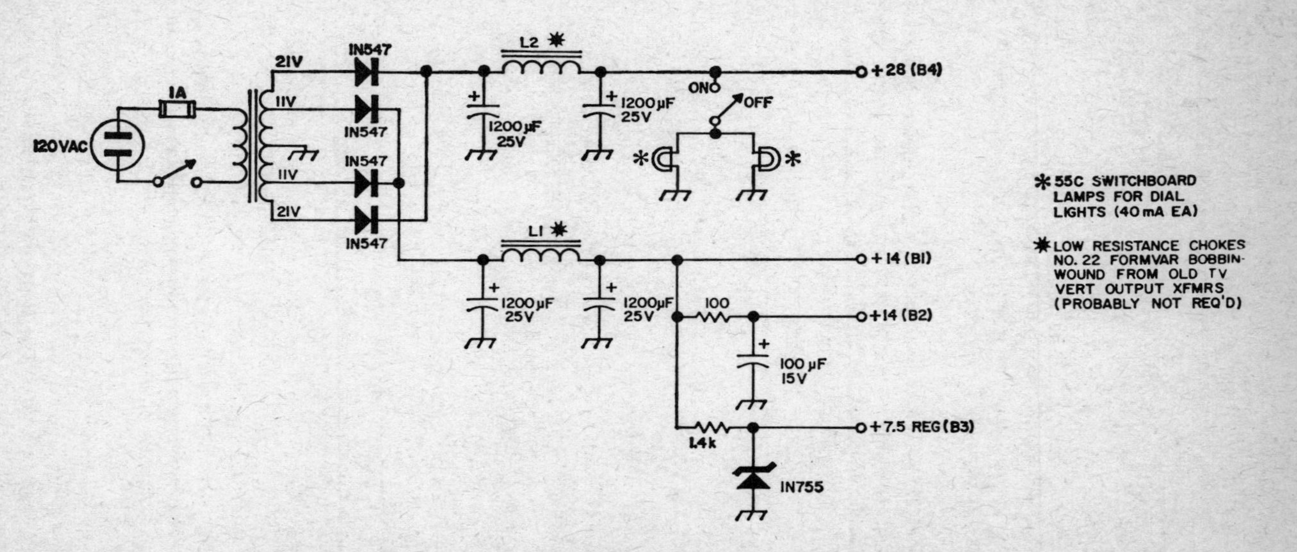

Fig. 2-3. AC power supply for the 2Q communications receiver.

available in a transistor junk box. Either PNP or NPN will work in most circuits, NPN being preferred in most cases for IF and oscillator transistors. Oscillator types should be those that have no internal connections to the case. The use of sockets for all transistors is highly recommended.

Construction

The receiver cabinet measures 8 1/2″ long × 6 1/2″ high × 6 1/2″ deep. The receiver is divided into a number of subassemblies mounted on a main chassis, made of 14-gauge aluminum. The subassemblies are of 21-gauge aluminum.

Only the 50 kHz IF amplifier and audio stages were built on the main chassis. The S-meter and AGC circuitry were mounted on the back of the S-meter. Figure 2-4 is a rough layout of the front panel.

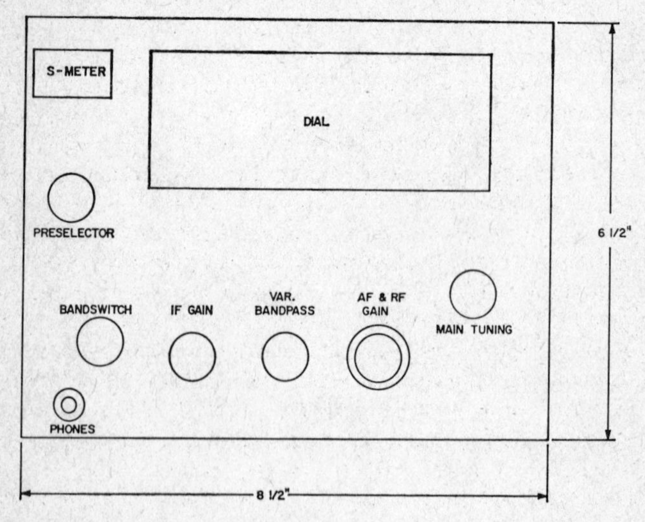

Fig. 2-4. Front panel layout used by W5ETT in the original model of the 2Q receiver.

The slide rule dial has a tuning rate of 45:1 or 45 turns of the tuning knob to cover 500 kHz. This gives at least 25 revolutions on the 40 and 20 meter bands. The mechanism consists of a weighted knob on a 1/4″ shaft driving a 2″ rubber tired wheel (Jenson #J1490-01) on a 1/4″ shaft driving a dial cord to a 3 1/2″ dial drum on the tuning capacitor. The dial scale was made on white paper (pasted to a piece of stiff cardboard) using a black ball point pen and a typewriter. The dial drum was made from a reinforced, nickle-plated lid from a peanut butter jar.

Conclusion

No wild claims shall be made for this receiver except to say it is a good homebrew receiver. Only 5 feet of wire strung up in the shack has been found necessary for good reception. Many hours were spent just listening and hearing signals that could never be heard with an old 14-tube homebrew receiver.

A SOLID STATE HIGH FREQUENCY REGENERATIVE RECEIVER

This receiver (see Fig. 2-5) tunes from 14 to 35 MHz, a range that covers three ham bands, short wave stations and public service frequencies. The detector uses the readily available Motorola HEP 55 NPN VHF RF transistor. The audio amplifier stages are a single RCA 3020 linear integrated circuit. This is an entertainment grade IC and is quite a bargain.

The receiver is built on a 4 × 6 × 2 in. aluminum chassis. The audio circuitry, with the exception of the audio choke, is mounted on a 2 × 3 in. piece of Vectorboard. The Vectorboard is mounted to the chassis on threaded stand-offs. The input audio choke L2 is just the primary of a small 100K to 95K audio transformer. The secondary is not used. Practically any small audio transformer having an impedance in one winding of from 10 to 100K can be used. An alternate approach is to use transformer coupling with a small 20K to 50K interstage transformer.

All of the RF components except the tuning capacitor are mounted below chassis. A transistor socket is used to permit experimentation with a variety of transistors. Actually a socket is a good idea unless you have had some experience in soldering to semiconductor leads.

You will notice that winding information is given for two different types of coils. The best bet is to use the small toroid cores which are carried by most major parts houses for about 10¢ each. If your local parts house doesn't carry them, use the alternate method. There isn't much difference in performance.

After wiring the circuit, carefully recheck your work. Remember, transistors, unlike tubes, are not forgiving about wiring errors. Now you are ready to hook up the antenna and apply power. Your best bet is to buy a battery holder that holds six D cells. While the idling current of the receiver is only 23 mA, peak currents are in the order to 50–60 mA. The cheap 9V transistor radio batteries aren't designed for this kind of use. Their internal resistance is quite high, and you will probably experience motorboating after short periods of use. The premium grade 9V batteries perform well, but their cost is high.

Connect the antenna. A 40 to 50 ft wire works well. Turn on the power and advance the regeneration control until a clean hiss is heard. Tune in a signal and alternately adjust the tuning and regeneration controls for best reception.

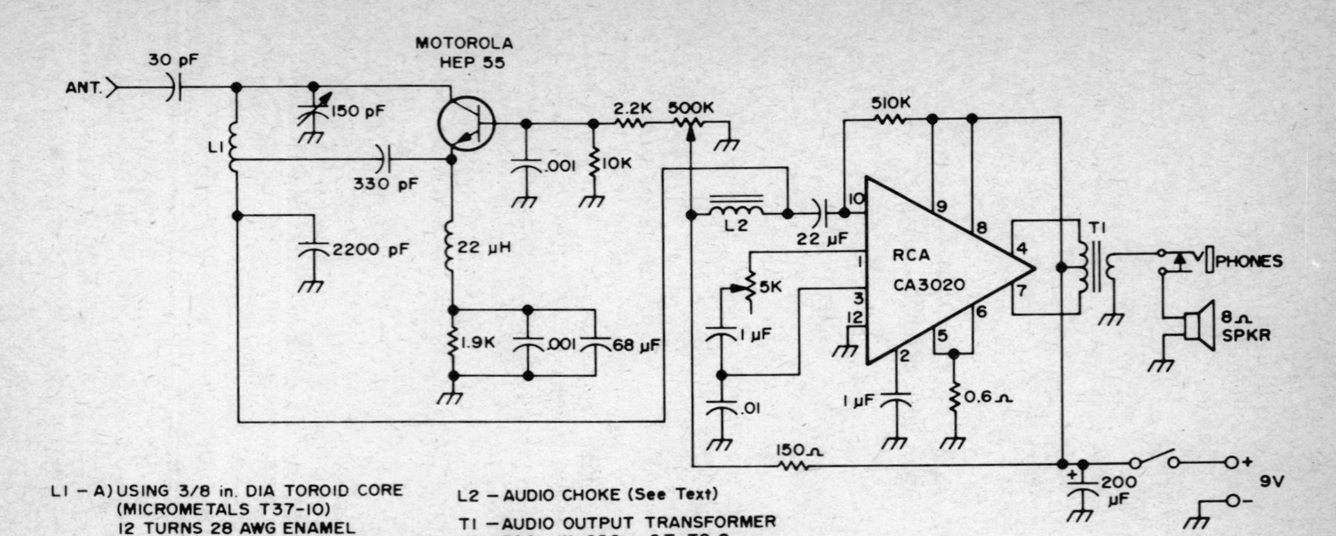

Fig. 2-5. Schematic of the regenerative receiver. L-1 A) using 3/8 in. toroid core (Micrometals T37-10) 12 turns No. 28 enamel tapped 3 turns from cold end; B) using No. 12 solid wire 8 turns 1/4 in. diameter tapped 2 turns from cold end; L-2 audio choke (see text); T-1 audio output transformer 500 mW 250Ω CT to 8Ω.

A little practice will give you the knack of it. Performance of this little receiver has been quite good, receiving signals from all over the world, not to mention the fun of listening to the local sheriff's department.

LISTEN IN ON A TWO-METER FM REPEATER

For those who don't have a two meter FM receiver and want to listen to the repeater (if you have one in your area), this little converter will do a very good job into an AM radio.

You will recognize this converter as being similar to police converters. (In fact, changing the crystal and repeaking the antenna coil should put a police converter in the ham band.)

The unit is built on a printed board 2 1/8 by 2 5/8 inches. The capacitors are all from a junk box and include small round tubular, round flat ceramics and small Mylar types.

Q1 and Q2 are 2N2996s, but 2N1141s and TIXM10s all work fine on three volts. Three volts was selected because this converter works into a radio that runs on two penlight cells. If you plan to use 6 or 9 volts, R1, R2, R3 and R4 will have to be changed to other values (found by experiment and measuring base voltages and collector currents). This unit using 2N2996s draws 8 1/2 mA at three volts.

After building, determine by GDO or receiver if the crystal is oscillating at three times its frequency, or if it is oscillating at crystal frequency, it will be ok at 3 times. Then place the converter near your BC receiver loop antenna and tune it to the approximate frequency (your IF in this case) or, using a signal generator or a signal from the repeater, adjust the converter loopstick for maximum signal or noise. With a signal picked up from repeater, adjust C1 for maximum.

You are receiving FM by slope detection, so tune your BC radio for the clearest reception. Some will be clearer than others, depending on how close to the frequency they are. You won't receive them as crystal clear as an FM receiver, but this will let you listen in.

Roughly, slope detection works like this (this is a quick and short explanation). Your 455 kHz IF passband looks something like Fig. 2-6 with the desired AM signal (when receiving AM) carrier at "x" or the top part of the curve while the FM signal should be somewhere around "y" (just high enough on the curve to give a signal through the IFs). Then, during FM modulation, the frequency swing can be between y and x, producing a varying signal strength at the detector. As the modulation approaches x on the curve, the IF is operating with maximum gain and when it is at a y, the IF gain is lower. This varying signal is detected by the AM detector (in other words FM modulation varies the gain of the BC IF as the signal rides up and down the curve and produces an amplitude voltage at the detector).

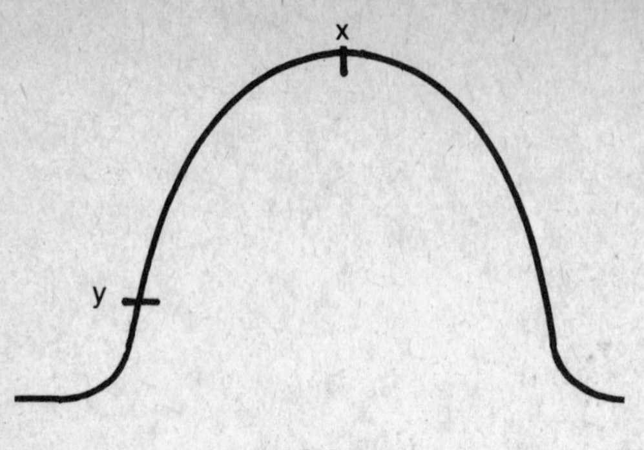

x = 455 KC peak

Fig. 2-6. IF passband used for slope detection of FM signal. Wideband FM will be muffled due to narrow frequency swing between y and x.

Some radios overload easier than others, so vary the distance between converter and BC radio for the best results (see Fig. 2-7).

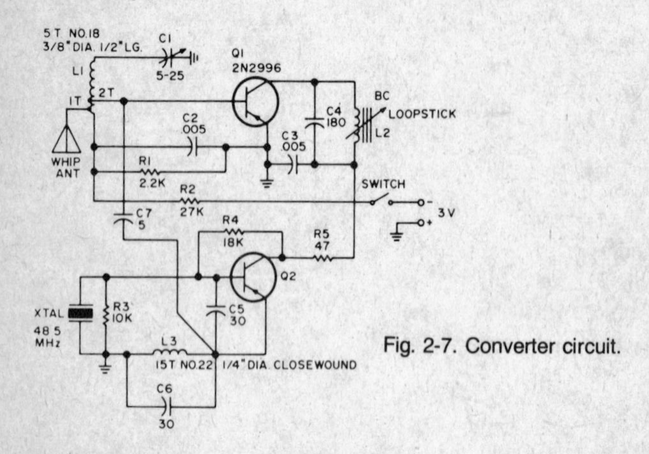

Fig. 2-7. Converter circuit.

This converter can be operated into a transistor 10.7 MHz IF. For those interested, the crystal can be 10.7 lower or higher than the received frequency. With a repeater output of 146.940 MHz, let's do a little math. Remember that we are using the third overtone crystal circuit as a tripler (that's why we divide by 3).

146.940 + 10.7 = 157.640 ÷ 3 = 52.5476 MHz = crystal on high side.

146.940 − 10.7 = 136.240 ÷ 3 = 45.4133 MHz = crystal on low side.

Either crystal will permit the 146.940 signal to come in at 10.7 MHz, the IF of most FM sets. The exact frequency of the crystal can be varied according to how much you can swing your IFs from 10.7 MHz.

Getting back to you who plan to use a BC radio. You will have to use this math to suit your crystals and also keep away from bc stations (pick a clear spot on the BC dial). This rig used a 48.716 crystal, giving an injection frequency of 146.148 MHz on the low side of the repeater frequency. 146.94 − 146.148 = 792 kHz.

For extreme ends of the BC band, using the xtal on the low side and the high side, you get:

146.940 − 1.600 = 145.340 ÷ 3 = 48.4466 MHz low side.
146.940 + 1.600 = 148.540 ÷ 3 = 49.5133 MHz high side.
146.940 − .550 = 146.390 ÷ 3 = 48.7966 MHz low side.
146.940 + .550 = 147.490 ÷ 3 = 49.1633 MHz high side.

A crystal between 48.7966 and 48.4466 will permit the repeater to come in between 550 kHz and 1600 kHz with the oscillator working on the low side of the incoming signal at the antenna, while a crystal between 49.1633 and 49.5133 will permit the repeater to come on the BC band between 550 and 1600 kHz with the oscillator working on high side of the signal at the antenna. So, in other words, if you have a third overtone crystal in your junk box between 48.4466 and 49.5133 you can put the repeater into theBCband!

CONVERTING THE AC/DC FOR WWV

Perhaps the single most complete piece of gear to invade the junkbox is the standard AC/DC five-tube table radio. Millions of these radios have been produced over the years, and they inevitably end up in the junk heap or are presented to the "ham" by good intentioned friends because they know "you like radios."

They go bad for any number of reasons, but mainly the problem is failure of tubes, electrolytic capacitors, or the output transformer. Other simple problems such as cabinet breakage or dial cord problems also tend to render the AC/DC useless to the owner.

The radio circuitry is almost always the same. An antenna loopstick or ferrite rod and coil forms the basis of the RF input circuit which is fed directly into a "converter." The antenna coil is tuned across the BC band by one section of the control grid and cathode of the converter and is tuned by the other section of the variable capacitor. When the two signals are mixed the resulting output of 455 kHz is produced, detected, and amplified. In most cases this converter tube is a 12BE6 and operates in conjunction with a 12BA6, 12AV6, 5OC5, 35W4. Not coincidentally, the total filament voltage in series is 115V (see Fig. 2-8).

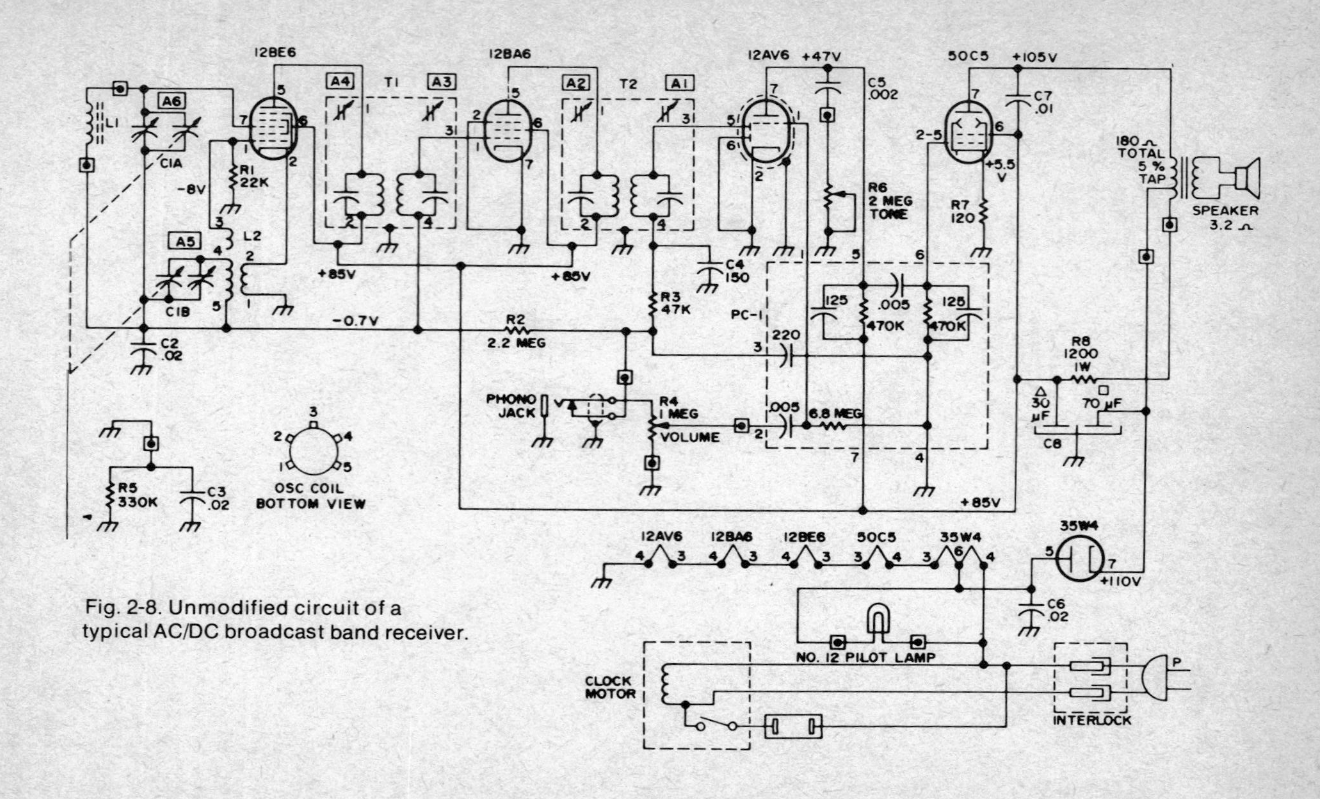

Fig. 2-8. Unmodified circuit of a typical AC/DC broadcast band receiver.

Converting the AC/DC to any fixed frequency between 1 and 30 MHz couldn't be simpler.

Remove the 365 pF variable, the ferrite coil, the oscillator coil, and all wires connecting to pins 7, 1, and 2 of the 12BE6. Next, remove all the coil windings from the oscillator coil form. This form is handy for the new oscillator coil and usually has mounting lugs and brackets already provided. When removing components, carefully trace all the ground wires to a single point. This is the grounding point that should be used, as it is usually isolated from the 115V AC line.

Using 26 AWG copper enamel wire, first wind a 5-turn link on the oscillator coil form connecting it to two unused lugs. Then scramble-wind the oscillator coil over the link and connect one lead to the ground lug of the link and the other to a third unused lug. Parallel a mica capacitor of between 50 and 200 pF and a 0–30 pF ceramic trimmer across the oscillator coil. Remount the coil and connect the link to the cathode (pin 2), and tie the oscillator coil to the control grid (pin 1) through a 47 pF capacitor. Insert a 22Ω, 1/2W resistor from pin 1 to the grounding point. Check the coil for resonance with a grid dip meter and then add 455 kHz for the indicating frequency.

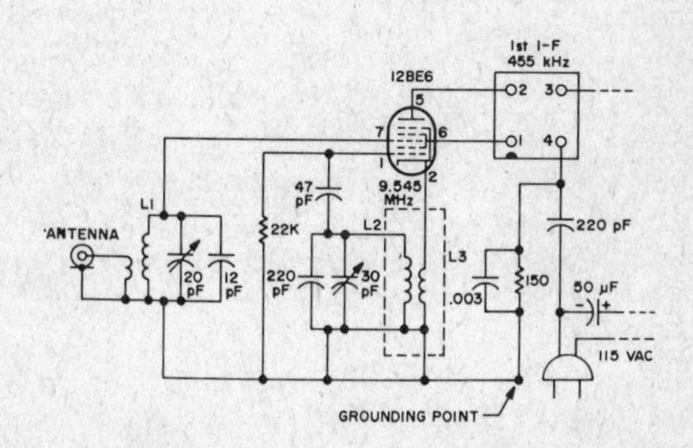

Fig. 2-9. Modified circuit of 12BE6 converter for WWV.

The RF section is all that remains; it consists of 30 turns on a 3/8 in. coil form paralleled by an APC 20 pF variable and a 12 pF mica capacitor, making the coil resonate between 11 and 9 MHz. This assembly can be mounted on the bracket used to support the old 365 pF variable capacitor. Figure 2-9 shows the revised circuit for the 10 MHz WWV receiver. Wiring is by no means critical and mounting of an assortment of parts can be left to the constructor's imagination. Care must be taken, however, to isolate the chassis ground from all components. If an external antenna is used do not ground either lead and make sure the "safe polarity" is established when plugging in the set.

Operation

Rough adjustment can be made with the grid dipper when putting on frequency. The antenna trimmer is used to peak the signal and the oscillator trimmer adjusts frequency.

The IF coils need not be touched as they still operate at 455 kHz. The new WWV monitor is quite stable and produces excellent quality, even during the daylight hours. In tuning the oscillator trimmer slightly up and down the band, equally pleasing results were realized. No difficulty in producing a fixed receiver for any desired frequency should result, as long as the oscillator coil and antenna coil are properly adjusted.

Some future modifications might involve replacing the 12AV6 detector tube with diodes and inserting a tube with equivalent filament current and voltage for a 455 kHz BFO. Also, the detector stage could be disconnected from the audio circuit and capacitively coupled to the speaker for a code practice oscillator. In all there are, and will continue to be, many uses for this common receiver.

FAIL-SAFE SWITCHING

Have you ever had a desire for a fail-safe method of turning on the filaments prior to turning on the transmitter high voltage supply? Or how about a guarantee that the bias power supply will always be turned on first, so that bias voltage would be present on the transmitting tubes prior to the time that the plate supply was turned on? Perhaps you need to turn on the final plate supply prior to turning on the modulator plate supply, in order to protect the modulation transformer. Operating the modulator without a load can cause excessively high voltages to appear across the modulation transformer windings. Result—shorted transformer.

At this point someone is bound to say, "Who uses a modulator in this day and age of Sideband?" Well for the SSB boys; you may want to work up a fail-safe arrangement to turn on the final plate supply that 4-1000 prior to turning on the screen supply. If you aren't careful—BLOOEY——the screen current will rise to excessive values and there goes another expensive final.

For most home constructed equipment, the only parts required are two DPST toggle switches.

The system can be carried a bit further and by using three each 4 pole switches any three load circuits can be turned on in a 1-2-3 order. No matter which switch is operated first, the no. 1 load is turned on. The second switch to be operated turns on the no. 2 load, and the last switch to be operated turns on the no. 3 load. At this point, operating any one of the switches to the "off" position will disconnect the no. 3 load. The second switch to be turned "off" will disconnect the no. 2 load, etc.

Other uses for this switching arrangement will no doubt come to mind, such as:

1. Always turn on the safelight in a photographic dark room prior to turning on the enlarger.
2. Always turn on the cooling fan ahead of the filament transformer when using air cooled tubes.
3. Turn on your CB receiver prior to turning on the transmitter. Many CBers leave their receivers on for monitoring purposes, and with the squelch operated, it may be difficult for non-technically minded persons—your mother-in-law, for instance—to know whether they turned on the transmitter or the receiver. Just explain to them to operate either switch to turn on the receiver.
4. How about turning on the RTTY power supplies prior to turning on the printer motor?

If the two or more pieces of equipment are widely separated, it might be best to mount the switches and provide outlets in a small Minibox. Be sure to label the outlets 1, 2 or 3 to indicate which outlet will energize first, second, or third. Plug the equipment into the Minibox and connect the Minibox cord to the electrical wall outlet.

You can also fool your friends with such a box. Plug two or three lamps, as the case may be, into the outlets and let them guess which switch controls which light.

BACK TO MOTHER EARTH THE EASY WAY

The importance of a good earth ground to an amateur radio station was iterated in QST (Aug 1970, p. 38). The point is an important one. Improper grounding can cause weird radiation patterns, severe energy losses, and in some cases a serious shock hazard to the operator. QST's solution to the grounding problem is an excellent one: a 10 ft length of galvanized iron rod, sledgehammer-driven into the ground.

My problem is that I have had difficulty finding 10 ft tall acquaintances willing to come over and wield the sledge.

The solution, it turns out, is simple. From your neighborhood hardware store, buy a length of 1/2 in. rigid copper pipe and a brass female hose connection to match. While you're there, pick up a bit of solid solder and a length of the heaviest wire you can find.

Use a torch to solder the fitting to one end of the pipe. Clean the pipe and fitting, tin the pipe all the way around with the solid solder, and put the preheated fitting in place. As it's only temporary, the joint needn't be "plumber approved."

The site for the ground rod should be chosen to give a short run to the shack (or in the middle for an array) and to have overhead clearance to wield the 10–20 ft pipe for insertion. It should be more than a foot or so from the basement wall to insure clearing the footings, and with luck, it should be within hose-reach of the nearest outdoor faucet. Hook up the hose, turn on the water full blast, and with the hose-end of the pipe in the air, drop the pipe straight into the ground. It will sink with a minimum of effort as the flowing water bores the hole ahead of it.

Judicious wiggling will clear most underground obstructions. If you hit one too big to pass, lift the pipe out and try again with a new hole. Leave about a foot of pipe sticking out of the ground.

Unhook the hose, let the water drain away, and unsolder the hose connection. The bus wire can now be soldered to the top of the pipe while the solder is still melted. Foot pressure on top of the pipe will now drop it even with the ground where it will pose less of a danger to kids and burglars sneaking around outside the house in the dark.

The ground bus should be routed to the shack by the most direct route, keeping in mind that it is an RF connection as well as a DC one. Many plush installations run the ground bus directly to a grounding bar, which may be fitted with pin sockets or binding posts, and runs across the back of the work table.

After the rod is installed, it may be some time until the soil resettles around the rod. This action can be hastened by gently flooding the area with water, as with a lawn sprinkler, and by tamping the area around the rod with a broomstick.

Checking the effectiveness of the final installation can present a problem, in that if you had a really good ground to begin with, you wouldn't have put in another. Probably the best reference is the house ground, used by the power and telephone companies, and connected (the green lead) to each box housing an electric outlet or switch (it says here). An ordinary ohmmeter is useless for this measurement. Stray electric fields can build up voltages that, with a high impedance instrument, can be positively frightening. Rather, connnect a low voltage source such as a variac-fed filament transformer in series with an AC ammeter, and hook this between the house ground and your ground. Monitor the applied voltage with a voltmeter and compute the resistance to your ground using $R = E/I$.

If the ground doesn't seem to be as effective as you'd like, there are three remedies. For the short term (perhaps longer with luck), soaking the area around the ground rod with water should increase subterranean conductivity. If this doesn't help, add salt. Copper sulfate is good if it is convenient, otherwise table salt is fine. Spread a pound or so around the rod and soak it in with water. The last resort is a longer rod (or another separate ground rod). To lengthen the rod, merely lift it out of the ground a bit, unsolder the bus lead and replace it with a coupling. Solder on another length of pipe fitted with the hose connector on the other end, and let 'er go.

Unfortunately, grounds, like everything else, deteriorate with age. If salt is used, it eventually leaches away, and often the pipe will build up a high resistance layer. Check the ground resistance once a year or so, when you "pull maintenance" on the rest of the gear in the shack. If you maintain the system this way, you too will be able to tell your friends that you are "well grounded" in amateur radio.

RAPID RECEIVER CONTROL SWITCHING

Amateurs who are really interested in improving their receiving setup for DX-hunting or other purposes usually end up with a receiver containing many modifications. These modifications, such as selectivity and preamplifier devices, may either be internally mounted in the receiver or contained in outboard enclosures. All the modifications may well prove their value in improving receiver performance, but one problem which usually develops is the control of the controls for the accessory circuits. The phrase "control of the controls" may sound a bit strange, but one of the greatest problems in DX-hunting with a receiver is to be able to concentrate on the tuning itself and not be distracted by having to look away to see the setting of other controls or, in fact, to operate other controls.

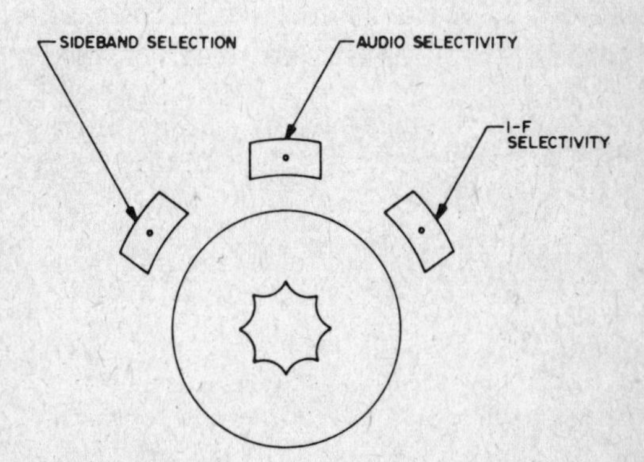

Fig. 2-10. Possible touch place arrangement on the front panel of a receiver around the main tuning knob.

The simple circuit described allows one to activate by touch control *any* receiver function which is *switch*-operated. The touch plates for different functions can be grouped around the receiver's main tuning knob, as shown in Fig. 2-10, so that one need not remove one's hand from the tuning knob in order to reach any touch plate. The choice of functions which one may wish to control in this manner is a matter of individual choice, but usually it will be such functions as IF bandwidth, audio selectivity, sideband selection, etc. Placing a finger on a touch plate will activate the function for which that touch circuit is wired for a time period of 5–10 seconds, then the circuit function will automatically revert to its original state. This method of operation was chosen so that manual override of the touch circuit would be possible at any time by using the regular control switch for a receiver function. Also, the time period chosen allows sufficient time to operate the regular control switch if it is desired to retain the function that was activated and yet the time period is short enough so that, if the circuit function chosen proves not to be useful, it will drop out before too much reception is lost or distorted.

Circuit

Figure 2-11 shows the simple circuit of the touch control unit. It consists of a basic SCR switching circuit utilizing a 3N60, or any similar SCR. Placing one's finger on the touch plate will provide enough pickup voltage at the cathode gate of the SCR so that the SCR will fire and the relay coil will be energized. Any relay can be used which has a 100–200Ω coil and does not require more than about 100 mA operating current. A 1 KΩ potentiometer placed between the anode gate and the power supply side of the relay coil can be used to regulate the sensitivity of the touch plate. Normally, the SCR when fired would continue to conduct indefinitely, unless a switch were provided to "unlatch" the SCR. To provide a means for the SCR to automatically stop conducting after a definite time period, power to the SCR is supplied via an RC network consisting of a 220Ω resistor and 1,000 μF capacitor. Once the SCR has fired it will continue to conduct until the charge on the capacitor decreases to the point where enough current does not flow through the SCR to allow it to

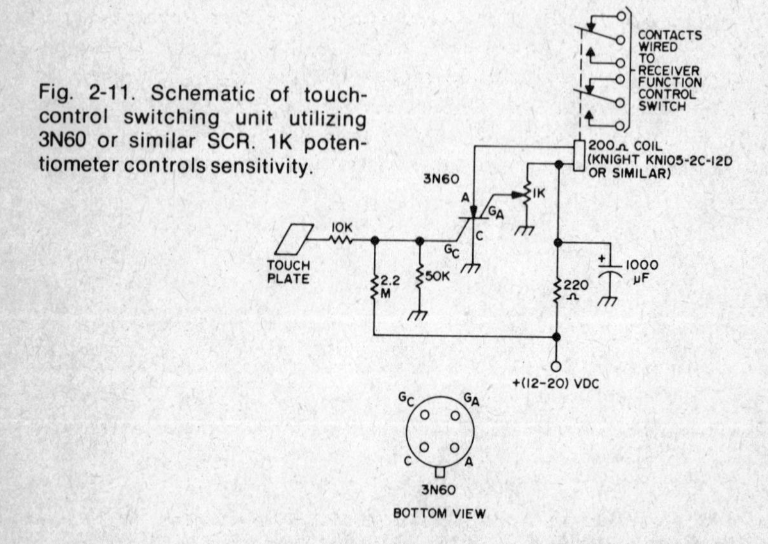

Fig. 2-11. Schematic of touch-control switching unit utilizing 3N60 or similar SCR. 1K potentiometer controls sensitivity.

remain in a conducting state. When this point is reached, the SCR will stop conducting. Depending upon the characteristics of the relay used, the relay will be de-energized shortly before or at the time the SCR stops conducting. Once the SCR stops conducting, the 1000 μF capacitor will again charge via the 200Ω resistor. The value of the components for the RC network can be chosen as desired for the time delay desired.

Construction

It is suggested that the circuit first be breadboarded so one can determine the proper components to use for the RC network for the time delay desired. The value of the components will vary depending upon the relay used and the supply voltage. The final unit can be assembled on a piece of perforated board stock and mounted in any convenient location in the receiver. The supply voltage need not come from a well-filtered source and can be taken from any available source in the receiver. If necessary, a simple halfway rectifier circuit from the filament line can be used.

The most difficult aspect of the construction of the unit will probably involve that for the touch plate itself, since it involved the question of to what degree one is willing to modify the front panel of a receiver. The area of the touch plate need be no larger than that of a penny, or dime coin. Probably the best way to try the circuit for its usefulness, in fact, is to use something like a penny coin with a wire soldered to it as the touch plate. The coin can be placed on the receiver front panel near the tuning knob by means of double-backed adhesive tape and the wire from the coin run under or above the front panel. A more permanent and far neater installation can be made for the touch plate, as shown in Fig. 2-12. Any piece of metal sheet about the area size of a penny (not necessarily a round shape) is glued to the receiver front panel with an insulated backing of tape or other material. An extremely small hole is then drilled through the metal plate and the front panel just large enough to pass size #30 or smaller enameled magnet wire. The end of the wire is

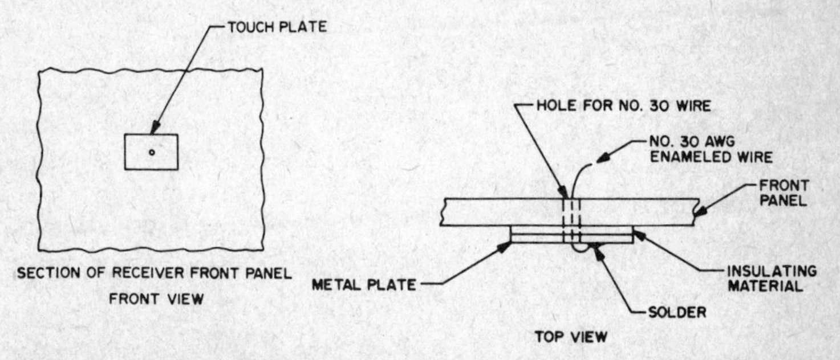

Fig. 2-12. Details of construction of a touch plate which can be mounted on a receiver front panel.

touch-soldered to the metal plate. In this way, if the metal plate is removed later, the very small drilled hole can easily be covered with touch-up paint, or other material.

The wiring of the relay contacts depends upon the control switch wiring in a given receiver. In most cases where simple on—off type control functions are involved, the relay contacts can simply be wired in parallel with the toggle switch contacts. In more complicated situations, as for instance when a multiple position rotary switch is used to set the receiver selectivity, the relay contacts must be wired across those positions of the switch for the two selectivity positions it is desired to alternate between using the touch switching. A study of the schematic for a specific receiver will reveal may useful ways in which the relay contacts can be wired into various control switches.

Summary

Touch switching of the most frequently used switch-type receiver controls can add a great deal of convenience to receiver tuning, especially under circumstances such as DX-hunting and contest work, where rapid control of various receiver functions is desired. Although touch switching is probably not the ultimate answer, it is at least a step in the direction of solving one of the major design faults with most receivers. That design fault, of course, is the grouping and nature of the controls for receiver functions not being really operator-engineered for rapid utilization.

THE BALL OF WAX—A CALIBRATOR

By now, the idea of integrated circuits as dividers in 100 kHz and 1 MHz crystal calibrators should be pretty well established. The acceptance of ECL and RTL integrated circuits by hams and experimenters is underlined by the fact that both of these logic families are now available in Motorola's HEP line, at most any radio parts store.

The calibrator to be described here is one that uses an admixture of bipolar transistors, a junction FET, MOS-Digital ICs, and MOSFET, and a linear IC. Such a unit could only be called a "ball of wax," because of the variety of components. In spite of the differences between the various components, they complement each other very well.

The finished calibrator provides the operator with a quite useful choice of frequency calibration marks: 200 kHz, 100 kHz, 50 kHz, and 25 kHz spaced

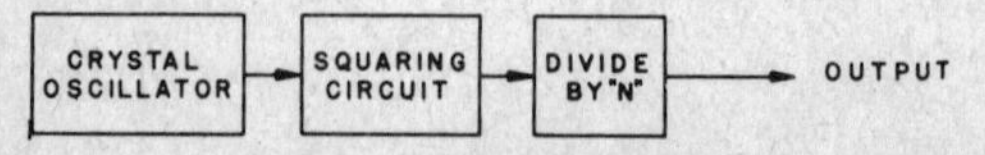

Fig. 2-13. Usual design circuit.

spectrum lines. This set of calibration markers is a convenient one for use with the general-coverage type of HF receiver.

The design philosophy of this calibrator is somewhat different than most others published in recent years. Usually, the crystal oscillator is followed by a squaring circuit, and *then* by dividers, as in Fig. 2-13. This is the method used in references 1 and 2. However, in the "The Ball of Wax," since MOS ICs were used as dividers (whose uppermost counting speed is only 500 kHz), the dividers were followed by a Schmitt trigger. This system is shown in Fig. 2-14. The Schmitt trigger circuit speeds up the waveform, assuring that the rise-time is fast enough to produce good, useable harmonics throughout the HF bands.

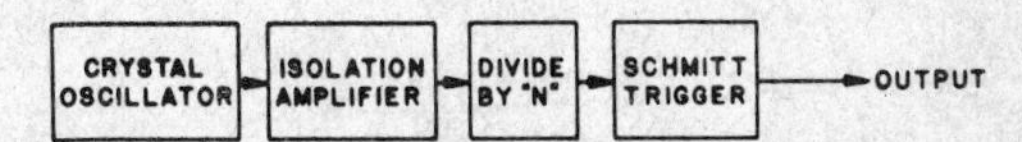

Fig. 2-14. Drawing of the Schmitt-Trigger system which is used to produce useable harmonics in the HF band.

The complete RF circuit of the calibrator is shown in Fig. 2-15. Although the circuit of Fig. 2-15 looks complicated, it contains less than $7.00 worth of semiconductors.

The crystal oscillator (Q1) is a Colpitts-type which operates the surplus FT241 crystal in the series mode. Most hams are accustomed to operating FT241 crystals (of the 400 to 500 kHz variety) in a parallel-mode oscillator. Apparently the 200 kHz FT241 crystal is the one exception in this holder style; so operate it series-mode! The 200 kHz crystal was obtained from Jan Crystals.

Following the crystal oscillator are two isolation amplifiers, Q2 and Q3. Q2 drives the MOS IC divider chain, and Q3 drives the Schmitt trigger when 200 kHz output is desired. Less elaborate isolation gave small variations in crystal oscillator frequency when one switched between 200 kHz and 100 kHz outputs.

The heart of the calibrator is, of course, in the three binary dividers: Hughes HRM-F/2 MOS integrated circuits. Unlike other ICs, these are packaged in TO-18 transistor cans with four leads. With one lead for power and one for common, that leaves the other two leads for input and output. If we'd look inside the F/2, we'd probably find a complete J-K Flip-flop, as offered in more complex members of Hughes' HRM family. However, in this simplified binary, only the "T" (toggle) and "Q" (output) are brought out of the can. The HRM-F/2 will operate up to 500 kHz and costs *less* than any RTL binary (even less than half of an MC790P dual-FF). Three HRM-F/2 ICs divide the 200 kHz output from Q2 down to 100 kHz, 50 kHz, and 25 kHz.

The Schmitt trigger is unusual in that it has an MOSFET in its input stage. This feature is used to prevent loading the output of any of the dividers of Q3. An ordinary Schmitt trigger, using two bipolar transistors, so heavily

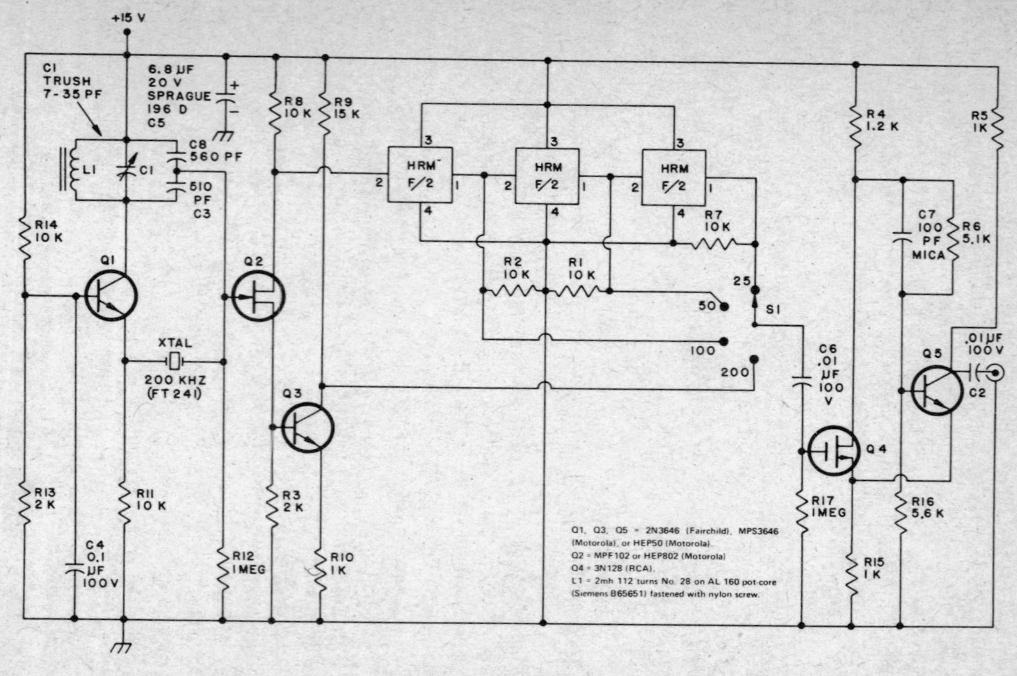

Fig. 2-15. Complete circuit of calibrator.

loads its input, that it is almost invariably driven via an emitter-follower. The use of a fast N-channel MOSFET gets around the input loading without sacrificing rise time.

If it is desired to operate the calibrator from its own line-operated power supply, a simple regulated supply is shown in Fig. 2-16. An integrated regulator made by Continental Devices Co. is used because of its low price and simplicity. The CMC 513-4 looks like an epoxy TO-5 transistor, having only 3 leads. Inside this package are all the components to make up a regulator: transistors, resistors, voltage reference, and even a thermistor to shut down the regulator when the temperature gets too high. The RF circuit of the calibrator only requires 15 mA, so the CMC 513-4 is more than adequate for this job. The Philbrick/Nexus 2105 IC regulator is apparently a similar device (at approximately the same price), and may also work here. A Triad F40x was used for the power transformer; it is capable of considerably more output current than needed, but smaller transformers are more expensive.

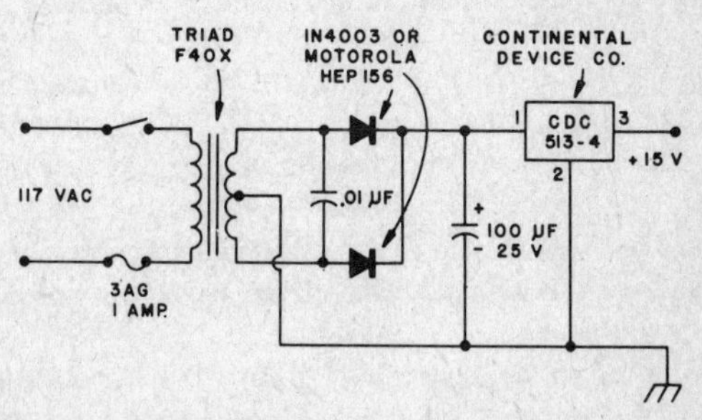

Fig. 2-16. A simple regulated power supply for the calibrator.

It was decided to put the RF circuitry and the power supply circuitry all on one etched circuit board. Since the power supply has so few components, little space is wasted if one decides not to build that portion. The board layout is shown in Fig. 2-17 and the parts layout in Fig. 2-18.

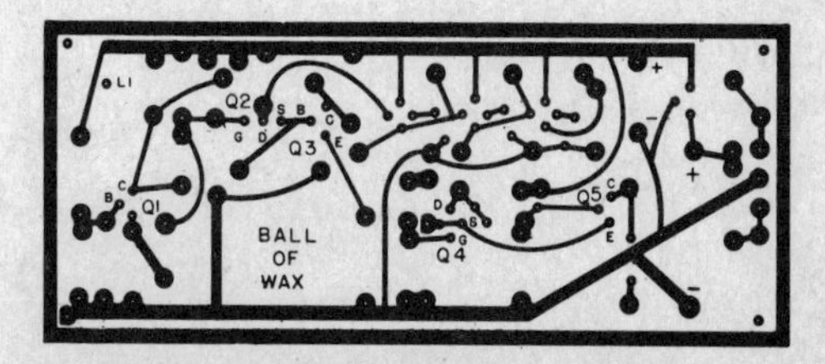

Fig. 2-17. Illustration of the circuit layout.

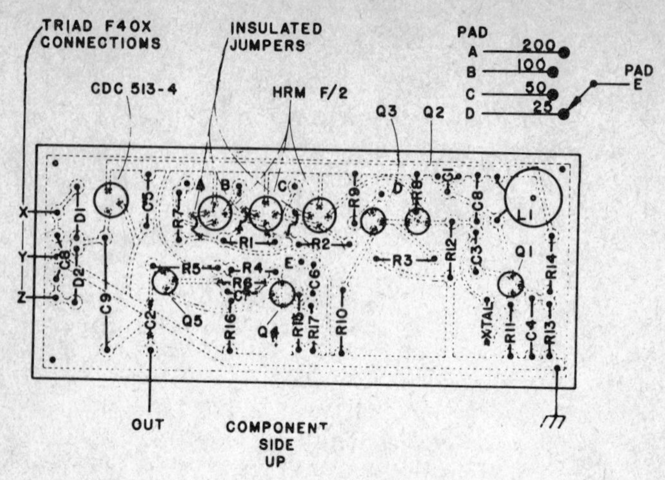

Fig. 2-18. Illustration of the parts layout.

There are a number of points about construction and materials that should be mentioned. L was handwound using a Siemens B65651 pot core. It was clamped together and held down on the etched circuit board with a nylon 4-40 screw. This core has an air gap in it *thru which the screw passes*; so don't use metal screws. The AL160 printed on the core would indicate a need of 112 turns for 2 mH, but 112 turns is nominal. (The AL number on a core is the number of nanohenries per turn-squared wound on that core.) The tuning capacitor, C1, is a Trush miniature ceramic type. Other types will work, but the board is laid out to fit this Trush model. The crystal socket is made from two pin-grips removed from an old octal socket. They are soldered through the board and bent 90° to accept the FT241 pins.

The procurement of parts, especially ferrites and semiconductors, may prove to be a bit more difficult than usual. The advantages offered by the modern components used in the "Ball of Wax," however, seemed to outweigh this increase in procurement difficulty.

RECEIVER FRONT END WITH AN RCA CA3103E

This little gem has two independent differential amplifiers inside, with its schematic shown in Fig. 2-19. Each of the six transistors has an FT in excess of 1 GHz, making this IC useful to 500 MHz. Special care has been taken in the internal chip layout to assure good freedom from reaction between the two independent amplifiers.

Since the true home-brewer is always interested in possible applications of what he spends his hard-earned money on, the following list of CA3102E uses is included: VHF amplifiers, mixers, multifunction combinations such as

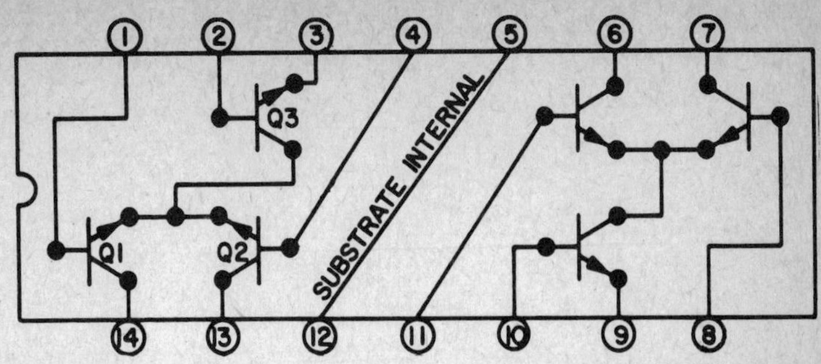

Fig. 2-19. Build a complete receiver front end with the RCA CA3102E integrated circuit.

RF/mixer/oscillator, converter/IF, cascode amplifiers for IF or diff amps, product detectors, doubly balanced modulators, balanced quadrature detectors, cascade limiters, synchronous detectors, balanced mixers, synthesizers, balanced cascode amplifiers, sense amplifiers (whew!) and others. The age of the VHF-UHF ICs is being opened up by these 1 GHz beauties.

Diff Amp and Cascode Details

Inasmuch as the LO is formed by a cascode amplifier, and the mixer section has an RF amplifier in it, a few words on the use of this chip in both modes is timely. Figure 2-21 shows one-half of this versatile little chip of 1 GHz capability in the diff amp mode, where Q1 is a grounded collector amp driving Q2, a common base amplifier. Q3 acts as a constant-current source. This is the configuration that is best if strong signals are present, as in metropolitan areas, or other areas where amateurs may be in close proximity to each other. This diff mode has slightly less gain than the cascode mode, but this need not be considered unless you are trying to minimize components, often a false concept. Two stages with lower gain, such as diff amps, are often much better and easier as well to build and line up than one of higher gain. Noise figure goes *down* with current, another reason for two stages.

Circuit

Referring to Fig. 2-20, the RF arrives at the tuned circuit L1-C2 from the input matching capacitor C1. A tap about one turn from the ground end goes to the base of Q1, a common collector amp. Q1 is emitter-coupled to Q2, a grounded base amp whose collector circuit has C3 and L3 for tuning. See RF section for details and component values.

Constant-Current Source

The makers of these IC RF circuits use a lot of transistors in these tiny chips for this sort of thing. Making them mainly for commercial users, they

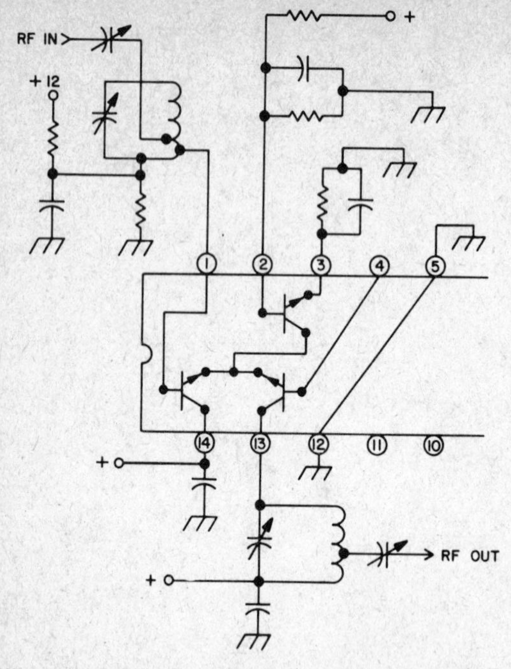

Fig. 2-20. Differential amplifier connection, RCA CA3102E (pin view). See Figs. 2-21 & 2-22 for representative values.

put in, for example, Q3, which does not appear to do anything at all. It does however do nice things if you look a little deeper. When the temperature varies, it tends to maintain a constant current on the amplifier and therefore lessens any detuning effect that might otherwise occur and other nuisance effects. Just use it as shown. It helps for a better circuit, especially when you can take it out in your car in the winter right out of a heated room or garage. After all, it only costs the maker a fraction of a cent to put it in there; so why should *you* worry?

Cascode Connection

Figure 2-21 shows connections for the cascode configuration, which has the highest gain but cannot handle strong signals like the diff amp. The RF goes through the series matching cap C1 (referring to Fig. 2-21), then to the tuned circuit L1-C2 and then to the low impedance base of Q3, a grounded emitter stage. This stage is cascode coupled to the emitter of Q1, and the signal goes out on the collector to the tuned output circuit C3-L2 and on out through C4. Q2 is not used in this mode, and simply floats for DC, with external connections 4 and 14 bypassed to ground for stability purposes. Keep L1 away from L2, and use a shield between them if needed. Put the shield more or less in line with pins 5 and 12, as shown in Fig. 2-21. More details in the RF section.

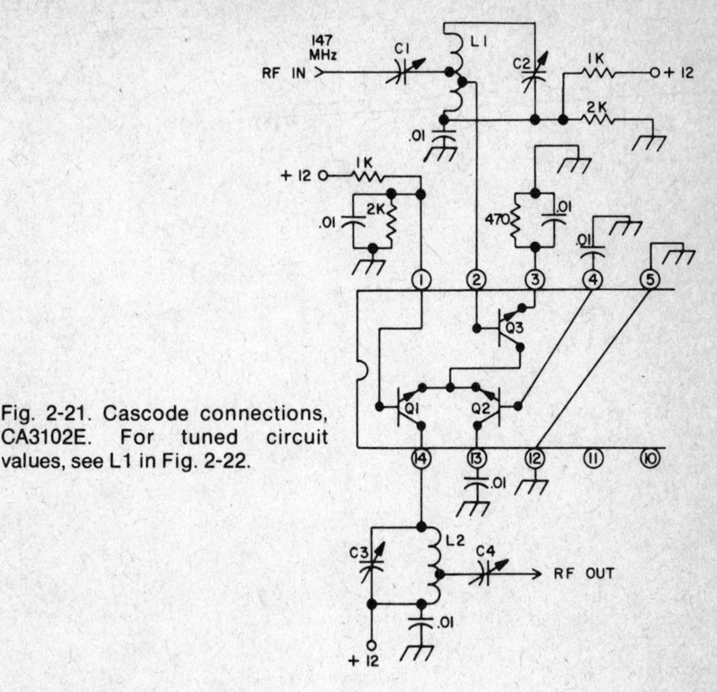

Fig. 2-21. Cascode connections, CA3102E. For tuned circuit values, see L1 in Fig. 2-22.

Internal Feedback and Stability

The compound (internally connected diff amp or cascode) connections of the diff amp or cascode are both much better for internal feedback (having less) than a single grounded emitter transistor, with the best in this respect being the cascode mode. The ratio of how much better may be as high as 1/140 at low frequencies, to perhaps 1/10 at 100 MHz. In the cascode circuit this may be as high as 1/135 to 1/1200. In any case the internal feedback is low enough so that no consideration of neutralization is needed. This reverse transconductance is sometimes labeled YR, in case you should meet it some dark night. Just be sure the *external* feedback does not clobber the nice internal feature. Pay attention to the details and notes in this section and it *won't* (shielding, bypassing, etc.).

AGC

The mixer gain can be easily controlled to a high degree in these versatile chips, up to 60 dB or more, by putting a negative voltage on the base circuit of

Q1 in the mixer circuit of Fig. 2-23. Remove R3 and put the negative going AGC voltage on the base at that point. You will more likely use the AGC on the RF amp, though. In that case the AGC voltage would be applied to the base return of Q3 through L1 in Fig. 2-21. A good balance of RF amp current (minimize) and gain, as well as mixer gain, should be sought.

Mixer

Figure 2-22 shows the half of the CA3102E used as an RF amp, mixer, and LO buffer-amplifier. Note the grounded line between pins 5 and 12.

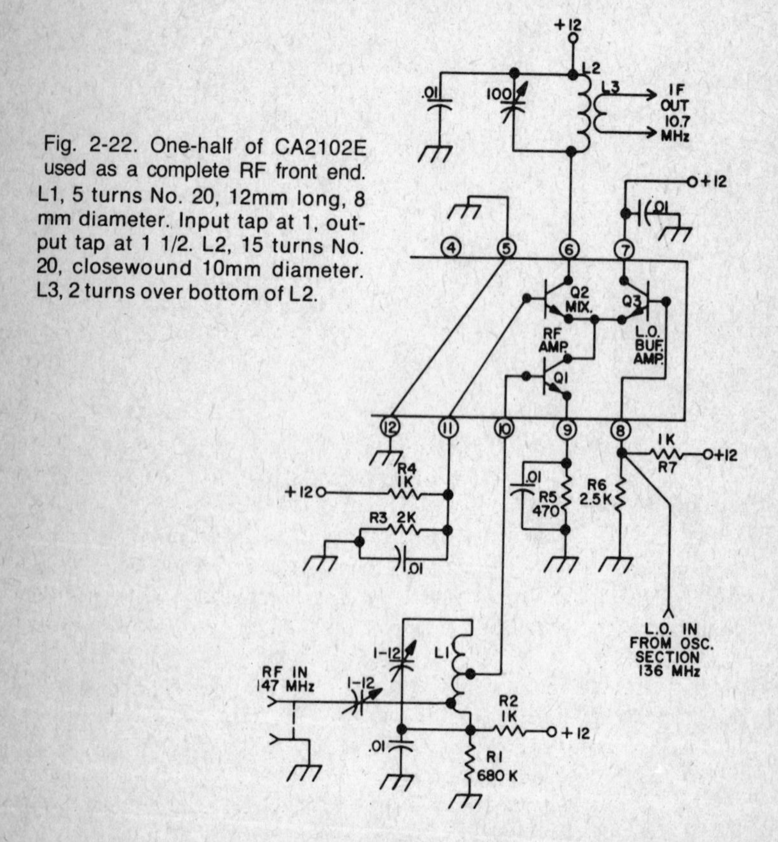

Fig. 2-22. One-half of CA2102E used as a complete RF front end. L1, 5 turns No. 20, 12mm long, 8 mm diameter. Input tap at 1, output tap at 1 1/2. L2, 15 turns No. 20, closewound 10mm diameter. L3, 2 turns over bottom of L2.

Q1 is a grounded emitter RF amplifier which takes the RF signal at 147 MHz in on L1 and direct couples it to Q2, a grounded base mixer stage. Q2's collector goes out at 10.7 MHz to the IF output transformer L2, link coupled by L3 to the output cable. Q3 is used as a buffer amplifier for the LO —the 136

MHz injection voltage coming from the other half of the CA3102E used as the crystal oscillator. This section is quite straightforward even though an RF amp and buffer-amp for the LO are included, and works well due to the 1 GHz transistors used.

Noise Figure

The noise figure of the CA3102E used as an RF amplifier is highly dependent on current, as indeed in most semiconductors, 2 or even 1 mA being good figures for RF amps. This noise figure runs about 4.6 dB at 200 MHz for the transistors used in the CA3102E.

There is a fairly simple rule to follow with these units, as already mentioned. Use of the lowest possible total current through Q1's emitter will in general give the lowest noise figure. This refers to the use of the CA3102E as RF amps also. Note the setup for tuning L1, the series cap bringing in the RF and the two tie-points serving as output connectors. This method has proven to be very efficient, both for building the first module and for layout of the PC boards later. If done correctly, each tie-point can serve as a place to drill the PC board for a lead hole, or component tie-in. It is particularly important in VHF and more and more so as you go up into 220 and 450 MHz, to furnish a layout man with exact placement detail. You can no longer leave it to him, at VHF and UHF, unless he knows a lot about RF work and particularly the IC under consideration. You should think this out carefully, as VHF and UHF ICs handle quite differently than AF, digital, or control circuits.

Base Bias

Due to the direct coupling of the various transistors, placing them in series DC-wise, attention should be paid to the base bias voltages. While in general the resistors may be as in Fig. 2-23, these should really be adjusted with signals flowing through for checking best noise figure, gain and current. There is also quite naturally a certain amount of interplay between R1, R3, R6 and the current controlling emitter resistor R5. The optimum values are not critical or touchy, but will be seen to be quite responsive to best values.

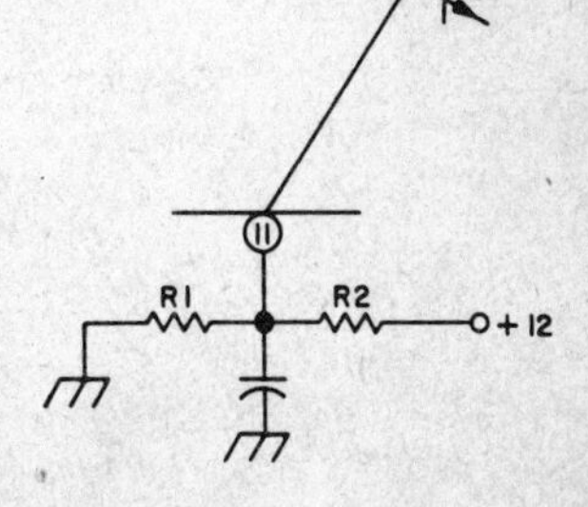

Fig. 2-23. Example of base bias in the CA3102E.

Generally, use a 5K pot at the R1, R3 and R6 places to start off with, and a 1K pot as R5. The proper values are soon determined with that arrangement, and you are not left guessing if you have the right values or not. Remember that monolithic photo-built transistors have quite a wide range of parameters at times.

The mixer was run with an outboard oscillator as well as the internal one, to see if there was any reaction between the two halves of the CA3102E when used as RF, mixer and oscillator, but none was found.

Crystal Oscillator on 136 Direct

Referring to Fig. 2-24, we see the RF signal from the crystal entering Q1 at the base. Q1 is the grounded emitter portion of the cascode amplifier forming the oscillator. Note the crystal used is cut for 136 MHz direct, without

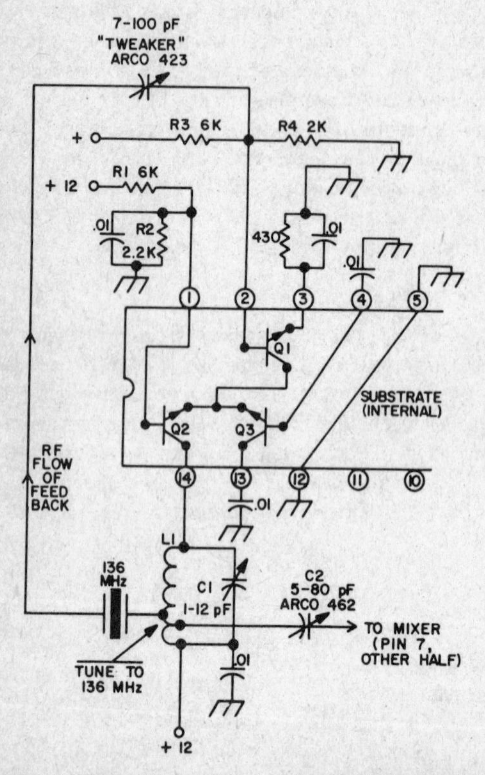

Fig. 2-24. Internal schematic, 1/2 RCA CA3102E, IC used in the cascode connection as a crystal oscillator. L1, 5 turns No. 20, 12mm long, 8mm diameter. Crystal tap at 1 1/2 turns, output tap at 1 1/2 turns.

multiplication of any kind. Some have expressed surprise at this, but it should be known that today control crystals are available up to 250 MHz direct, not only for two meters. Such crystals are commonly referred to as "overtone," which smacks of obscurantism, the real functioning of that precious little piece of "glass" being as shown in Fig. 2-25. This gives you a rough idea of what is going on inside that little tin can. You can also see that only odd numbers can be used. If you try to use the even number combination in Fig. 2-25 you will of course get no AC out of it. Never forget that a crystal operates on sound waves and that only certain numbers can be used. If this sounds to you like quantum mechanics, good for you, you're getting warm! A piezoelectric chip, if cut right, gives off electricity, positive on one side and negative on the other when compressed, and that is good old DC if you hold it that way! Madame Curie's brother used to demonstrate this nearly one hundred years ago by hitting a piece with a hammer and demonstrating to the entire hall a spark therefrom. What you need first is that the oscillator be running, and if it's a good oscillator, it will be as soon as you throw the switch. It has to run, it's a law of nature. So, as shown in Fig. 2-23, the oscillator voltage enters the base of Q1 on pin 2, a grounded emitter amplifier. This is internally connected (direct coupled) to the emitter of Q2, a grounded base amplifier, the two-transistor compound forming a cascode amplifier good to 450 MHz. Here it is used at 136 MHz. The output of Q2 goes to a tuned circuit L1-C1, tuned also to 136 MHz. In

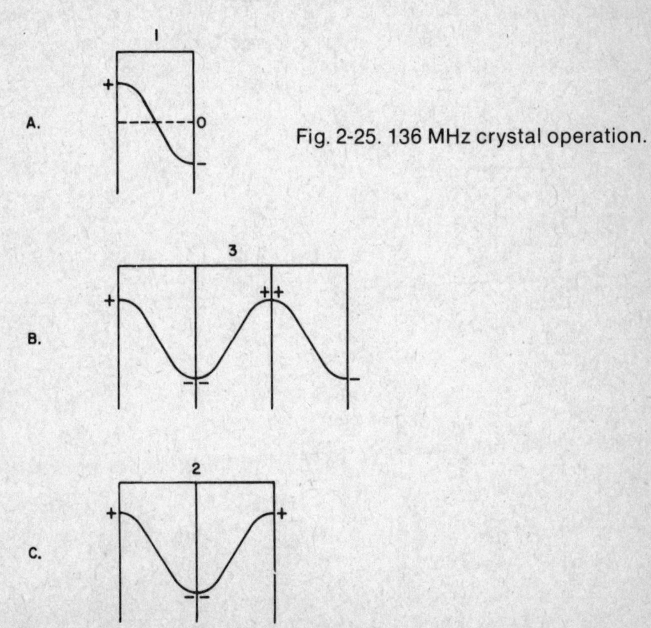

Fig. 2-25. 136 MHz crystal operation.

this circuit, L1 has two taps, one for the output going to C2, and then to the mixer in the other half of the IC. The full gain of the cascode Q1 and Q2 is used and needed. Crystals on 136 MHz are not to be taken lightly. They must be matched as to AC impedance fairly well into and out of the cascode. The input, base of Q1, is sort of a natural, as the base impedance is quite low and the matching can be helped by adjustment of both the base bias and the emitter bias.

The output match of Q2 into the crystal is done by tapping down on L1, and is also seen to be low in impedance. It is also possible to couple into the crystal by a two turn link, wound in the same direction as L1, and placed over or inside of L1. Tune up such an oscillator with a 5K pot as R2, and another as R4. These are not at all touchy but you can make the oscillator run better that way. To check out and tune up—not only for frequency but also for good starting, absence of squegging, noisy hysteresis jumps either in frequency or power and proper base biases—a tuned diode detector is very handy. A one or two turn link around the cold end of L1 with a cable into a diode through-line cavity does the trick. You must know if it is really on 136 MHz. The figure 2-26 shows a helpful setup to do this. Set up the tuned diode detector and check for good, smooth, quiet power out, using both meters to measure and AF to listen. Then bring in wire, as shown, close to or around the diode tuner center conductor as shown in Fig. 2-26. Tuning the signal generator very slowly over the region 135 to 137 MHz, you should be able to hear the desired heterodyne resulting from the crystal oscillator beating with the signal generator on 136 MHz. If you do not, something of course is wrong. Do *not* use a sensitive receiver for these tests.

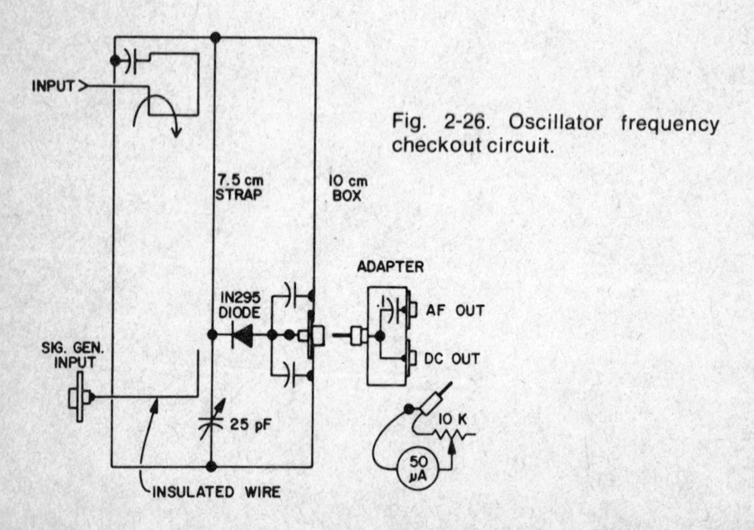

Fig. 2-26. Oscillator frequency checkout circuit.

Test operation should include good, clean tuning of C1 (Fig. 2-24) with the oscillator coming on and off with a nice "plunk" in the AF speaker indicated in Fig. 2-26. The meter should also come up and drop off with the proper action. Marginal operation is to be avoided like poison. For example, the oscillator may be *just* working, and the next time it won't come on at all.

If you build this IC front end as shown, it will work and in good style. The operation is stable and secure—not marginal. You can put any RF stage in front, or use another CA3102E as the RF stage.

Chapter 3
SSB, DSB, AM and Voice Communications

THE FET COMPRESSOR

The addition of a speech compressor to a rig is a worthwhile project. A speech compressor will not only increase talk power but will also reduce flat topping. The simple circuit described here in Fig. 3-1 uses two Motorola HEP 801 field effect transistors. The entire unit can be built for approximately $15 or less, depending on the size of your junk box.

Features

The compressor has an overall gain of unity. Therefore, the output of the circuit is comparable to the output of the microphone connected to it. This is accomplished by the voltage divider R3 and R4 in the output. If for some reason more output is required, the value of R4 may be increased provided the value of R3 is decreased by an equal amount. In other words, the sum of R3 and R4 must remain approximately 70.2 ohms.

Because of the inherent high impedance of the HEP 801, the input impedance of the compressor is determined almost entirely by R5. The value of R5, 470K , is not critical and may be changed to match the output impedance of any high Z microphone. However, this resistance will provide a good match for most crystal, ceramic or dynamic microphones. At high signal levels, a 6 dB input change will result in only 1 dB change in output. The circuit also has an inherent threshold voltage of approximately 10 mV due to the characteristics and operating point of the FETs.

Circuit Operation

The compressor is composed of two audio amplifiers, an AGC rectifier and an AGC circuit. The output from the microphone is amplified by Q1 and then

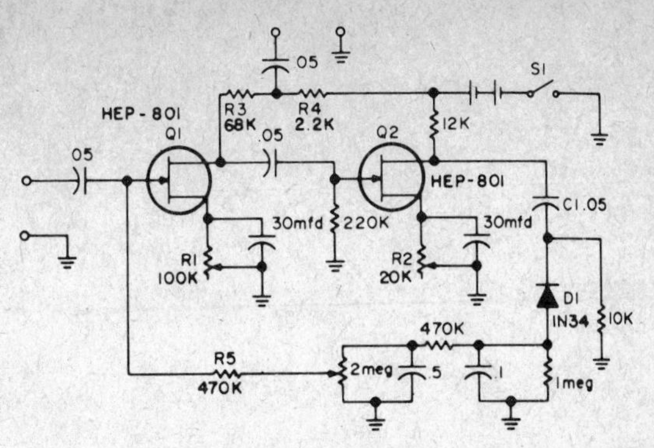

Fig. 3-1. Audio compressor schematic.

capacitively coupled to Q2 where the signal is amplified further. The output from Q2 is coupled through C1 to the AGC rectifier, D1, and is rectified. The rectified audio is then filtered and fed through the AGC circuit and applied to the gate of Q1 as negative bias.

A small input signal will develop very little AGC voltage and have little or no influence on the circuit gain. A large input signal will develop much more AGC voltage. This large voltage will reduce the transconductance of Q1, thus lowering the circuit gain considerably and resulting in an output that is only a few dB greater than that of a small input signal.

Construction

The compressor was built on a 2 1/2- by 4 1/4-inch piece of perforated Bakelite which was then mounted in a Minibox. The size of the unit can be reduced through the use of miniature components.

The FET used as Q1 must have a low pinch off voltage. To determine which transistor has the lowest pinch off voltage, the source and gate leads should be temporarily connected together. The resistance from the drain to the source is then measured with a VOM or VTVM. Whichever FET shows the highest resistance reading should then be used as Q1.

The parts layout of the circuit is not critical if proper construction methods are followed. In the original unit three pieces of bus bar were run across the board and used as tie points for the B+, ground and AGC line.

Adjustments

The resistors R1 and R2 should be adjusted to give maximum output. This is done by placing the compression level at minimum and then measuring the voltage across R6. The value of these resistors is critical and pots had to be used. Fixed resistors were used at first but failed because the nominal values were not accurate enough.

When these adjustments are complete, the compressor may be inserted in the microphone line. With the compression level at minimum, adjust the audio gain on the transmitter to a normal level. Compression can then be raised to the desired level. The audio gain can now be raised again until the output meter peaks at the same reading as it normally did.

Operation

Two 9-volt transistor radio batteries were used to supply the B+ to the circuit. The compressor draws only one mil so battery life should be very reasonable.

Besides increasing the talk power of your rig and reducing flat topping, the circuit has many other advantages, depending on the operating practices of the builder. For example, the man who runs phone patches will no longer have to continually reach for the audio gain when the caller changes the level of his speech.

ADDING dBs TO THE AUDIO COMPRESSOR

Simple audio compression units can easily improve the apparent effective signal strength of any SSB transmitter by several dB under difficult weak signal conditions. Simple audio compression is not as effective as RF speech clippers as many tests have demonstrated. However, with frequency response shaping and "soft" clipping added to an audio compressor, the latter can be made almost as effective as much more elaborate speech processing methods. This section describes how various simple accessory circuits can be added to any exisiting audio compressor which will considerably improve its effectiveness. The circuits which are added all operate at audio frequencies so no complicated construction is required. The cost is minimal when one considers that they can produce several dB more effective signal strength under poor signal conditions. This is especially true when one considers what the cost would be to increase one's signal strength by 3 dB by conventional means. That means doubling transmitter power, a directive antenna array, etc.

Frequency Response Shaping

If one uses any type of compressor/preamp, it operates on the basis that audio levels beyond a certain threshold activate a gain reduction circuit which reduces the gain of some circuit early within the compressor/preamp unit. Input signals below the threshold are amplified fuller and those exceeding the threshold activate the gain reduction circuitry so that the wide variations in input signal levels are compressed into a much smaller range of output signal levels. Although many compressors have a stated frequency response which includes only the 300 to 3000 Hz range, this restricted frequency response is not shaped sharply before the gain reduction circuitry is reached. The result is that if one uses a wide response microphone, the gain reduction circuitry is often activated by low frequency audio/signals which are not passed anyway by the SSB generation circuits in the transmitter. The result is that the audio compressor's action is partially wasted on responding to audio signals in a range which is later rejected and are not useful for voice intelligibility. The result is at least a partial waste of the compressor's effectiveness.

The rather simple solution to this situation is to frequency shape the audio signal input to the compressor before any audio compression action is started. One often hears about some type of microphone which seems to have a particularly effective response or audio "punch." Using such a microphone before a good compressor often makes several dB difference in signal effectiveness. However, this type of effect can be duplicated using almost any simple type of microphone if a frequency shaping preamplifier is used before any existing audio compressor! A suitable circuit is shown in Fig. 3-2. A low noise FET preamplifier is used to provide initial gain for a high impedance microphone input (low impedance via a matching transformer). This stage is

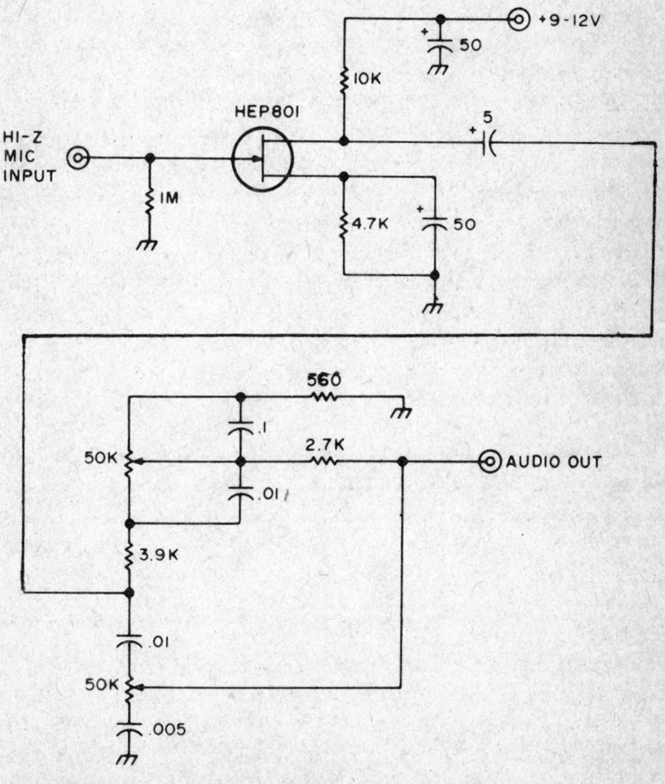

Fig. 3-2. Shaping circuitry to be added ahead of existing compressor.

followed by a low and high rolloff circuit which can provide about a 15 dB boost or rolloff to frequencies centered on approximately 1 kHz and extending both higher and lower than this center frequency. By the adjustment of both potentiometers, one can just about duplicate the sound of any commercially available communications microphone as it is heard after the audio compressor. It cannot, of course, compensate for the directional pickup characteristic of a microphone which may be of consideration when a location with high background noise is in question. But it can compensate to a very great degree for individual voice characteristics when used with a given microphone and this is the main advantage of using the circuit in conjuction with a given audio compressor. This initial frequency response shaping is useful by itself but particularly worthwhile when combined with "soft" clipping and further frequency response shaping as described next.

Soft Clipping

Soft clipping refers to the type of limiting action that takes place when a sine wave signal is fed into a clipping circuit, such as a diode pair, but where the sine wave is not cut off along a flat line on its positive and negative excursions. Rather, the sine wave is rounded off abruptly and a far less harmonic rich output is produced. "Soft" clipping is not well defined and where a diode provides soft clipping by its basic characteristics, it may be regarded as a poor diode for clipping purposes where a very sharp, hard clipping characteristic is desired. Also, one can produce soft clipping by driving a diode pair through some resistance so the diode operates over the portion of its characteristic where it has a rapid current/voltage change characteristic. The diodes operate as a form of a variable resistor element, one diode in the pair responding to positive going voltages and the other to negative excursions.

The later type of soft clipping is used in the circuit of Fig. 3-3, which is meant to be placed at the output of an existing compressor. Most compressors provide more than enough output voltage, usually several volts, to drive the clipper circuit. If the compressor has an output level control, it should be set to provide full output for initial adjustment. The clipping can be adjusted by means of the variable resistor in series with the diodes. The clipping itself is not meant to be in action constantly but only as a further adjunct to the basic compressor action. Peaks which the compressor doesn't handle are acted upon by the clipper and the overall average to peak ratio of the processed audio signal is increased *without* significantly more distortion.

The output level of the clipper is very low and it should be followed immediately by a good low noise amplifier. This action, in fact, is one of the most important considerations in making the circuit effective. As shown in Fig. 3-3, the clipper is followed by a low noise FET voltage amplifier stage which has a broadband, flat frequency response. The voltage gain is sufficient to provide for the loss in the filter in the drain output of the FET stage and still have the final output level more than sufficient to drive the audio input for any transmitter.

The filter following the FET stage is a carefully designed double unit which provides extremely sharp attenuation of frequencies higher than about 3000 Hz and continues to provide excellent harmonic attenuation up to 30 kHz or more. If one can obtain the inductors (commercial types are available in the correct values), this type of filter is highly recommended because of its excellent harmonic filtering capabilities. Unfortunately, miniature audio chokes of good quality are not very inexpensive, unless obtained from surplus sources. Therefore, Fig. 3-4 shows an alternate type filter using only one choke which is only slightly less effective. Audio chokes which provide a Q of 60 at least at a few thousand cycles should be used. One side of the audio transformer winding cannot normally substitute for a choke even though it may have the correct inductance. The increased resistance of such windings and other characteristics deliberately engineered to provide a broad frequency response in the transformer result in a very low Q for the windings. A few transistor transformer windings the author measured had Qs of 1–2!

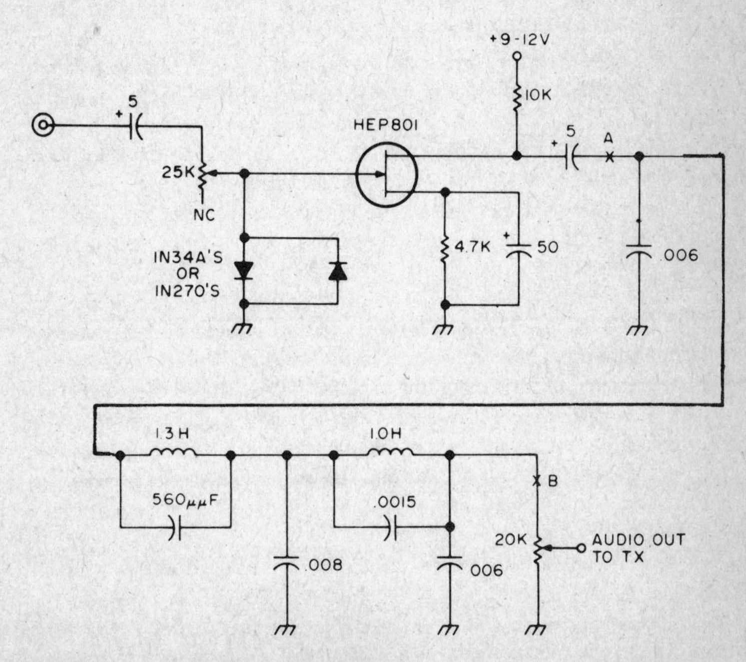

Fig. 3-3. Clipper and filter for use at output of existing audio compressor.

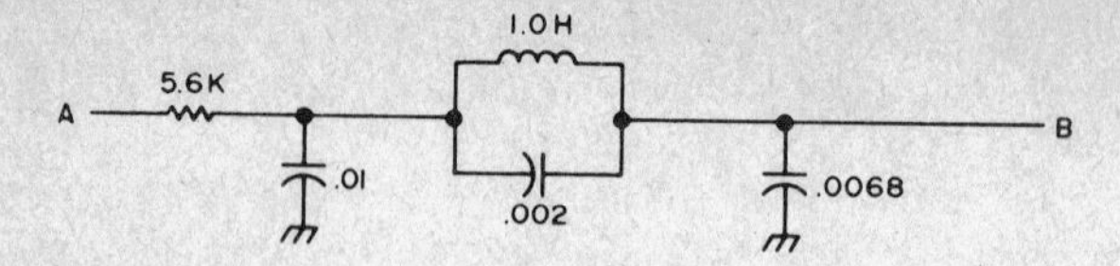

Fig. 3-4. Alternate filter design to be connected between points A and B in Fig. 2 instead of filter requiring two inductors.

Construction and Adjustment

The auxiliary circuitry described is best constructed on perforated board stock and placed in the same enclosure as the audio compressor with which it is used. Two commercial type audio chokes are used.

The adjustment process may seem a bit confusing especially since so many potentiometers are present, including those present in the basic compressor. However, approached step by step the adjustment process is basically simple. Set up the compressor first as it is used normally for best results. Then, connect only the circuitry of Fig. 3-2 to the input of the compressor. Adjust the rolloff potentiometers in different combinations for the most effective audio sound. This can be done by running the compressor output through any good audio amplifier and listening to it over some headphones. However, one can become confused by this method when listening to one's own voice. A far better procedure is to use an over the air check with a local station. The other station should, however, constantly reduce the RF and not the AF gain on the station receiver to simulate a barely readable DX signal. One is interested in intelligibility under poor conditions and not fidelity at this point. The audio input level to the compressor (part of the compressor) may also have to be adjusted to prevent over driving the compressor. If problems with RF feedback are encountered, use an RFC of 1 to 2 mH in the input lead from the microphone to the circuitry of Fig. 3-2.

Then, the circuitry of Fig. 3-3 with either filter is connected after the compressor being used. A similar type of adjustment process is gone though by adjusting both the output level control on the compressor, if it has one, and the variable resistor in series with the clipping diodes. The correct adjustment point is a compromise between distortion generated and improvement in audio effectiveness. If the compressor is only intended to be used for DX contacts, adjustment should be made under simulated weak signal reception conditions. A final adjustment can be made then by going back to the rolloff potentiometers in the input circuitry.

When one considers the expense involved in building the accessory circuits described to obtain more effectiveness on DX contacts as opposed to increasing power or antenna gain, the circuitry described represents very good value. If the type of equipment you are using works better when a conventional audio compressor is added it will work somewhat better when these accessory circuits are used.

AN IC AUDIO PROCESSOR

The unit makes use of two ICs—an SL630 amplifier and an SL620 automatic gain control unit. The two ICs were designed to mate together for a speech processing function and, therefore, should be used together even if one is tempted to replace the relatively simple amplifier IC by another type of IC.

The complete schematic is shown in Fig. 3-5. The unit is designed for both a low impedance input and output and a matching transformer is necessary for use with a high-impedance microphone. Usually, no transformer is needed on the output even if it is connected to a high-impedance audio input on a transmitter because of the amplitude of the output voltage. The input provides either for a balanced or an unbalanced type of microphone input. The former can be quite useful since by having both microphone leads ungrounded, many problem with noise pickup, RF pickup, etc., on the microphone leads are automatically avoided. If the usual type of unbalanced input is used, however, it is connected to pin 5 of the SL630 via a 1 μ F capacitor. Pin 6 is then left unconnected. The capacitor between pins 3 and 4 of the SL630 provides a high-frequency rolloff characteristic. The values shown provide a rolloff starting at 3 kHz, but this can be changed, if desired, by experimenting with the capacitor value.

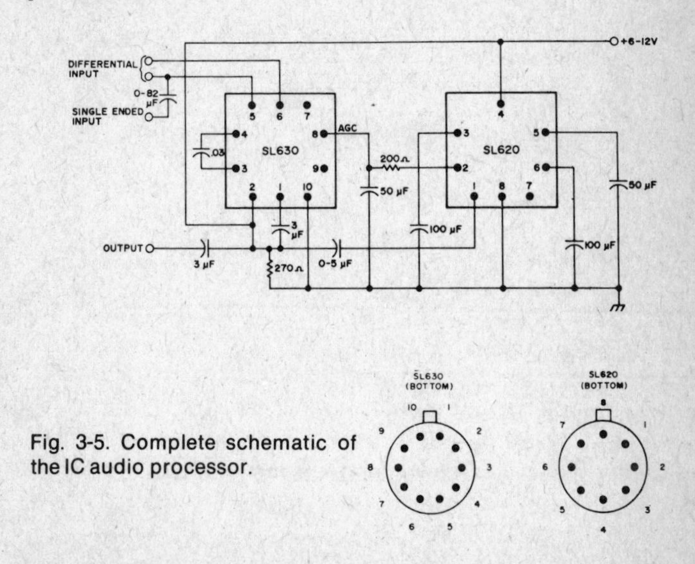

Fig. 3-5. Complete schematic of the IC audio processor.

The output is taken from pin 1 of the SL630 via the two 3 μF coupling capacitors. Part of this output is coupled to pin 1 of the SL620 IC via the 0.5 μF capacitor. The SL620 IC uses this voltage to generate an AGC voltage which is eventually available at pin 2 of the SL620. From there it is coupled to pin 8 of the SL630 to control the gain of the latter IC. What these two little ICs can accomplish is shown nicely by Fig. 3-6. The upper curve shows a varying speech input, while the second curve shows how the audio output appears after processing. Notice that when the speech input either rises rapidly or falls rapidly the output remains essentially constant. Noise bursts, because of their much shorter time duration, are recognized separately by the unit. The noise burst shown occurring during speech, although much higher in amplitude than the speech level, produces practically no increase in output. An automatic squelch feature is also provided. When there is a pause in speech, the output is disabled to prevent the background noise buildup common to most simple compressors. The pause time before the output is disabled is about 1 second and can be changed, if desired, by varying the value of the capacitor from pin 6 of the SL620 to ground.

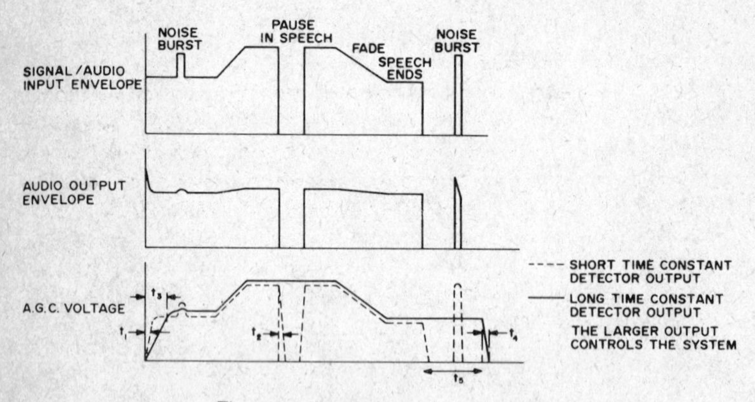

Fig. 3-6. Illustration of the input/output and AGC voltage characteristics of the processor.

The control range of the unit is illustrated by Fig. 3-7. Only a very slight change in AGC voltage is necessary to control the output over a 60 dB range. In practice, the input can change over a 35 dB range and the output level will remain between 70 and 87 mV.

Construction

The parts layout is in no way critical and is just a matter of convenience. The units will operate equally well with a 6 or 12V DC supply and draw up to about 15 mA. The operating voltage can be borrowed from a well filtered point in a transmitter or a battery supply used. In the latter case, a usual 9V battery is ideal. There are no controls to the unit, so it may be housed easily inside a transmitter, avoiding any location near high RF fields.

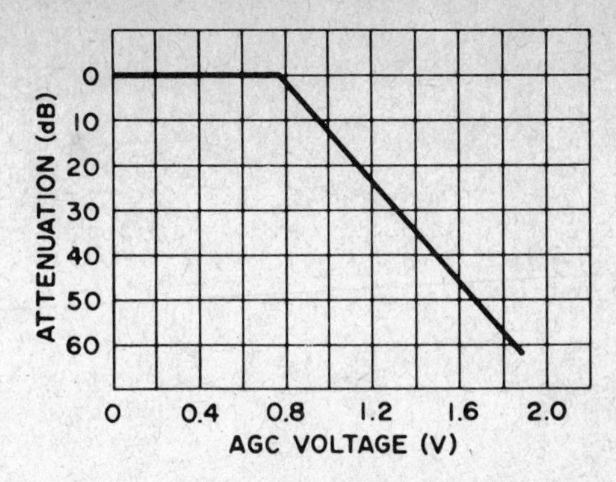

Fig. 3-7. Control range of the unit exceeds 60 dB. The graph illustrates the control of the input signal by the SL630 with AGC voltage supplied by the SL620.

Operating Results

When comparing operation to several types of conventional audio compressors, in every case, the unit described exhibited a much smoother compression action without pops, clicks, etc. It simply sounded more like the type of quality speech processing found on commercial circuits rather than the usual harsh, noisy compressor type action, which usually has to be disabled for local contacts because the inherent distortion then becomes so noticeable. The squelch and noise immunity features also added a great deal to the cleanliness of the speech output and should be particularly useful in a mobile type situation where a great deal of extraneous background noise can exist.

A TRANSISTORIZED 10 METER DSB TRANSMITTER

The DSB transmitter described in Fig. 3-8 uses crystal control and runs 1 watt or so, which can be used barefoot or to drive an RF amplifier.

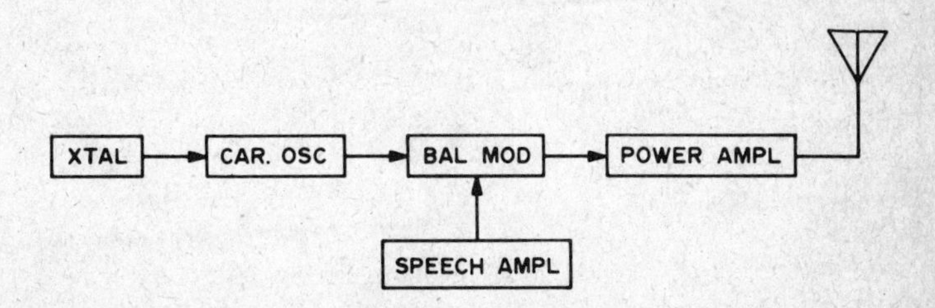

Fig. 3-8. This block diagram shows the simplicity of the homebrew 10M DBS transmitter. The ten-watt unit can be used to drive a low-power linear or, for QRP fun, it can be used barefoot.

The unit should be assembled in a small Minibox or built on perfboard or PC board and then installed in a small Minibox, as stray hand capacitances can upset the carrier balance.

The transmitter consists of speech amp, carrier oscillator, balanced modulator, and PA stages.

The amount of carrier suppression available with a diode-type balanced modulator is − 40 dB.

Care should be taken in selecting the diode pair. Check the forward resistance of several diodes with your VOM until you find two with the same or nearly the same forward resistance. Germanium diodes were used in this unit which were in a grab-bag pack of 50 for $1 from Poly-Paks. The diodes should read at least 10:1 (forward-to-reverse resistance ratio). Figure 3-9 shows the DSB schematic.

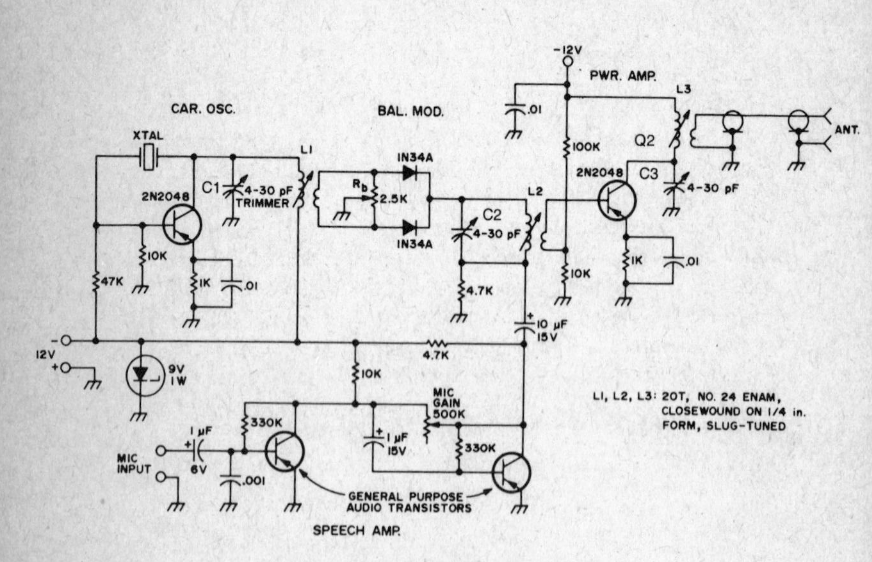

Fig. 3-9. 10 meter DSB transmitter schematic.

Capacitors C1, C2, and C3 are 4-30 pF variable trimmers. R_b is a potentiometer. Coils L1, L2, and L3 are 1/4 inch slug-tuned types removed from a TV PC board. Many types of these coil forms were tried and, although the ferrite slugs vary in Q, all coils resonated at 10 meters by juggling the slugs and tuning capacitor settings.

After checking the transmitter for operation by listening for the signal on a receiver, peak L1 and C1 for maximum S-meter strength. Then null out the carrier (minimum S-meter indication) with pot R_b. L2 and C2 may also be varied to null out the carrier. With the carrier nulled out, speak into the microphone with the mike gain control half-open. You should be able to hear the double-sideband signal in your receiver. Next adjust L3 and C3 for maximum output.

Observing a scope with the signal tuned in on 10 meters makes balancing the carrier simple. If you don't have a scope, monitoring your S-meter is sufficient.

The carrier suppression of this transmitter is sufficient to put it way below noise level on the 10 meter band.

The carrier balancing adjustments should be carried out with the rig installed in its Minibox cover on. Small holes should be drilled in the appropriate places for adjusting the coils and trimmers. Changing crystals may upset the carrier balance and you will have to make the balance adjustments again.

Today's receivers can receive a DSB signal with no difficulty. Most of the time, operators will not be aware you are using DSB. This rig could easily drive a 6146B which would give you about 100W PEP.

This transmitter can also be used on any other band by changing the coils and crystal. To do this, "borrow" coil data from other sections for the band you want. A homebrew VFO could also be built. It could even be made into a walkie-talkie or hidden in the glove compartment of a car and would be just the thing for talk-in at hamfests.

You can squeeze more power from Q2 by reducing the resistor values shown or by applying more voltage to the collector (not to exceed the rated voltage for the particular transistor you use). However, you could also drive Q2 into a nonlinear operating condition. For DSB, as in SSB, you must operate the PA in its linear operating zone.

By turning the R_b balance pot to either side, you can use low-level AM.

THE ICMITTER

This transmitter was designed with the QRP fiend in mind. It uses inexpensive RCA type CA3000 integrated circuits. As crystal oscillators they will work to about 10 MHz, and if you're lucky you just might be able to use them at 20 meters. This transmitter will work on 160, 80, 40, and maybe 20. This circuit will put out AM or CW at the flick of a switch.

The schematic in Fig. 3-10 should be self-explanatory. If you build on Vectorboard, use sockets. As a matter of interest, the 2N3904 is for AGC. Capacitor C1 should be chosen for good AGC action—try 0.05 or 0.1 μF. Transformer T1 is a swamped 10 to 5 kΩ matching transformer. The values shown for it give a fairly good match. The C2-L1 network should resonate at the crystal frequency. Tap L1 wherever you get the best match. The AM-CW function is switched by a 3-pole double-throw wafer or toggle switch. For AM work, use a 50 ohm mike (or something close).

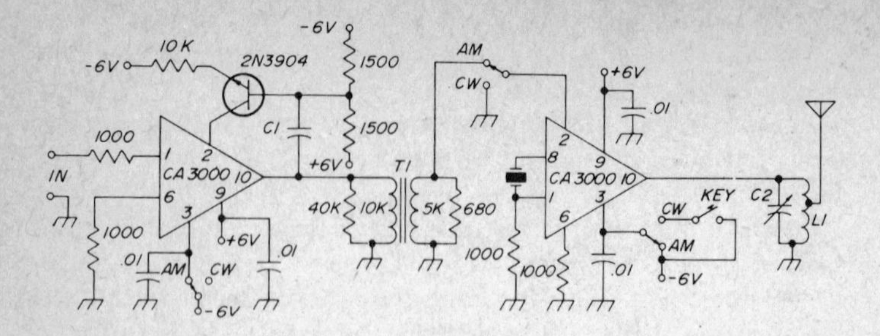

Fig. 3-10. See text for C1, C2, L1, T1. For CW you can also switch out the transistor if you want. Instead of keying the voltage, you could key the output. If you don't want AGC, simply ground pin 2 of the first IC.

These circuits are by no means completely original. Only partly. The were lifted from RCA publication ICAN5030. Also from File 121. Both are available from RCA for the asking. These papers provide much useful data on the CA3000.

ANOTHER IC TRANSMITTER

In these days of rising costs on practically everything, it is indeed a pleasure to construct and operate an efficient piece of radio equipment for pennies, and that is the reason for this circuit.

Circuit Description

The RF section of the transmitter, which consists of a crystal controlled Motorola HEP 53 oscillator followed by an RCA 2N4427 RF power amplifier running class C, develops about 1.25 watts output at 28 MHz.

The oscillator stage is a Colpitts type, providing excellent frequency stability with respect to supply voltage and temperature and delivers over 100 milliwatts to the input of the RF amplifier stage. Figure 3-11 shows the schematic.

The power amplifier stage uses a class C common emitter configuration and is modulated through the collector circuit.

A pi-network is used in the output resonant circuit to provide a measure of harmonic suppression. A double pi-network was originally used, but was later charged to a single pi. A Drake low-pass filter is used for additional harmonic suppression, as solid state finals are noted for putting out many harmonics of quite healthy levels.

Other types of transistors may be substituted for the particular ones used here, but remember that the final amplifier impedance might be different. This would have to be taken into consideration for modulator impedance matching purposes.

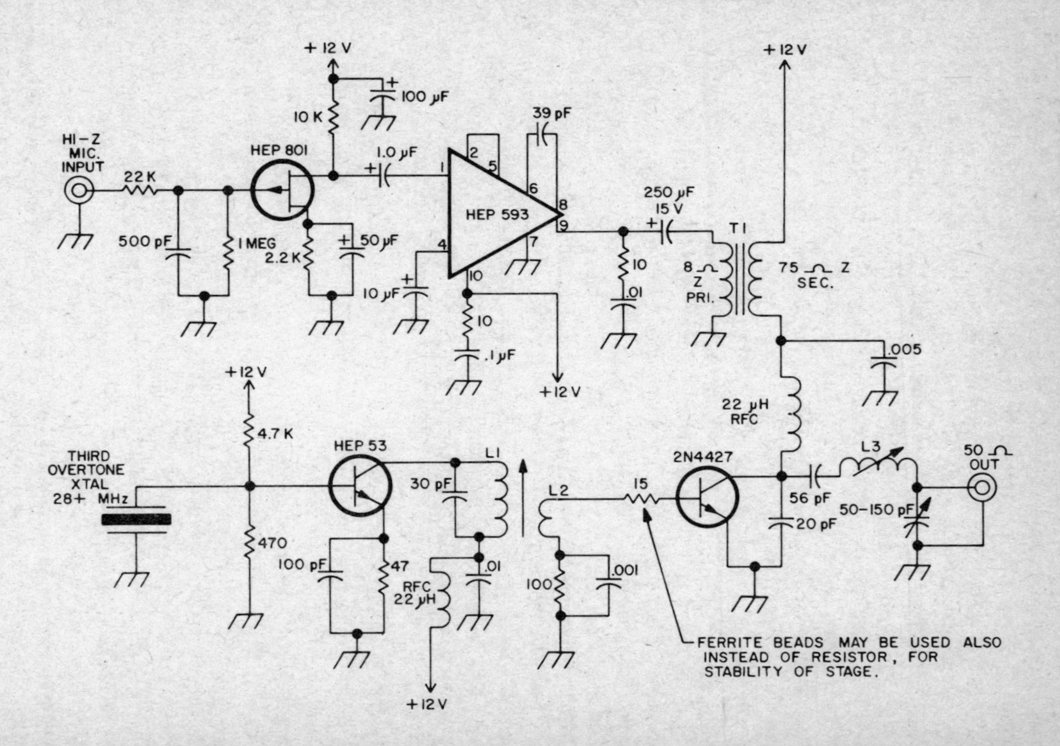

Fig. 3-11. L1-13 turns No. 28 P.E. CW, L2 2 turns link on cold end, L3 13 turns No. 28 P.E. C.W. All coils close wound on 6 mm diameter slug-tuned ceramic forms.

The modulator section starts out with an FET microphone amplifier (HEP 801) to obtain a high impedance input for the crystal microphone used. The output stage is a HEP 593, a 1W output IC having an 8 ohm output impedance.

Other audio amplifiers may be used here, such as the Amperex TAA-300, and these IC modules are preferred to standard audio boards for their compactness that lends itself readily to miniaturization.

The output of the audio stage is fed to an 8 ohm input modulation transformer that steps this up to the required modulating impedance, being about 75 ohms in my particular case.

An output transformer with a primary impedance of 75—80 ohms and a secondary of 8 ohms is used reverse connected as a modulation transformer.

If you elect to wind your own transformer, then a few words are in order here to assure good modulation results.

Be sure that the core that you use to wind this transformer has enough iron in it so that it cannot saturate during modulation, and also that there is ample space for the windings. The wire sizes used for transistor modulation transformers are larger than for comparable tube devices due to the current demands of the transistor and thus needs more room on the bobbin.

Good results can be had using an old 5W output transformer with 70 turns of No. 26 enamelled wire for the 8 ohm primary and about 210 turns of the same size wire for the 75 ohm secondary.

The final in most units draws about 180 mA. The supply voltage is regulated at 13.5V so the final impedance is about 75 ohms (collector voltage divided by collector current). The transmitter runs about 2W input and readily delivers 1W plus output.

Construction

This little transmitter is constructed in a small Minibox measuring approximately 10 × 5 × 4 cm, and a top cover is fabricated from an aluminum front grille from an old transistor receiver with the sides folded to form a shield and the seams cemented. When dry, the corners are filed smooth and given a coat of metallic green spray paint for a pleasing appearance.

The ends of the box are used for mounting the receptacles for the DC power input and microphone and the RF output to the antenna. Two phone jack types are used for microphone and power input and a BNC type connector is used for the antenna to readily accommodate coax cable.

The coil forms used are ceramic surplus, just over 6 mm in diameter and tuned with a powdered iron slug. All coils are close wound with No. 28 enamelled wire and given a coat of clear household cement.

Tuning Up

The tune up procedure is very simple and there should be no difficulty in obtaining proper output providing that all parts are good and the circuit has been wired correctly.

Connect some sort of dummy load to the transmitter output, such as a 51 ohm resistor (carbon) and a diode and voltmeter, or use a QRP wattmeter as I do.

Starting with the oscillator slug and with power applied, adjust the slug with an alignment tool for output indication on the meter, and turn the circuit on and off several times to make sure there is reliable starting of the oscillator each time. Then adjust the amplifier slugs and the pi-network capacitor for maximum output. Be sure to use heat sinks on the RF amplifier and on the audio amplifier in the interests of cool operation and efficiency.

For the final touch-up in the alignment procedure, install the low-pass filter in the line from transmitter to the dummy load and repeak all stages for maximum. The harmonic content will now be at the lowest level and will not influence the output reading.

Transistor output stages are not as tolerant of high standing wave ratios as tube circuits, so be sure that your SWR is kept to a low level at all times

Results

On the air results are most gratifying. Using a variety of antennas, modulation reports were very good indeed.

Using a 1/4λ whip attached to the side of the house about fifty to sixty contacts were made from both coasts and Canada with reports ranging from S-zero to S-9 plus. Always noted was Q-5 copy, attesting to the modulation capability.

With a 2-element beam, South American contacts were made as well as Central America and even the Windward Islands on a CQ!

STRAIGHTFORWARD SSB FOR 6 METERS

There have been many different circuits put forward over the years for getting an SSB signal on 6 meters. They range from the simple transverter to the full-fledged single band affair. This unit was designed to be simple yet effective; thus the PTT and single conversion with option of spot frequency injection or VXO. The 2E26 final provides about 30 watts PEP input; however, substitution of a 6146 with suitable changes in power supply voltages would almost triple the power input.

The Circuit

The block diagram of Fig. 3-12 illustrates the overall layout of the exciter. The 7360 oscillator balanced modulator is a conventional circuit in which the cathode, grid and screen form the oscillator circuit while the plates and deflection electrodes are utilized in the balanced modulator circuitry. The DSB signal produced in the 7360 tank circuit is passed through the 9 MHz filter at which point one of the sidebands is removed. After amplification in the 6AU6 IF amplifier, the 9 MHz SSB energy is applied to the 7360 balanced mixer. 41

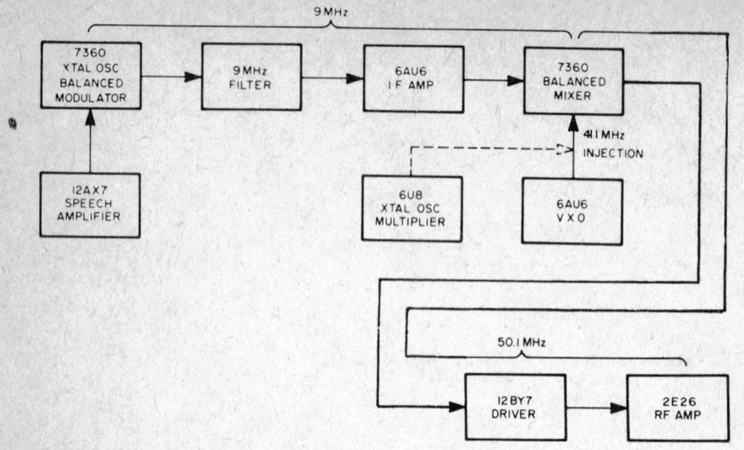

Fig. 3-12. Block diagram of the RF section. Single conversion path is illustrated.

MHz energy fed to the grid of the 7360 mixer is cancelled out in its tank circuit. The 5 pF butterfly capacitor is brought to the front panel as the "mixer tune" control. CW operation is provided through cathode keying of the mixer stage and has proved to be quite satisfactory.

The 12BY7 (Fig. 3-16) functions as a straight-through class A amplifier whose plate circuit trimmer appears on the front panel as the "driver tune" control. Link coupling to the grid of the 2E26 is used to further attenuate any undesired feedthrough. Neutralization of the 12BY7 was not found necessary because of careful shielding across the tube socket below the chassis. With suitable pin connection changes, a 6CL6 would serve as a good substitute for the 12BY7. The 2E26 final operates in class AB_1 in a conventional circuit employing capacitive neutralization.

Power Supply and Control Circuits

The power supply (Fig. 3-13) may consist of any common "TV" type transformer functioning in a dual voltage arrangement; the final's plate voltage being provided by bridge rectification, and the lower plate voltage by full-wave rectification through two legs of the bridge. Bias voltage for the final amplifier is derived from a reverse-connected 6 volt filament transformer connected to the used 5 volt filament winding of the power transformer. Transmit operation is provided by grounding the floating side of the 6 volt DC relay. This may be accomplished with a push-to-talk microphone or by an auxiliary "transmit" switch. When not transmitting, plate voltage is removed

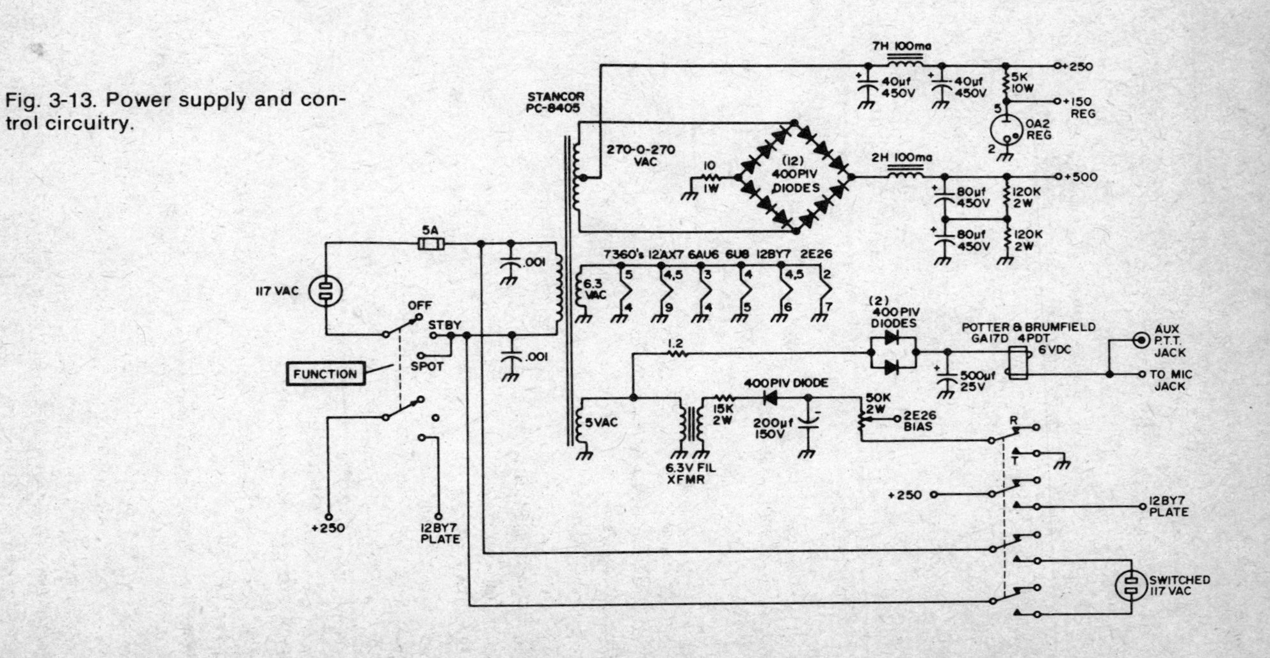

Fig. 3-13. Power supply and control circuitry.

from the 12BY7 to allow it to run cooler. The 2E26 bias is increased during standby by ungrounding one end of the bias adjust potentiometer.

Construction

As detailed in Fig. 3-14 the 7360 sideband generator and the 12AX7 speech amplifier are located next to the front panel. Consequently, the three controls and the jack associated with these stages can be mounted directly on the front panel without excessive lead length. The main chassis is assembled from aluminum, while the shielding is cut from thin brass sheet. Brass or copper partitions are placed across the IF amplifier, mixer and drive tube sockets. All ground connections can be made directly to these partitions eliminating the need for ground lugs. Two of the partitions, also serve as mounting for the mixer and driver tune controls. The 2E26 final is completely enclosed in a compartment at the rear of the chassis. A suitable enclosure can often be salvaged from the high voltage "cage" of older style TV receivers. Extensions to the front panel for the mixer, driver and final amplifier controls were made from 1/4 inch brass tubing. Short lengths of tight-fitting rubber tubing were used to couple the extensions to the capacitor shafts.

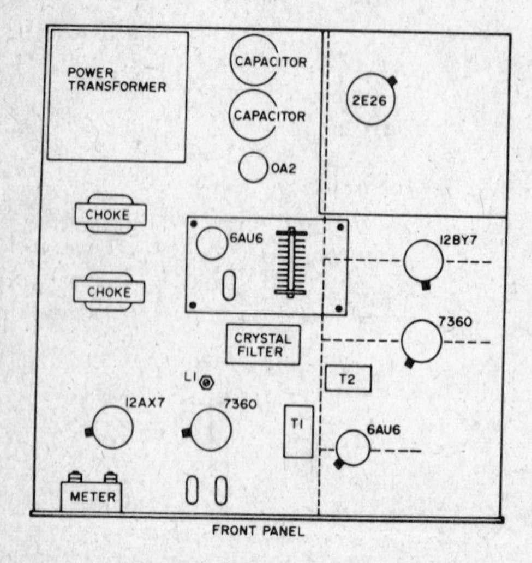

Fig. 3-14. Component layout viewed from above. Dashed lines indicate partition placement below chassis. The tab at each socket indicates position of pin 1.

There are two small subassemblies used in the construction. The bridge rectifier diodes are assembled on a phenolic board which is bolted vertically underneath the chassis. Small holes are drilled in the board; the leads of adjacent diodes are placed through a hole, bent over, clipped, and then soldered together. If desired, 270K resistors and .002 disk ceramics may be paralleled with the diodes to give voltage equalization and transient protection. The components associated with the RF output meter are mounted on a terminal strip which is bolted inside the 2E26 compartment near the antenna jack. A shielded lead carries the rectified voltage to the meter on the panel.

The 41.1 MHz injection (Fig. 3-15 or 3-17) is supplied by a VXO assembly built on a small plate and mounted above the main chassis. Certain of the VXO components such as the crystal socket and tuning capacitor are mounted 1/4 to 1/2 inch from the plate to minimize capacitance to ground.

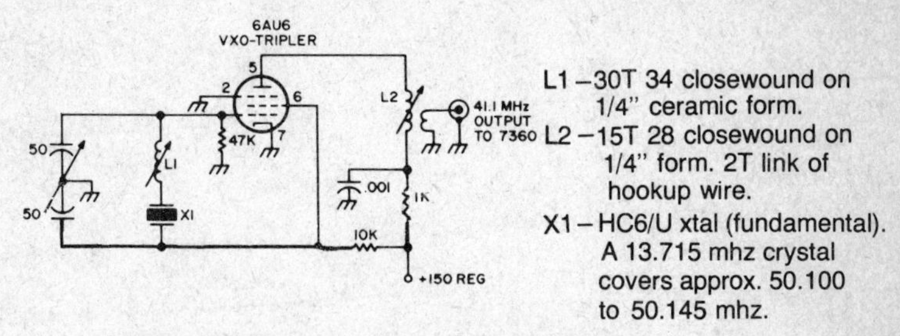

Fig. 3-15. Schematic diagram of the VXO tripler.

The usual VHF wiring techniques must be observed in order to produce stable TVI-free operation. All filament leads should be shielded and .001 disk ceramics should be placed from the hot filament lead to ground at each tube socket. In the higher frequency RF stages, lead length should be kept short. Grounding is done as directly as possible and it is desired to have only one or two common ground points for each stage. All plate voltage leads to the mixer, driver, and final are routed through the main partition with feedthrough capacitors.

Special Components

The 10.7 MHz IF transformers used in the 6AU6 amplifier are an older style of transformer using slug-tuned windings. They are more easily adapted than the smaller "K-Tran" type. In T_1 the primary winding was removed and replaced with a 3 turn link. The secondary of T_1 is moved down to 9 MHz by the addition of a mica capacitor. In a similar fashion the primary and secondary of T_2 are lowered in frequency.

Some will argue that L_1 should be bifilar wound. Both a conventional and bifilar winding were tried and no significant difference in carrier suppression was noted. The differential capacitor associated with L_1 is constructed by taking a regular air trimmer, setting its rotor at half mesh and after inserting cardboard wedges between the stator plates, cutting through the stator with a

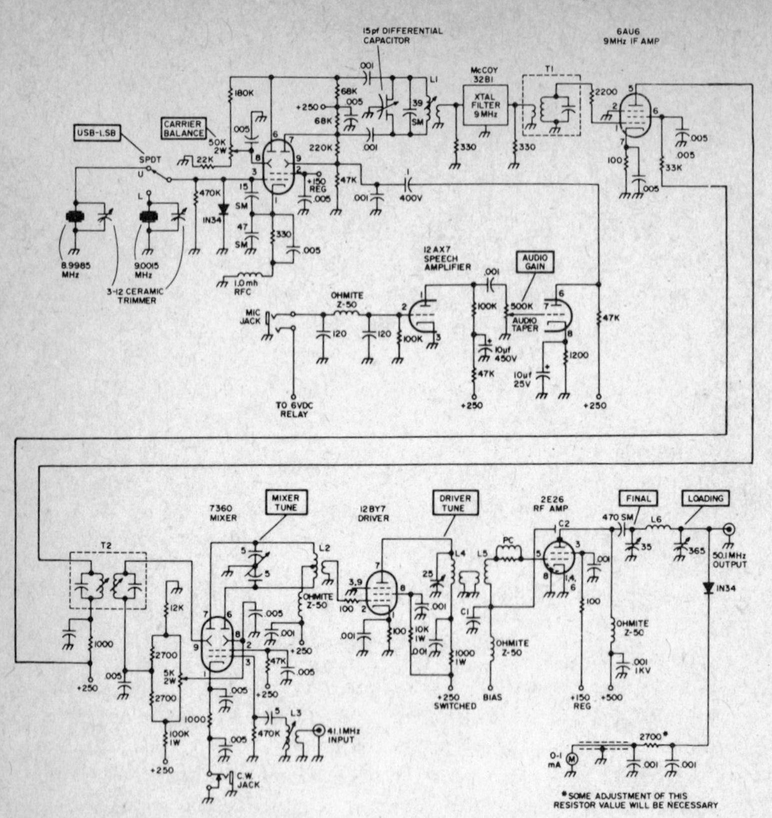

Fig. 3-16. Circuitry of 50 MHZ exciter. Controls brought to the front panel are shown in boxes.

fine tooth hacksaw blade. Of course this capacitor is available as a regular item if you can obtain one.

The 2E26 final is stabilized with capacitive neutralization. Capacitor C_1 should be mica and will be in the 330 to 560 pF range while C_2 is simply a stiff wire passed up through the chassis into the final amplifier compartment and placed near the tube plate.

Tuneup and Adjustment

It is assumed that the constructor will have checked all coils for approximate resonance (with the tubes in their sockets,incidentally). Applying line voltage and setting the function switch in the standby position will place plate voltage on all stages up to the 12BY7 driver. A VTVM with an RF probe is almost mandatory for tuneup. Place this probe at the 6AU6 grid pin and adjust the carrier balance control to secure a reading. With the differential capacitor set at center, adjust L_1 and T_1 for a peak reading. The differential capacitor is then adjusted for a null. Set the ceramic trimmers across the carrier oscillator crystals at minimum setting and while switching from USB and LSB positions check to see that the VTVM reading remains about the same. This indicates that the crystals are centered reasonably well on the filter curve. Proceeding next to the 7360 mixer, place the probe on pin 9 and adjust both windings of T_2 for a maximum reading. Then with 41.1 MHz energy applied check pin 3 for a reading. It should be about 1 volt with L_3 adjusted to resonance. The next step is to remove the 6AU6 IF tube to prevent 9 MHz energy from reaching the 7360. With the probe at the center-tap of L_2 adjust the 5K 2W potentiometer for minimum 41.1 MHz feedthrough.

Apply voltage to the 12BY7 with a temporary jumper and with the probe at pin 5 of the 2E26, peak L_2 and L_4 and adjust L_5 by the "squeeze" method. Moving the probe to the plate of the 2E26 adjust C_1 and/or C_2 for minimum feedthrough of 50 MHz signal.

A fairly satisfactory alignment of the carrier oscillator crystals can be accomplished as follows: with the RF probe at pin 9 of the 7360 mixer and with some carrier inserted increase capacity across the 9.0015 MHz crystal in order to move it down into the filter passband. Note the VTVM reading and then adjust the trimmer to yield a reading approximately three-quarters of the passband reading. Set the 8.9985 MHz trimmer to give a similar reading.

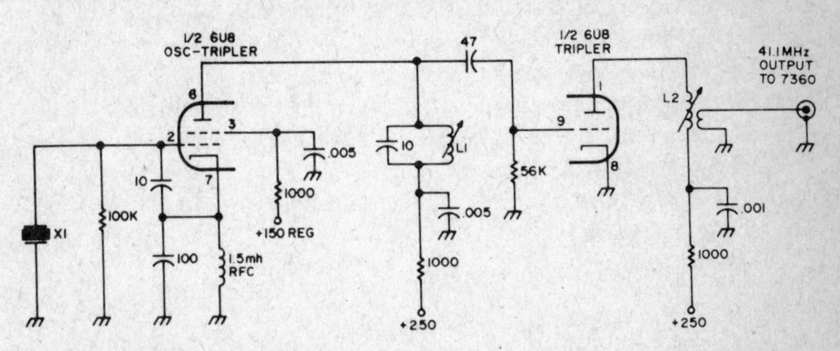

L1 – 26T 28 Closewound on 1/4" form.
L2 – 15T 28 Closewound on 1/4" form. Link 2T of hookup wire.
X1 – FT243 surplus xtal. A 6.850 MHz crystal will give a spot frequency of 50.100 MHz.

Fig. 3-17. Diagram for the oscillator tripler.

Afterthoughts

Some builders will wish to modify portions of the circuitry to fit their likings and their junk boxes. One improvement would be a four position function switch to allow for a "manual" operation position. In the control section, by use of the negative bias voltage to disable the low-level RF states, one could use a relay with fewer contacts. Furthermore, only one side of the AC line need be opened for the switched 117 VAC.

A BETTER BALANCED MODULATOR

Balanced modulators have been a vital part of amateur transmitters ever since development of interest in single-sideband. Although improvements have been made from time to time, most circuits still require resistive and capacitive balancing by means of adjustable elements. Unfortunately, the long-term and temperature stability of this approach is dependent on the characteristics of the pots and trimmers used. Temperature effects can be minimized by use of compensating elements but the whole process now becomes very involved. Circuit unbalance means not only carrier leak but greater distortion for a given audio input. The purpose of this section is to provide an introduction to a circuit capable of providing the same order of carrier rejection as conventional modulators with much better temperature and time stability. It can do this with absolutely no initial or routine maintenance type adjustments.

The circuit shown in Fig. 3-18 can be recognized as a ring modulator whose operation is adequately described in the literature. T1 and T2 are usually wound in trifilar form on toroidal ferrite cores since this provides close coupling and more uniformity than other types of construction. Randomly selected diodes will provide about 15 or 20 dB of carrier suppression below the double-sideband output. This figure can be increased to about 30 dB at IF frequencies with careful matching of the diodes. It is soon evident to the experimenter that any further improvement can only be obtained by the addition of balancing adjustments. The reason for this threshold is a basic limitation of the trifilar winding coupled with the method used for driving the transformer. Due to the single-ended input, one side of the secondary has more capacity to ground than the other. This unbalanced capacity has increasing effects as the operating frequency is raised.

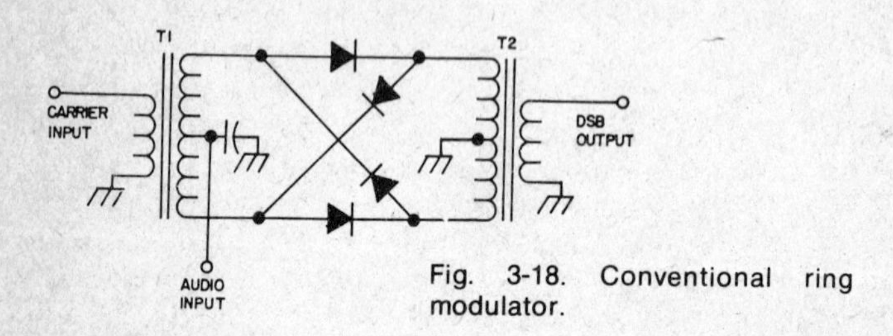

Fig. 3-18. Conventional ring modulator.

The easiest way to neutralize the effect of this unbalanced capacity is to isolate the return side of the primary from ground. This can be done by the arrangement shown in Fig. 3-19. T1 is a bifilar wound transformer with unity turns ratio which serves to isolate the primary return to T2. It may not be immediately obvious, but there is no loss of power in this transformation. The requirements that must be met are that the turns ratio be unity and the coefficient of coupling must approach unity. The mathematics used to prove this is not presented here since it is not common knowledge to most amateurs. The above requirements can be realized by the use of a bifilar winding on a toroid core.

The complete schematic of a practical modulator for use in the range of about 2 to 20 MHz is shown in Fig. 3-19. No tuning is used since T1 through T4 are broadband in nature with the impedance levels shown. When terminated with a 50 ohm load at the output, the impedance looking into the carrier input is 50 ohms. With 1 volt RMS of carrier injection, the impedance at the audio input is approximately 12 ohms. From the impedance levels it can be seen that this circuit is suited to transistorized circuitry but can be adapted for tubes.

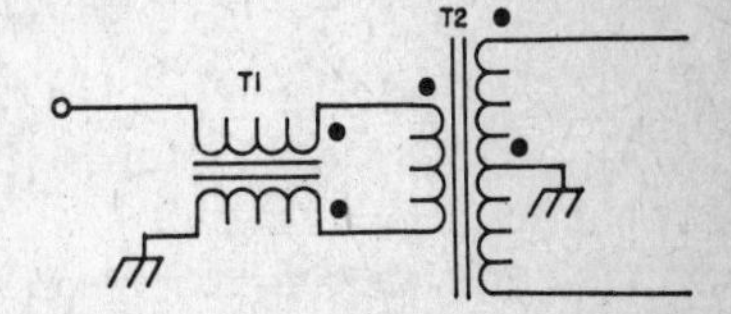

Fig. 3-19. Method of reducing effect of unbalanced capacity to ground in T2. Dot markings are explained in Fig. 3-21.

Measurements were made at frequencies of 3 and 9 MHz, corresponding to the usual range of IF frequencies used in amateur equipment. Optimum carrier injection at both frequencies is about 1 volt RMS. Optimum audio level at this point is about 200 millivolts RMS total (140 millivolts per tone for a two-tone signal). At 3 MHz, these operating conditions result in a double-side-band output of 137 millivolts RMS with the carrier suppressed by 52 dB below either tone. Intermodulation distortion is better than 50 dB down. At 9 MHz, DSB output is 130 millivolts RMS with a carrier suppression of 45 dB and IM rejection of better than 50 dB. At 30 MHz these figures will be degraded by a few dB. High frequency performance can be improved by using fewer turns on the transformers. A higher carrier level will result in lower distortion but less carrier suppression. Less carrier injection will have the opposite effect. The figures given above seem to be a good compromise.

It should be evident that the circuit can be stuffed into a very small space and hidden on a printed circuit board or in a corner of a chassis since no access to it is needed. An unclad epoxy board should be used to minimize unbalanced coupling to ground. The circuit layout should be symmetrical (see Fig. 3-21). Actually, the schematic gives a good physical layout and was used in the experimental unit. T1 and T2 should be placed at electrical right angles to eliminate mutual coupling. The same goes for T3 and T4. The close proximity of the cores makes this necessary as will become evident to anyone who experiments with the position of the cores while monitoring the carrier suppression.

The bifilar winding is formed by taking two pieces of wire, each of sufficient length to wind the required number of turns on the core, and twisting them together to form a twisted pair. This composite wire is then wound evenly around the core. The trifilar winding is exactly the same, except a third piece of wire is added to form a twisted triple. Referring to Fig. 3-20, 20 turns bifilar wound actually means 10 turns of a twisted pair. Similarly, 30 turns trifiler wound means 10 turns of a twisted triple.

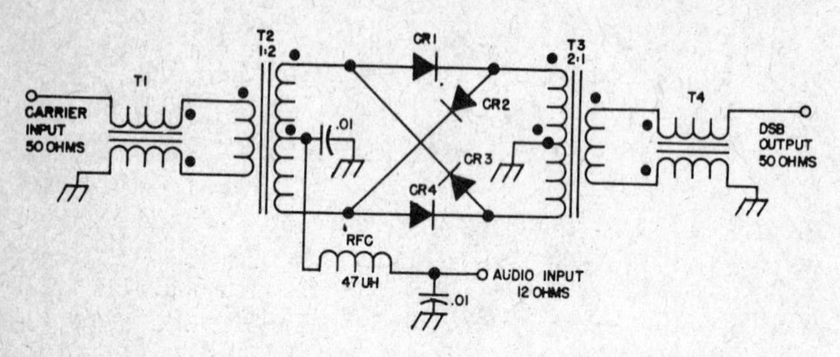

Fig. 3-20. Schematic of the wideband balanced modulator. The dots indicate winding polarities. Bifilar and trifilar winding is discussed in the text. Coil data: T1, T4-20 turns No. 32 enamel bifilar wound on 0.23 inch o.d. Ferrox cube toroid core (3D3 type ferrite). T2. T3- 30 turns No. 32 enamel trifilar wound on same core as T1 and T4. CR1, CR4, hot carrier or high speed silicon diodes.

The diodes should be matched by choosing those with the closest forward resistances on the X1 ohms scale of a VTVM. Those with lower forward resistances are better from an efficiency standpoint. Most any silicon diode will have much higher back resistance than is necessary for proper operation. The zero-basis capacity is a more important consideration and should be no more than a few picofarads. The lower the better. Germanium diodes should not be used since their characteristics are not suitable for use in this circuit as regards efficiency and distortion. Hot carrier diodes have proven themselves more desirable at the higher frequencies. Another good choice for the diodes would be the RCA CA3019 which is an integrated circuit diode array. The advantage of using a unit like this is the excellent matching and temperature tracking of the diodes.

The modulator can be unbalanced by inserting a variable DC voltage of 0 to about ± 1.5 volts at the audio input. This should be fed through a choke of 200 mH or more to avoid distrubing the audio frequencies. It is also worthwhile to mention that the load resistance at the output terminal should be kept in the vicinity of 50 to 100 ohms for proper operation with the circuit constants shown. Probably the most reliable way of obtaining a constant 50 ohms is by shunting the output with a 50 ohm resistor and using a buffer amplifier.

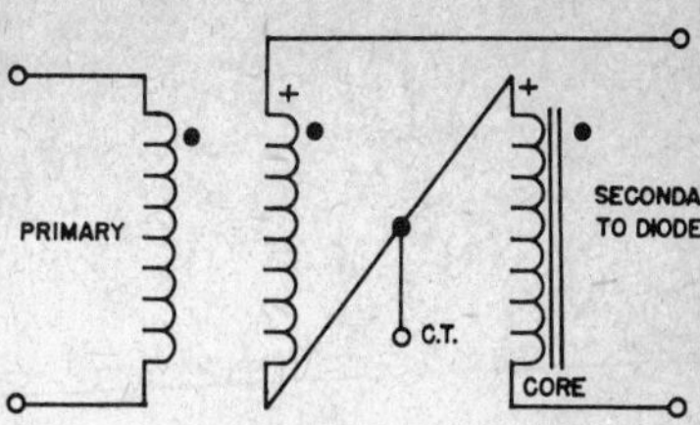

Fig. 3-21. Detailed illustration of the trifilar transformers, showing hookup of the three windings. The dots indicate that a current entering the primary in the direction shown will induce positive voltages at the ends of the secondary windings marked by the dots.

This circuit is not limited to use as a balanced modulator. It functions very well as a low distortion product detector by adding a step-up transformer at the audio terminal to bring the impedance level up to 10 or 20K ohms. With an IF input of about 100 millivolts RMS, a few volts of audio will be produced at the secondary of the matching transformer. This can be fed directly into a power output stage using a pentode. A lower step-up from the detector should be used for a transistor output stage. The dynamic range of this detector is on the order of 130 dB. The addition of a third trifilar transformer makes the circuit useful as a well-balanced mixer over the range of 2 to 10 MHz as shown in Fig. 3-22. Dynamic range is about 130 dB with suitable diodes (hot carrier or very high speed switching types). Of course, a filter is needed after the mixer to remove the unwanted sideband. A voltage-controlled attenuator can be realized through the use of a variable DC voltage applied to the audio terminal. The IF signal to be controlled is fed into either of the other ports and is taken out from the port which is left. Signal input at the RF ports should not exceed a few hundred millivolts to avoid distortion. A little thought will reveal many other applications of this circuit.

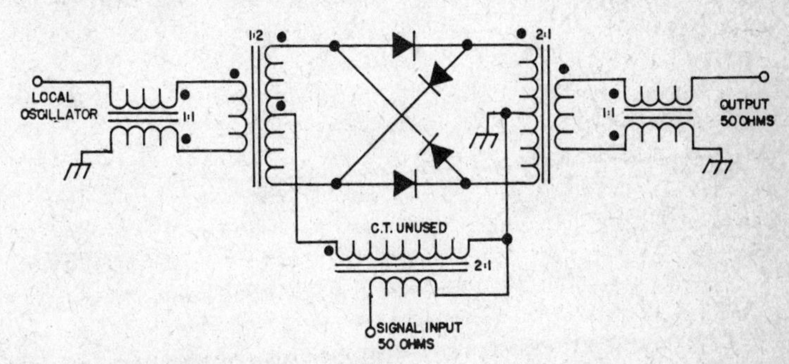

Fig. 3-22. Wideband balanced mixer. Transformers and diodes are the same as in Fig. 3. Signal levels should be the same as for balanced modulator service.

A STABLE VFO FOR SSB

One of the main problems in building SSB equipment is the need for a really stable VFO. Most commercial units are mechanical nightmares which have various types of temperature compensation. Adequately compensating these type VFOs is often beyond the means of the average ham and his limited test equipment. Even so, these best commercial units only say that eventually they will settle down to a 100 cycle drift per some time or other. The three main causes of drift or frequency change in the usual order of difficutly are: 1. Heat, 2. Mechanical Stability, 3. Voltage variations. Therefore, to have a stable VFO, simply eliminate these three items. It really isn't that hard.

Instead of isolating the tuned circuit from any heat producing sources, isolate the whole circuit from any heat source. That is, put the whole thing on a separate chassis. Next, use transistors at very low power levels. Finally mount the transistors in and the components on a 1/8 inch thick aluminum chassis. The thirty milliwatts of input power is dissipated into about 2 pounds of aluminum. Thus room temperature prevails.

The "wobbliest" part in any variable capacitor VFO is, of course, the variable capacitor. Using a ruggedized capacitor with small, thick plates and double end bearings is essential. The capacitor must be firmly mounted, thus the 1/8 inch thick chassis again. The coil and *other* components must be rigid also.

Finally, since very little power is used and that at a low voltage, batteries are ideal as a stable power source.

Construction

Figure 3-23 shows the schematic. Silicon transistors are used because they tend to change characteristics much less than germanium units with temperature changes. Since the input capacitance of the 2N2219 is only 20 pF, much less capacitive swamping is needed. In larger capacitors, the actual capacitance change per degree of temperature change can be larger than that of the transistor. Since the 2N2219 is a high gain, high frequency transistor, it can be very loosely coupled to the tuned circuit and still function well. A much less expensive 2N697 or other transistor could probably be used by increasing the coupling somewhat. All fixed capacitors are silver mica units for best stability. The variable capacitor is a Johnson Type R ruggedized unit with wide (.071) spacing. The actual value isn't too important since C_2 and C_3 are used to set the value of capacitance across L_1. The series-parallel combination of C_2 and C_3 reduce the range of C_1 to the value needed, and more important, make any changes in C_1 due to heat or vibration much less noticeable. L_1 is a coil from an ARC/5 unit. Be sure to carefully clean out all extraneous windings and make sure no turns are shorted.

The VFO is constructed on chassis which is bent from a single piece of 1/8 inch thick aluminum. This is the basis of the exceptional mechanical rigidity. A second piece of 1/8 inch aluminum forms the front panel and holds the Millen 10039 dial. These must be fastened rigidly together.

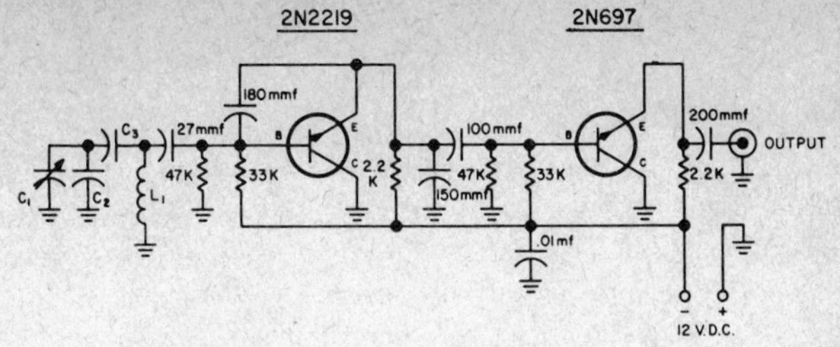

C1 Johnson Type R, ruggedized variable .071 spacing 3" long.
C2 82 mmfd
C3 200 mmfd
L1 21½ turns of #16 wire on 1¼" dia. ceramic form.

The variable capacitor is not only bolted to the chassis by the mounting feet, but spacers are glued in under the ends of the ceramic supports to take up any torque forces. A large hole was cut in the chassis and the ceramic coil also glued in place. The front panel extends above the dial so that the case for the unit does not get too close to the "hot" end of the coil and thus lower the Q. All components are wired to ceramic insulators or standoffs. All components are also glued to the chassis for maximum rigidity. The transistors are mounted upside down in press fit holes in the chassis and also glued in place. This "glue" is Ross epoxy weld which is available in most dime stores. Most two tube epoxy glues will work well.

The output of the unit is about 6 volts peak to peak of nice clean sine wave. If the output is not good looking, juggle the values of the 180 and 150 pF swamping capacitors. This particular unit covers from 4.95 to 5.6 MC for use with a 9 MC crystal filter. When in its case (a box built of 1/8 inch aluminum naturally) and when the unit has reached room temperature, the total change in frequency around the nominal value should be less than 25 cycles, with reasonably constant room temperature. Don't try to prove it by your receiver; it took a Hewlett-Packard 5243L frequency counter to prove the VFO even changed. There is a little more drift right at turn-on but nothing like those tube units. Since the whole thing only draws 4 milliamps, there is no reason why it can't be left on continuously. Size D cells or lantern batteries will last practically for their shelf life.

Chapter 4
RF Power Amplifiers

2 KW PEP BUILDING BLOCK LINEAR

The amplifier described in this section is a particular case in point. It illustrates a particularly simple construction technique and illustrates more than just the construction of a specific linear. By selection of the number and types of tubes used, an amplifier of anywhere from the 200 to 2000W PEP level can be constructed. The cost is very low on a watts to dollar ratio and can easily approach 20W of input power per dollar of material cost.

Planning the Linear

If one looks at the overall diagram of any linear, it presents a rather complex picture; that is, perhaps not so much in terms of electrical functions of the various parts but in terms of mechanical construction when one starts to consider in *detail* how the amplifier will be built and how to arrange all the components on a chassis, do the mechanical work to mount the various components, etc. It is usually at this point that the home brewer starts to falter and forgets about the entire project as being too complex and, particularly, too risky. The risk factor comes into play since if one assembles all the necessary parts—tubes, cabinet, chassis, transformer, etc., and then the amplifier project doesn't work out, one will be left with a fair amount of money expended for no real return.

The risk in construction can be almost completely eliminated if one looks at the construction of the amplifier from a different point of view rather than as a complete whole. As shown in Fig. 4-1 the amplifier can really be considered to be made up of three main "sub-blocks"—a power supply, a pi-network output circuit and a sub-block containing the actual tube circuitry with the antenna switching. The power supply sub-block should really present no problem. Any number of power supply circuits are available which have been thoroughly proven in operation and one can even purchase relatively inexpensive high power capacity power supplies on the surplus or used equipment market. Any amateur should be capable of building a simple bridge type rectifier power supply. The pi-network output circuit sub-block should present even less of a problem. Pi-network circuits for single band or multiple band use using conventional components or using toroid coils have been thoroughly described in many handbooks and articles. With a bit of patience, anyone can find the correct tuning conditions and component values to use with a pi-network circuit on any given band.

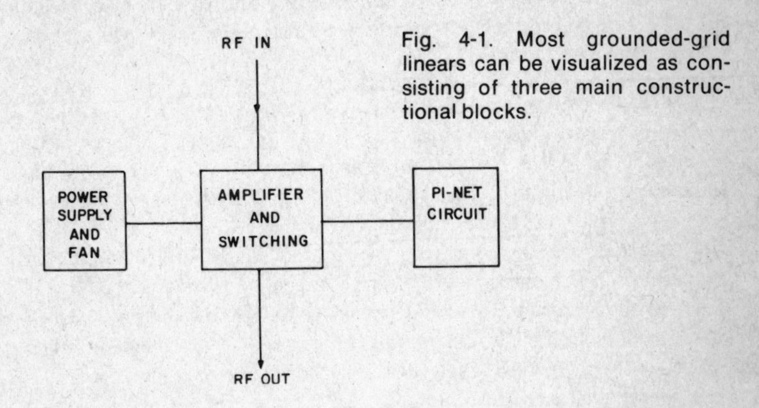

Fig. 4-1. Most grounded-grid linears can be visualized as consisting of three main constructional blocks.

However, it is the tube circuitry or actual amplifier sub-block that is at the crux of the whole linear construction project both in an electrical and in a mechanical sense. If this sub-block can be electrically complete so that the *only* other electrical wiring that is necessary is the power supply and the pi-network circuitry *and* if this amplifier sub-block can be easy to construct, almost 70 to 90% of the work in building a linear will work. To take a very conservative approach, one need invest initially only in the components necessary to build the amplifier sub-block without risking a total investment in all the linear components that might not be used should one decide to give up the project.

A 6KD6 Amplifier Sub-Block

This article describes in detail the construction of a linear amplifier sub-block designed for use with one to five 6KD6 tubes. Other tubes can be used as well, using the same construction ideas. Five 6KD6 tubes can provide a maximum input of 2 kW PEP on SSB. The rock-bottom price found for these tubes was $1.50 each. They are available new from almost any discount tube dealer for around $5.00 each. Wherever these tubes, or other TV sweep type tubes for use in a linear are purchased, be sure to obtain tubes of the same manufacture. This applies also in case a tube needs to be replaced.

The question of distortion products might be raised when beginning to describe any linear amplifier using TV type sweep tubes. Rather than get involved in this subject, which is not the purpose of this book, it is sufficient to say that the linear described will have 3rd order intermodulation products of −25 to −30 dB. Also, the purpose of this book is not so much to emphasize the use of a particular tube type as to emphasize a particular construction technique. If extremely low intermodulation products are an overriding consideration, one can easily use regular transmitting tubes in place of the 6KD6s. 572Bs, for instance, would lend themselves nicely to the type of construction presented here.

Sub-Block Circuitry

If one extracts from a linear amplifier all the circuitry which can be combined to make it as complete as possible except for the power supply and output circuit, the result will be the circuit of Fig. 4-2. In this case, the antenna switching circuitry has been included in the sub-block. The antenna relay used is a triple pole type so it can also switch a 12V zener regulator in during transmit to bias the tubes properly. During receive the tubes are cut off by the higher unregulated bias voltage to prevent noise generation. The sub-block contains within itself all the external connections needed except the power line connection for the power supply (assuming the power supply will be in the same overall enclosure as the amplifier). The pi-network output circuit need only be connected between the plate circuit and the feed-through insulator which connects to the antenna relay. The power supply requirements to the amplifier sub-block are: a high voltage from 800 to 900V (at 200 mA peak for each tube used), a filament voltage of 6.3V (at 2.8A for each tube used) and a bias voltage (unregulated) from −25 to −100V at 50 mA. The bias voltage is not critical as long as it is more than 25V so the tubes are completely cut-off during receive. The series resistor for the 12V zener will have to be adjusted according to the bias voltage so that the zener regulates properly without undue heat dissipation. The above are the only external power supply requirements, for the antenna relay is powered by the bias supply line. A monitoring circuit for relative power output is included in the sub-block which would be used to drive a 1 mA meter via a 25K series connected variable resistor. These components would be mounted on the front panel of the overall linear enclosure. This meter plus an ammeter (0—1 or 0—1.5A) connected in series with the plate supply (the ground lead, if possible, for safety) will suffice for metering of the amplifier. A cooling fan is necessary if the amplifier is to be used at its maximum input rating. This can be simply connected across the primary of the power transformer so it is activated at the same time as the

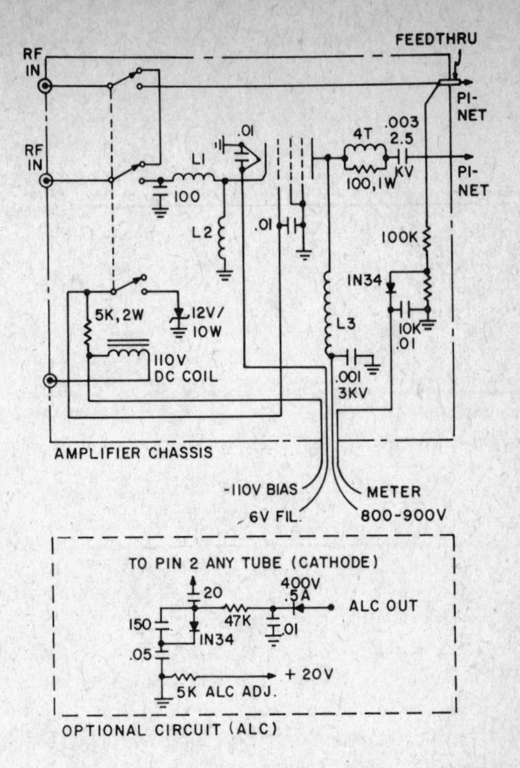

Fig. 4-2. Amplifier sub-chassis diagram. Up to 5 tubes can be wired in parallel. L1, L2 and L3 are discussed in text.

power supply is turned on. A 115V fan which is popular is a Barber Coleman D YAF 761-110 with their AYFA-403 fan blades. The larger electronic mail order houses can supply this item. Other suitable surplus type fans can also be used, but there must be a good flow of air (about 100 cfm). The 6KD6 tubes run hot and I have seen several removed from TV sets where under normal operating conditions the glass envelopes have started to distort due to heat problems.

The external connections for the amplifier sub-block consist of the usual RF input and output connectors, a relay connection for switching the linear to transmit via the exciter, and an ALC connector for use with the exciter. The latter is an optional feature which can be used if the ALC voltages needed by the exciter from the driven linear are known.

The actual circuitry of the amplifier sub-block is extremely straightforward. The 6KD6 tubes operate class AB in a grounded grid configuration. The tubes are not individually current balanced and this is not necessary as long as the tubes used are of the same manufacture. The input circuit to the cathodes of the tubes contains a low-pass filter (L1 and the 100 pF capacitor). This filter helps to prevent further amplification of harmonics from the exciter used and improves the overall distortion rating of the linear. L2 is an RF choke which will pass the cathode current of the tubes while still keeping the cathodes above RF ground. The choke is quite easy to construct and consists simply of close wound turns of #24 wire on a 1.25 cm by 7 cm long ceramic or high temperature plastic form. L3 is the plate RF choke. One has the choice of using one of two simple approaches to construct a suitable RF plate choke. The choke used is wound on a 9.5 × 2 cm ceramic form with #24 wire. It consists of five sections, each section separated by about 1 cm. The top section consists of 11 turns and the following sections are 16, 27, 33 and 43 turns. A somewhat simpler RF choke can be made if desired, but then two chokes have to be used in series. The first choke is wound full with #24 wire except for 0.5 cm at the top and bottom. The second choke is an Ohmite Z-50. The .001 μF plate bypass capacitor is used after the Z-50 choke, not between the two chokes.

Construction

The amplifier sub-block is fully contained on a standard 5 × 7 × 2 in. aluminum chassis. The tube sockets are arranged in a semicircular fashion with the RF plate choke in the center. The circle is arranged so the center of the end tube sockets are about 2.5 cm away from the side of the chassis on either side. This will permit the tubes to be separated enough to allow good air passage while not allowing the RF leads to become too long. The RF plate choke is about 2.5 cm away from the edge of the chassis also. The tube sockets are oriented so that the number two pin (cathode) on the socket always faces the center of the semi-circle. At the rear of the chassis, the connectors used are a SO-239 for the RF input, phono jacks for the relay control and ALC output (if used), and another SO-239 for the RF output. The ALC adjustment potentiometer is also located on the rear panel of the chassis. The use of single hole mounting SO-239 connectors and phono connectors greatly simplifies the mechanical work involved.

At the underside of the chassis the antenna transfer relay is mounted on one side wall above a tube socket and close to the front panel. There is no need to mount the relay in any exact position but it should be mounted so the length of the wire between the RF output connector to the pi-network is as short as possible. This is to minimize coupling between the output line and the rest of the amplifier circuitry. A short piece of coax is used to go from the relay to the input connector. The zener diode for bias regulation may be mounted in any convenient location near the relay. The cathode of the diode is directly bolted to the chassis, as the chassis functions as a heat sink for the diode.

The wiring of the cathode circuitry of the tubes deserves mention. This wiring should be kept as short as possible and an insulated terminal post is mounted for this purpose in the center of the ring of tube sockets. The post can be held in place by the same screw which supports the RF plate choke on the top of the chassis. The cathode choke, L2, is mounted on the side wall of the chassis opposite the relay. The low pass filter coil, L1, is wired between the relay contact and the cathode terminal post in the center. Also the 100 pF capacitor associated with the filter is wired in beneath L1 and utilizes the bottom of the center terminal post for ground. You can also use the ground post of the small terminal strip located in the middle of the inside front panel. This latter terminal strip supports the few components for the relative RF output level circuitry. The 100K resistor in this circuit can be seen going between the relay and the terminal strip. Each cathode pin of the tube sockets is connected with an equal length lead to the center terminal post. 0.5 cm wide copper material is preferred, but large size wire such as # 12 or larger can also be used. Wire the tube sockets as fully as possible before mounting them, then complete the wiring of the sockets on the chassis. Mount the rest of the components on the chassis and then complete the final wiring. The whole process may sound a bit involved, but actually once the chassis has been drilled and the components assembled the whole wiring can be completed in an afternoon.

All the pins on the tube sockets are wired to ground via the ground lugs on the sockets before mounting the sockets on the chassis except pins 2 (cathodes), 12 (filaments), 5 and 9 (grids) and 7 (internal tube connections). The two .01 bypass capacitors associated with each socket are wired in at the same time. Once mounted on the chassis, the filament pins (12) and grid (9) are wired together from socket to socket. The wiring to the power supply leaves the chassis via a grommet on the side wall opposite the relay.

The mechanical working of the chassis can be done completely with hand tools if desired. The only large holes which require work are the ones for the tube sockets and the SO-239 connectors. These can be made with a punch or with a nibbling tool and round file. There is little sense in investing in a punch if one is not likely to use it again, but a nibbling tool can be used for a variety of chassis work.

Pi-Network Circuits

A number of pi-network circuits have been tried which will work properly with the amplifier. If the linear is to be used on only one or two bands, the pi-network circuitry becomes quite simple and one need only experiment with the component values that produce the highest power output. Extract the single band component values from the multiband circuits shown in Fig. 4-3. Under properly loaded conditions and with 800—900V on the plates, the CW plate current will run 200—220 mA per tube used. For the full five tube circuit, the plate current for CW can be loaded up to between 1. and 1.1A. The key-down periods must be kept to 10 seconds or less.

If an 80 to 10 meter multiband pi-network output circuit is desired, Fig. 4-3 presents two good possibilities. The first circuit has an advantage in that it requires only a conventional 2 pole, 5 position switch for all of the capacitor and coil switching functions. The second circuit avoids the use of auxiliary fixed capacitors for the lower frequency bands but the plate tuning capacitor shaft requires an insulated shaft coupling and mounting above ground. In any case, the final coil tap positions should be experimentally peaked up for maximum power output on each band. The plate tuning capacitor in either circuit should have 1000—1500V spacing while the output loading capacitor need only be a broadcast band type. The fixed capacitors should be mica or ceramic types rated from 1 to 3 kV. Ceramic transmitting type capacitors such as the Centralab 850S are perferred if one can find them surplus at a reasonable price.

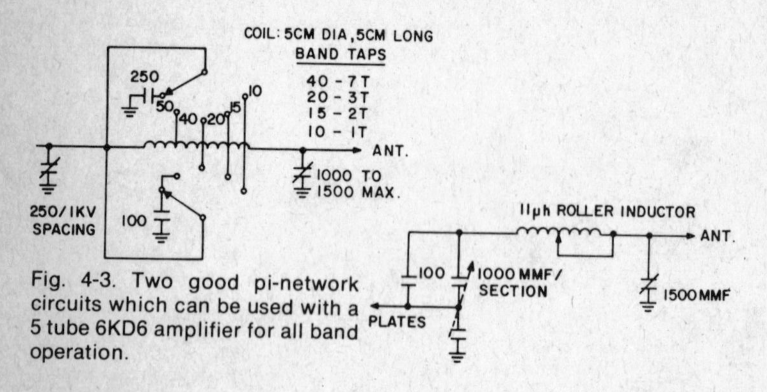

Fig. 4-3. Two good pi-network circuits which can be used with a 5 tube 6KD6 amplifier for all band operation.

Final Enclosure Assembly

Exactly how one wishes to finish up an amplifier in an enclosure depends to a good degree on how fancy an appearance is wanted, so this part of the project is pretty well left up to the desires of the individual builder. Certainly, if the power supply is ready, the amplifier chassis finished and the parts for the output circuit secured, the grouping of these items in an overall enclosure is not a big problem.

Although not one of the fanciest approaches, but certainly a simple and adequate one which has been used for a number of amplifiers, is standard large size metal utility cabinets (Bud CU or AU series, Par-Metal MC series) as the overall linear enclosure. These enclosures are inexpensive ($5 to $6) and come in black wrinkle steel finish or bare aluminum. The black steel cabinets are attractive in that they need no further painting. These enclosures are typically square looking with removable top and bottom covers. The mounting of the amplifier sub-chassis in such an enclosure is particularly simple since the connectors on the back of the amplifier sub-chassis can be used to secure the sub-chassis to the enclosure. Figure 4-4 shows how the total linear might be assembled in such an enclosure. For ventilation purposes, a series of holes should be drilled in the two side walls of the enclosure which are in the path of the air flow produced by the cooling fan used.

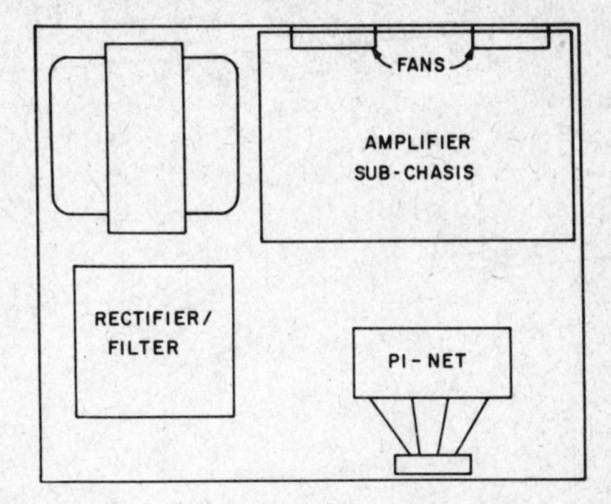

Fig. 4-4. There are many ways the amplifier subchassis can fit into an overall enclosure to fit almost any taste. Build the sub-chassis first and then integrate it into an overall enclosure.

THE TINY TIM LINEAR

The *Tiny Tim's* linear is only 3 in. high, 12 in. wide, 10 in. deep, and will tune 80—10 meters. It also includes a built-in antenna relay and an RF output indicator. This compact linear will run 500W DC input on SSB with 1500V on the plates of the two 811As in grounded grid.

Tiny Tim Circuit

The 811As are used because they do not need a bias supply and are not expensive for the power they deliver. They also work well following a low-power exciter, and require as little as 20W driving power on all bands. The circuit is nothing startling, but a proven one squeezed down for simplicity and compactness. Many parts can be scavenged from the junkbox, TV, and old radios. The filaments are kept above ground with a homebrew choke and driven directly from coax input from the exciter. The plate choke is wound on a plastic rod and held in place at the ends. The plate tuning capacitor is from an old ARC-5 transmitter, and the antenna tuning capacitor is a two-gang type from an old radio with the stators wired in parallel. The tank coil is smaller in diameter than the garden variety used in most, but the Q is correct and works well. The extra contact on the tank switch can be used on 80 meters to add

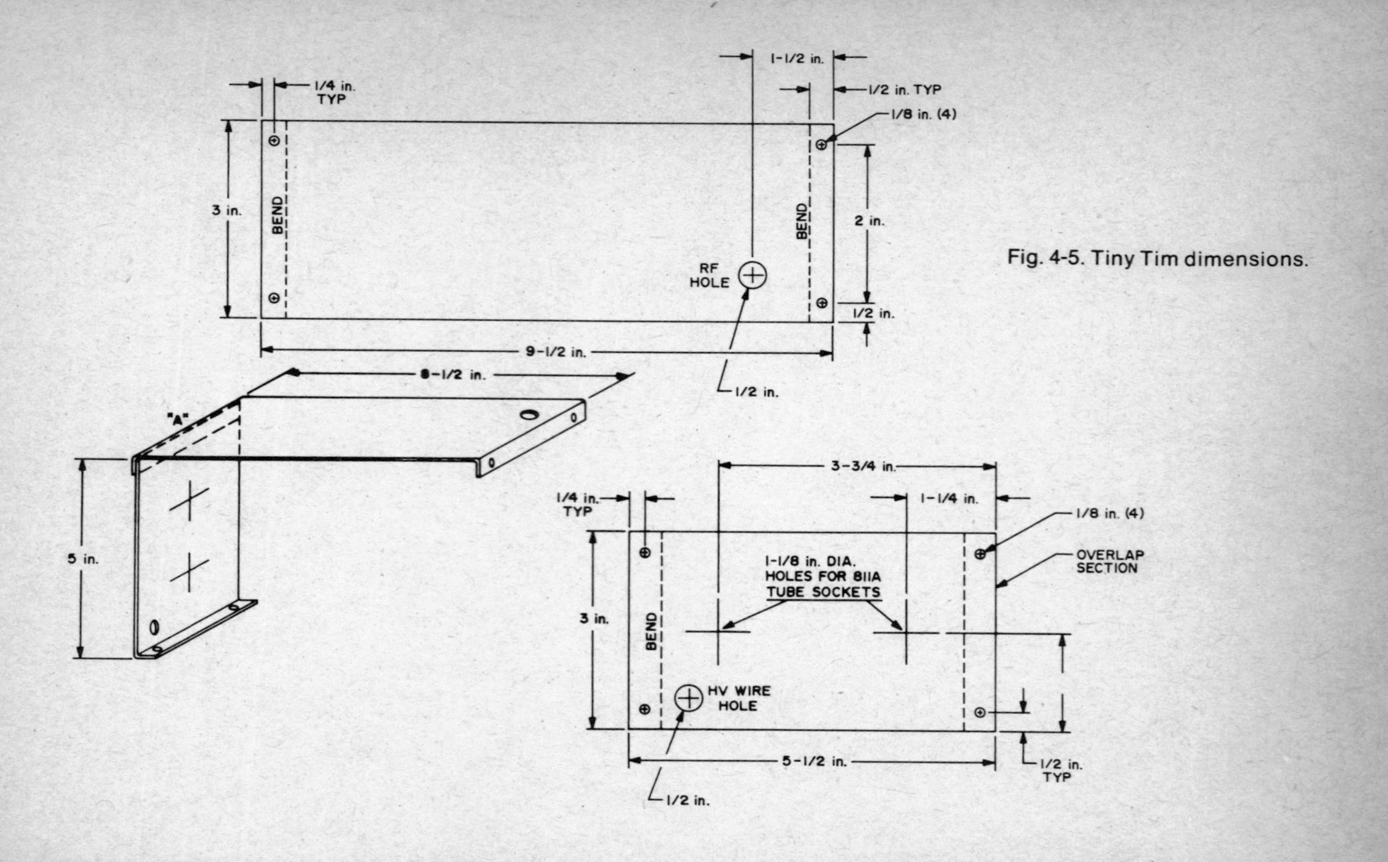

Fig. 4-5. Tiny Tim dimensions.

extra capacitance in cases where the load resistance happens to be unusually low or reactive.

The antenna relay is a four-pole double-throw, but a 5A three-pole is all that is needed. The third pole contacts are used to lift the filament centertap from ground on standby. This keeps the transformer much cooler. For those using *Tiny Tim* with the SBE 33, you will find 10V DC on the octal male plug at pin 7 for controlling the antenna relay, but be sure and use a two prong plug that is insulated from the linear chassis. Consideration should be given to the relay control voltage, if used with other units than the SBE. The output meter provides the necessary indication for loading the amplifier, and the 25 kΩ pot keeps the reading in the proper scale, along with a push-pull switch on the back to control the antenna relay for tuning just the exciter, or turning off the linear with the AC power switch still on.

Almost any of the power supplies shown with linears of this plate voltage requirements will work. An old husky TV transformer and a surplus capacitor with some bargain high voltage diodes works well. No choke is needed in the supply; this helps make the unit easier to hide.

Tiny Tim Construction

Tiny Tim is built inside a 3 × 12 × 10 in. aluminum chassis with the top cut out leaving a half-inch ledge the same as on the bottom (see Fig. 4-5). Allow extra diagonal metal in the corners for mounting the rubber feet. Cut two 3 in. strips from the leftover scrap for the RF section divider, and assemble as in illustration. Care should be taken on construction of the divider as the tube tolerance is quite close.

It is important that the filament pins (large holes) on the tube socket should face each other on assembling to prevent filament sag. Although the tube filaments are shown in series for a 12.6V transformer, they could be paralleled for 6.3V.

High-voltage blocking capacitor C1 (Fig. 4-6) was taken from an old TV high voltage cage and screwed into a half-inch standoff insulator.

Plate choke RFC1 was constructed of plastic rod cut to 5 in. in length, tapped at both ends for mounting screws, and also tapped at the sides for the start and finish position of the winding.

The three sections of the tank coil were first cut to the right number of turns, then soldered to the end turn of each coil, and held together, where the plastic sections meet, with epoxy cement. Centering of the tank coil in the chassis is important to its operation. The coil is held at the center by a standoff insulator secured to the RF divider, but care should be taken when soldering to the turns of the coil that you don't short out the next turn.

The filament choke is made by winding two side-by-side (bifilar) lengths of 14-gage wire on a 0.5 in. ferrite core cut to 5 1/2 in. Leave enough length on the ends of the wire to reach to the power plug and filaments. Make one wrap of

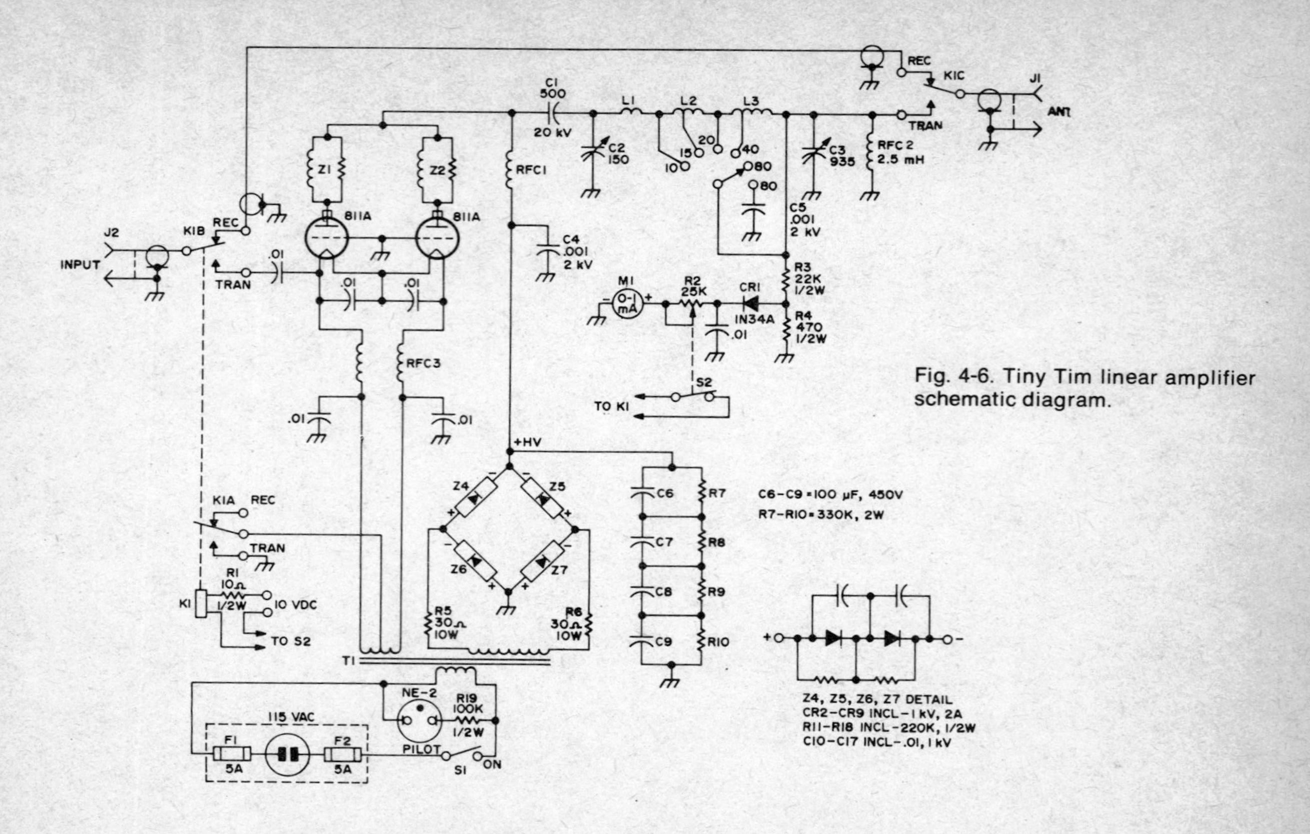

Fig. 4-6. Tiny Tim linear amplifier schematic diagram.

I1 — Ne-2 panel mount neon lamp
J1,J2 — Coax connector, chassis mounting (SO-239)
F1,F2 — 5A fuse, fuse-in-plug assembly
Z1,Z2 — Parasitic suppressor, 100Ω 2W carbon resistor assembled inside 2 1/2 turn coil of No. 16 tinned, 1/2 in. dia, 3/4 in. long

Capacitors

C1 — 500 pF 20kV TV-type high voltage (Allied 43A5599)
C2 — 150 pF var. (Johnson 250E20)
C3 — 935 pF var. (Allied 43A3528)
C4, C5 — .001 μF 2kV ceramic (Allied 43D6290)
C6—C9 — 100 μF 450V electrolytic (Allied 43A4547)
C10—C17 — .01 μF 1kV disk ceramic
CR1 — 1N34A
CR2—CR9 — Silicon diodes - 1kV 2A each

Note: All capacitors 600V ceramic unless otherwise listed

S1 — Single-pole single-throw toggle (Olson SW412)
S2 — Push-pull SPST switch on back of R2
S3 — Single-pole, 6-pos. ceramic rotary (Centralab 2501)
K1 — Relay 4-pole, double-throw, 5A contacts, 6V coil (Olson SW468, Potter & Brumfield GA23671)
RFC1 — 90 μH, 500 mA 4 in. closewound No. 26 Formvar on a 5 in. long, 3/4 in. dia solid plastic rod
RFC2 — 2.5 mH rf choke
RFC3 — 28 bifilar turns No. 14 Formvar or Nyclad wire closewound on 1/2 in. dia, 5 1/2 in. long ferrite rod (Lafayette 32R6103)
L1 - 4 turns, No. 10 wire, 1 3/4 in. inside dia, 1/4 in. spacing
L2 — 8 turns, 2 in. dia, 4 tpi (AirDux 1604)
L3 — 21 turns, 1 3/4 in. dia 8 tpi (AirDux 1408)
COIL BAND TAPS — All taps measured from end of C2

10m 2 1/2 turns
15m 4 1/2 turns
20m 7 1/2 turns
40m 17 1/2 turns
80m all of coil

M1 — Dc milliammeter, 0—1 (Allied 52A-7614)

Resistors

R1 — 10Ω 1/2W
R2 — 25kΩ pot with push-pull switch (Lafayette 33E14838)
R3 — 22kΩ 1/2W
R4 — 470Ω 1/2 W
R5,R6 — 30Ω 10W
R7—R10 — 330 kΩ 2W
R11—R18 — 220 kΩ 1/2W

Parts

T1 — TV power transformer 350 mA (with 4A fil)

Fig. 4-6. Parts list.

plastic over the core before starting your winding. A couple of plastic straps screwed to two standoff insulators hold the choke firmly to the side. RFC2 is mounted on the same side at the end of antenna tuning capacitor C3.

The metering rectifier unit is all mounted on a 2 × 2 in. Vectorboard, and held to the back by the meter lugs (Fig. 4-7).

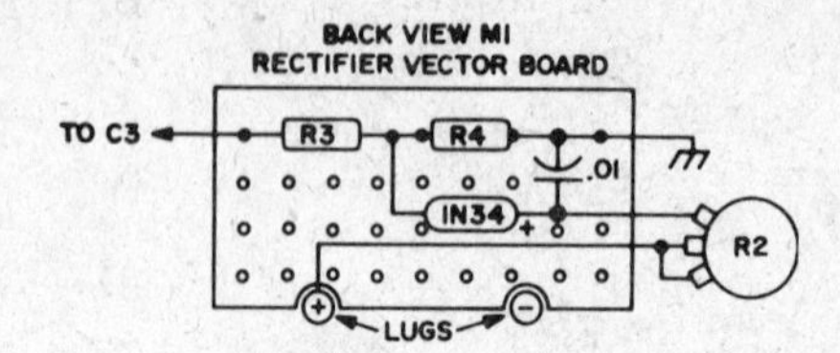

Fig. 4-7. Meter rectifier board.

The three wires from the AC power switch are run in shield under the chassis ledge to the power plug on back. High-voltage capacitor C4 and the filament bypass capacitors are mounted at the power plug.

The top and bottom covers of the amplifier are cut from a sheet of perforated aluminum (do-it-yourself type) available in most hardware stores, and held in place with metal screws.

The transformer is fastened on its side inside of a 3 × 5 × 13 in. chassis with the filter capacitor secured at the other end. Placing a piece of plastic under the capacitor helps to reduce any possible flashover. There is plenty of room left for the standoff insulators and resistors.

Fasten a piece of pegboard over the capacitor area, leaving a 2 in. space from the transformer for the barrier terminal strip. On top of the pegboard is fastened a 2 × 3 × 5 in. Minibox that holds the rectifier diodes on two terminal strips with the leads going out the bottom. The shielded rubber covered 8 ft. power cord passes all the way through the cover, with the leads screwed to the terminal tie strip. Another piece of pegboard over the terminal block area will completely eliminate the possibility of accidental shock.

A male type TV AC connector, mounted flush on the side of the chassis, is a safe quick way to be sure your power is completely disconnected.

Half-inch rubber feet are mounted at the four corners with metal screws, the same as on the linear.

Check all wiring at the relay and power plug with an ohmmeter for possible shorts before applying the smoke test; it may save a few parts.

Keep in mind that although *Tiny Tim* is small, the knockdown voltage is big and you might not get up.

Tuneup Procedure

Load your exciter into a 52Ω dummy load with linear amplifier relay switch S2 off. Note that the antenna relay is still active with S2 on, even when the linear AC power switch is off. After normal exciter loading, turn on antenna relay switch S2, and center the meter output control. Switch your exciter momentarily to tune, and swing plate tuning capacitor C2 for the most output indication on the meter. Repeat with antenna loading capacitor C3, and again with C2 until you have peaked the output with the meter set in the upper scale.

To check for proper operation of the tank circuit for efficiency on each band, antenna loading capacitor C3 stator plates should be considerably less than fully meshed. If not, this can easily be corrected by changing the band tap, and adding one more turn to the tank coil.

After loading the linear on your favorite band, carefully loosen the setscrews on the turning knobs C2, C3, and the output meter control knobs; adjust so that the knob markers are all vertical. This will give a reference for future tuneups.

4-1000A GROUNDED GRID LINEAR

Simplified module construction incorporated with up-to-date features including high plate dissipation make this linear amplifier an excellent choice for reliable contest, rag chewing or SSTV applications (see Fig. 4-8).

Construction

The following tools were used in the construction of the amplifier: metal munching tool, pop rivet gun, electric drill, various hand tools and a good soldering gun.

The main chassis 43.18 × 43.18 × 10.16 cm (standard) and the front and back of the amplifier are constructed from 43.18 × 43.18 × 7.62 cm (standard) and 33.02 × 17.78 × 7.62 cm (standard) chassis, respectively. The front and back are mounted to the main chassis by pop rivets around the perimeter. The cover is manufactured by hand bending a 48.26 × 121.92 × .16 cm (standard) sheet of aluminum to tightly fit chassis assembly and is held in place by sheet metal screws at 3 cm intervals.

Bending can be accomplished with two pieces of angle iron. The material to be bent is clamped between the angle irons using C-clamps and a vise. Use a piece of flat wood as a protector between the hammer and the material to be bent to avoid unsightly hammer marks. A commercial metal bending brake will no doubt do a better job in far less time provided you have or can borrow one.

The air is exhausted by mounting home air vent assemblies on the sides of the cover and perforated aluminum sheeting over holes on the top and back. A total of six holes are covered by the perforated aluminum.

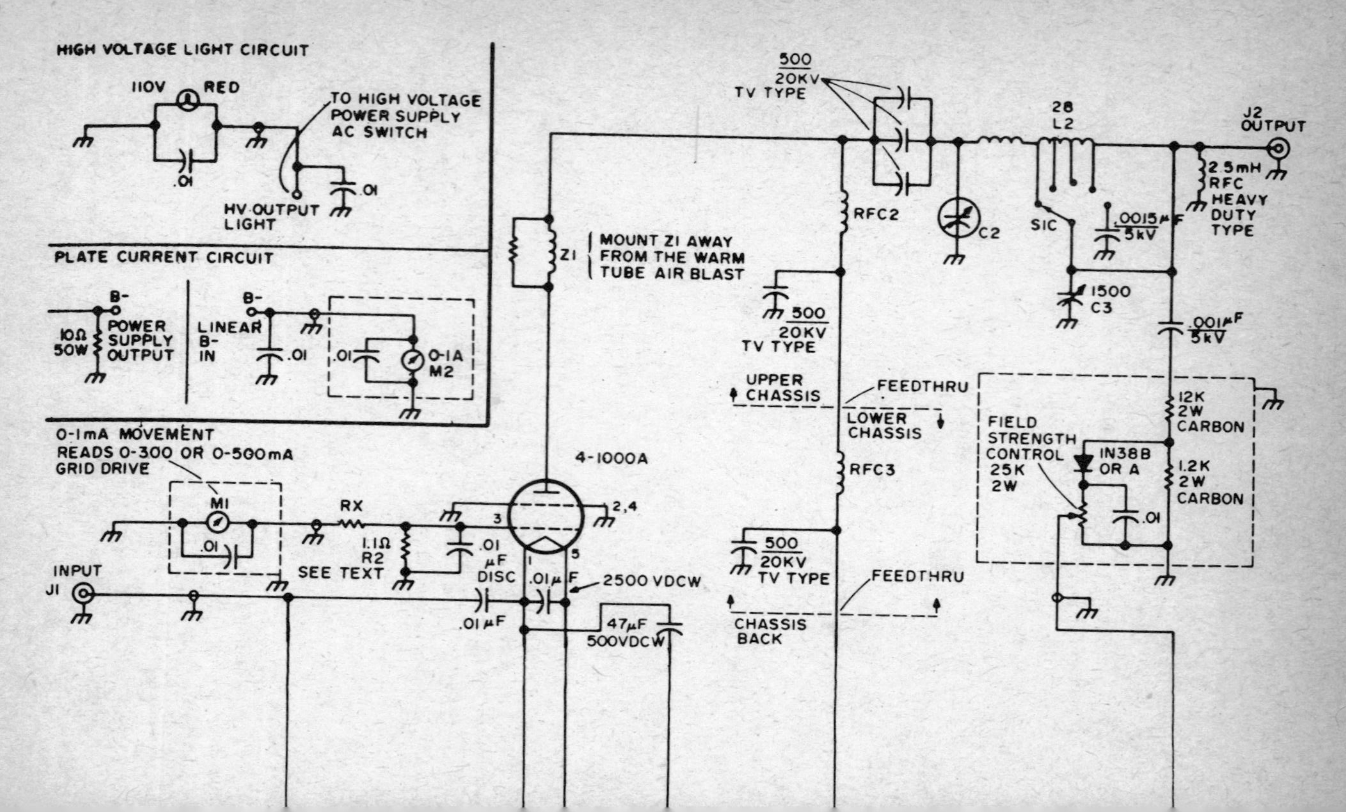
HIGH VOLTAGE LIGHT CIRCUIT
110V
RED
.01
TO HIGH VOLTAGE POWER SUPPLY AC SWITCH
.01
HV OUTPUT LIGHT
PLATE CURRENT CIRCUIT
B-
10Ω 50W
POWER SUPPLY OUTPUT
B-
LINEAR B- IN
.01
.01
0-1A M2
0-1mA MOVEMENT READS 0-300 OR 0-500mA GRID DRIVE
M1
.01
INPUT J1
RX
1.1Ω R2 SEE TEXT
.01 μF DISC
.01μF
.01μF
2500 VDCW
47μF 500VDCW
3
1
5
2,4
4-1000A
Z1
MOUNT Z1 AWAY FROM THE WARM TUBE AIR BLAST
500 20KV TV TYPE
RFC2
500 20KV TV TYPE
UPPER CHASSIS
FEEDTHRU
LOWER CHASSIS
RFC3
500 20KV TV TYPE
FEEDTHRU
CHASSIS BACK
C2
28 L2
S1C
.0015μF 5kV
1500 C3
.001μF 5kV
J2 OUTPUT
2.5mH RFC HEAVY DUTY TYPE
12K 2W CARBON
1.2K 2W CARBON
FIELD STRENGTH CONTROL 25K 2W
1N38B OR A
.01

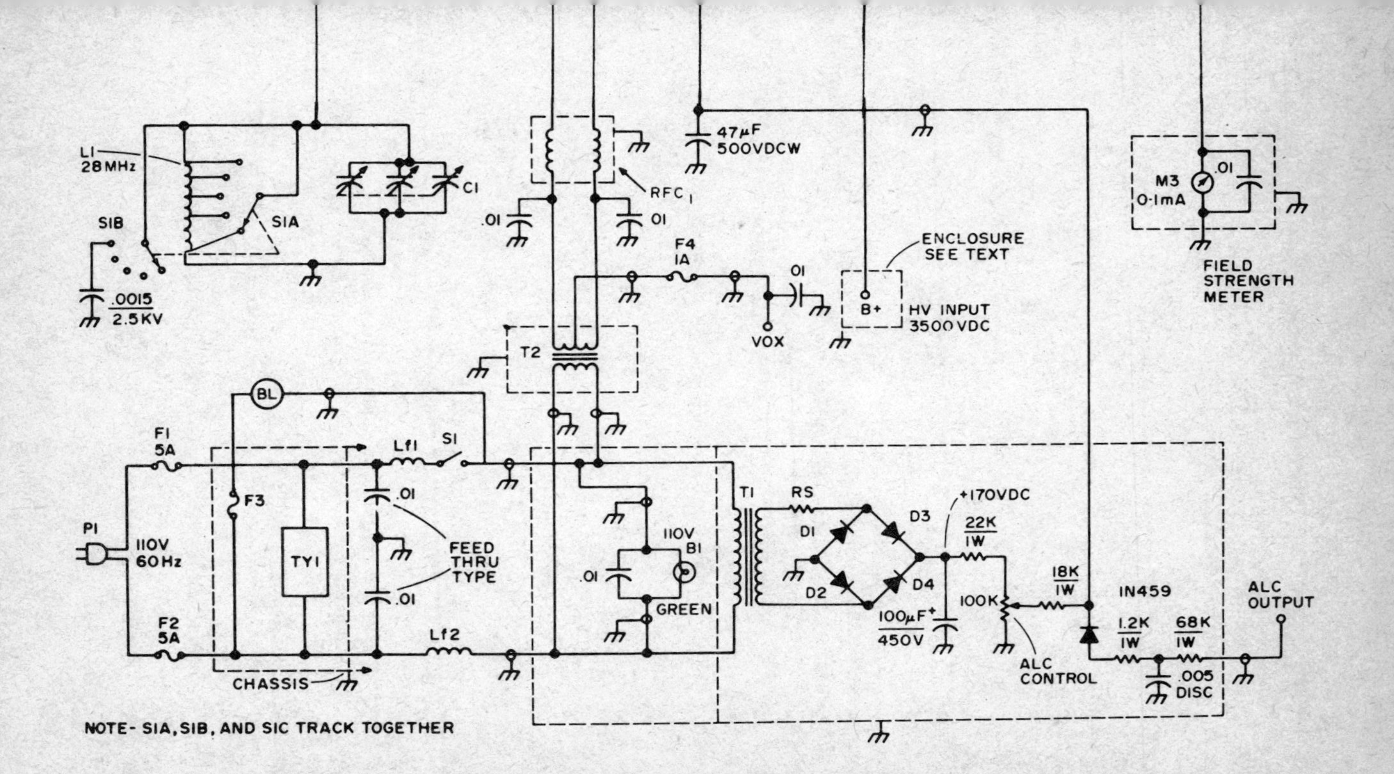

Fig. 4-8. Schematic diagram of the modern 4-1000A, Class B, grounded grid, linear amplifier.

To achieve proper shielding, the screen in the air vent is removed and replaced with perforated aluminum eave trough screen, available at most hardware stores. When properly installed it provides excellent shielding but it is too weak to be installed without a frame. Attach the air vents a minimum of every 2 cm around the perimeter of your cover to provide adequate shielding. Small aluminum pop rivets really come in handy here but make sure each rivet is fitted tightly. Before attaching aluminum to aluminum rough each contact surface with extra fine sand paper to assure a good electrical connection. Also, each chassis is electrically connected together by lengths of copper braid.

The front of the amplifier contains a relative output meter, plate current meter and grid current meter. The high voltage reading is taken from a voltmeter mounted directly in the power supply.

The three controls on the front in the lower left are cathode tuning, relative output level and ALC level. Also included is a turns counter in the upper right for the vacuum variable capacitor and a vernier dial for the loading capacitor. The band switch is located in the middle and ganged to the cathode circuit via two right angle drives, providing single switch band switching.

The bottom plate is made from .32 cm (standard) aluminum or steel with one caster at each corner. It should be held in place by sheet metal screws placed a distance of 3 cm to insure adequate shielding.

To provide safety and stability, many sub-chassis are used. The ALC and relative output circuit are shielded in the final enclosure. The sub-chassis are relatively inexpensive if purchased, but could also be hand formed.

The ALC and relative output circuits are enclosed in aluminum Miniboxes. The plate and grid current meters are enclosed in a 12.7 × 25.4 × 7.62 cm (standard) chassis with steel or aluminum bottom plate attached. The relative output meter is shielded and provides low measured leakage with no enclosure. The loading capacitor has an added frame enclosure but it is not critical in the design. The rear of the amplifier is designed with safety in mind. The B+ and B− connections are in a 12.7 × 10.16 × 7.62 cm (standard) chassis with two grommetted holes in the bottom. High voltage cables should have a minimum rating of two to three times the voltage expected to be encountered. The blower is fused and the blower solder connections and fuses are located in a small Minibox. The AC line is terminated in a small sub-chassis and at this point the thyrector attenuates line transits providing protection to your solid state devices. The AC is fed into the bottom chassis through two feedthrough capacitors and is fused in the plug with two fuses, one on each side of the line. This places all but one fuse external and readily available. Located on the back are three shielded banana outlets for vox, ALC and the high voltage light. Located below the banana plugs is a heavy-duty ground connector *which should be utilized to reduce possibilities of electrical shock.* The RF is fed directly into the lower compartment and output is taken directly from the upper chassis to eliminate the problem of feedback. Shielding between input and output circuits of a grounded grid amplifier reduces the possibility of parasitic oscillations. The large opening in the rear of the amplifier where air enters the pressurized bottom chassis from the blower must be adequately shielded. An aluminum screen mounted over the hole and bolted in place at a minimum of every 2 cm insures good electrical connection between the screen and chassis. Be sure to clean the screen occasionally as it will clog up with dust, decreasing air volume. Place the blower a good distance away from the tube socket and seal any undesirable air leakage points in the bottom chassis with silicone rubber sealer to provide a good pressurized system. Cutting the lower flange off the tube socket will also provide better unrestricted air flow if you use the standard SK-510 socket.

To protect the operator, metal shafts that protruded out the front are equipped with insulated shaft couplings. All shafts at ground potential are connected directly to ground via flexible copper cable. This is done as double protection even when the shafts are already at ground potential, as in the case of the loading capacitor. Take great care at these connections to guarantee electrical and mechanical strength.

Wiring

The wiring of the amplifier is straightforward. If you are contemplating construction and have not built a linear before, I would recommend the reading of construction techniques, as applied to RF amplifiers, in the handbooks—particularly the section on preventing radiation from the transmitter.

The input circuit is a tuned cathode circuit. The 4-1000 requires a substantial amount of drive; therefore a cathode tuned circuit is a must. This reduces drive requirements and also provides improved distortion products. When constructing this circuit, keep in mind that 10 meters may be a problem in respect to drive—therefore align coil and capacitor combination to provide short connecting leads on ten. Silver plating the cathode coil and leads also provides measurable improvement on ten.

Grid drive is monitored with a 0-1 mA meter utilized as a millivolt meter. *Be sure to calibrate meter before soldering the screen.* As the screen is operated in parallel with the control grid it must be disconnected from ground so that you are reading control grid current only. To calibrate the meter you must determine the proper value of the series resistor Rx. This is found by placing a regular milliammeter with a scale of 200 mA or more from the vox terminal to ground. Carefully apply excitation with no plate voltage and substitute resistors at Rx until both meters have the same deflection at 100 mA. One meter required 82Ω but this is variable depending on characteristics.

Another meter may require a larger or smaller value. As the 4-1000 has no plate voltage at this time, the control grid dissipation can be easily exceeded. Therefore be extremely careful and work quickly during periods of excitation.

The plate current is measured by the meter being shunted across a 10Ω resistor in the negative high voltage lead. The resistor was placed at the power supply and the negative terminal of the supply must not be grounded except through the resistor.

The screens are connected to ground via a .64 cm flexible copper ground strap which passes through the slot of the socket directly from ground to the screen pin. Both pins are grounded in the same manner. Keep leads short! All power and metering leads are shielded and bypassed according to good construction procedures.

The entire plate circuitry is silver plated and the connecting output lead from the coil assembly is of silver plated .65 cm (standard) copper tubing. The plate and vacuum capacitor lead are made from silver plated flexible 1.27 cm copper ground strap.

Silver plating is accomplished with a small electroplating unit powered by flashlight batteries. It is simple and requires no special skill. There are different units available from various electronic outlets with prices starting at a few dollars.

Adjustment

Before applying any potentials recheck all wiring. Set the sensitivity of the output indicator to minimum, that is, maximum resistance with the slider of the 25K top at ground. Connect a dummy load of 52 to 72Ω. Select the proper band with S1 and apply plate voltage. Apply a small amount of excitation. Peak up C1 for maximum reading. Resonate the output circuit with C3, making sure the loading capacitor is set at full capacity. Adjust C3 and C2 to increase the plate current to the input power level you desire while increasing drive to 120 to 150 mA. You will probably note that the RF output meter and plate current meter conflict somewhat. Maximum output does not always occur at the point of resonance. Don't let this alarm you, adjust your drive and loading to a point where for a given drive, the output drops off slightly at your desired input. This is the point where you will generally achieve the best linearity. Once you have established your tuning for a band, record turns and loading for future reference.

To utilize the ALC properly, connect the exciter to the linear via the ALC connector on the rear of the amplifier. Once connected you may adjust your ALC to achieve the maximum limitations you desire. The ALC section, if properly utilized, can increase the average power output while maintaining the linear within the amateur power-input limit, providing that extra punch to communicate under adverse conditions.

Operation

The operation of the amplifier has been excellent. It provides adequate efficiency and it is built with long-term reliability in mind. The multi-band linear circuit compared most favorably with a single band configuration on 80 meters which would be expected. Comparison of harmonic levels with commercial ham linears provided data that equalled or exceeded the commercial designs tested. In respect to tube operation, the 4-1000 will loaf at the 1 kW DC level, whereas some commercial circuits are operating the finals at or beyond maximum level which could lead to unpredictable reliability. The 4-1000 can be damaged, however. The control dissipation is limited to 25 watts—therefore keep an eye on the grid drive. Do not exceed 200 mA drive, and *never* apply full excitation without any plate voltage.

Technical Summary 4-1000A

The 4-1000 proved not to be guilty of rampant TVI. The circuit described was compared with commercial amplifiers at the same power level, 1 kW DC, and it provided equal or better results. Circuits designed, constructed and operated according to good engineering practice should provide excellent results. Persons experiencing severe TVI problems with the 4-1000 should review their design and operating parameters for possible errors. This is not to discount problems due to rectification or overload. However, these problems are not the fault of the linear.

Drive Requirements. The 4-1000 does require a fair amount of drive. Most commercial exciters should drive the amplifier to 1 kW DC up to 15 meters, however.

Ten meter drive requirements may be a problem for two basic reasons: (1) Some exciters fare poorly when it comes to output on 10. They provide substantially less output on 10 as compared with 15 meters. A check of your exciter's output efficiency would be advised before selection of this tube for 10 meters. (2) When building the multi-band input cathode circuit, much energy can be lost in this configuration. Normally the majority of loss is at the highest frequency of operation.

Parasitic Oscillations. If the 4-1000 circuit is wired according to good engineering practice with proper preventive measures taken, you should have no unusual problem with parasitics. Before you place your amplifier on the air, check it for parasitics on all bands. When first constructed, a parasitic may well occur but once the amplifier has been stabilized, it should provide excellent results. Parasitics are generally the result of the layout and wiring and not the tube per se. Keep in mind that it is indeed a lucky ham who builds a linear with any type tube who has a completely stable amplifier to begin with.

Component Modifications. To prevent the possibility of high voltage breakdown due to moisture at the base of the B&W 800, a fiber screw replaces the metal one supplied. Also, the B&W 800 is not mounted directly on the chassis. A 2.54 x 5.08 x .32 cm (standard) piece of plexiglass is cut and the choke mounted to it. The plexiglass is mounted 1.27 cm above the chassis on two 1.27cm insulated stand-offs.

An extra switch position is required for the B&W 850A so that the .0015 μF mica capacitor can be switched in on 80 meters. An extra contact for constructing the switch is available from Barker & Williamson in Bristol, Pennsylvania.

A note on the B&W 850A. The B&W 850A multi-band inductance does an adequate job of impedance transformation considering it covers 80 through 10 meters. It does not, however, provide the optimum results that a well-designed single band linear can provide.

In essence, if you are a band hopper and prefer convenience, the B&W 850 A will fill the bill. But if you are basically a one-band man, critical about efficiency, you would probably be happier omitting the B&W 850A, saving a few bucks and going single band by replacing L2 with a single band inductance. For further details see pi-network design in the various handbooks and journals.

A SIX METER 1 KILOWATT PEP LINEAR

This linear consists of a pair of 4 × 150A/7034 or 4CX250B tubes in parallel in a passive grid circuit. The plate voltage is 2000V, plate current is 500 mA peak (see Fig. 4-9).

The grid circuit load is dependent upon the amount of drive available. A 50 ohm, 50 watt Globar may be mounted internally for rigs in the 100 watt PEP output class or a Cantenna may be connected to the input through a coax "tee" connector. Much lower drive levels may be used by increasing the resistance at this point. A noninductive (carbon) resistor of sufficient wattage rating will allow the amplifier to be driven by five watts or less. The plate circuit is shunt fed through a home-brew or Ohmite Z-50 choke. Matching the load is accomplished by tapping down on the tank coil with a clip. Screen overcurrent protection is afforded by a #327 light bulb which acts both as an indicator of screen current and as a fuse.

Purists may frown on some of the preceding, however there are good reasons for it. The passive grid circuit will accommodate any reasonable drive level and while wasteful at high levels it should be remembered that linears are for use only when required to maintain communication. It also eliminates neutralization as well as several parts. The shunt fed tank also makes for simpler, less expensive, and easier to adjust equipment. The bulb in the screen circuit functions as a protection relay and meter as well, but costs only a small fraction as much.

The screen power supply consists of a light to medium duty replacement or TV transformer, with full-wave rectifier and capacitor input filter. This supply is zener regulated at 300 to 350 volts. One half of the same winding is half-wave rectified and regulated (in the linear) at 100 volts negative for operating bias. The same supply furnishes cutoff bias in the receive condition. Filament voltage is obtained by phasing the 6.3 and 5 volt transformer windings in

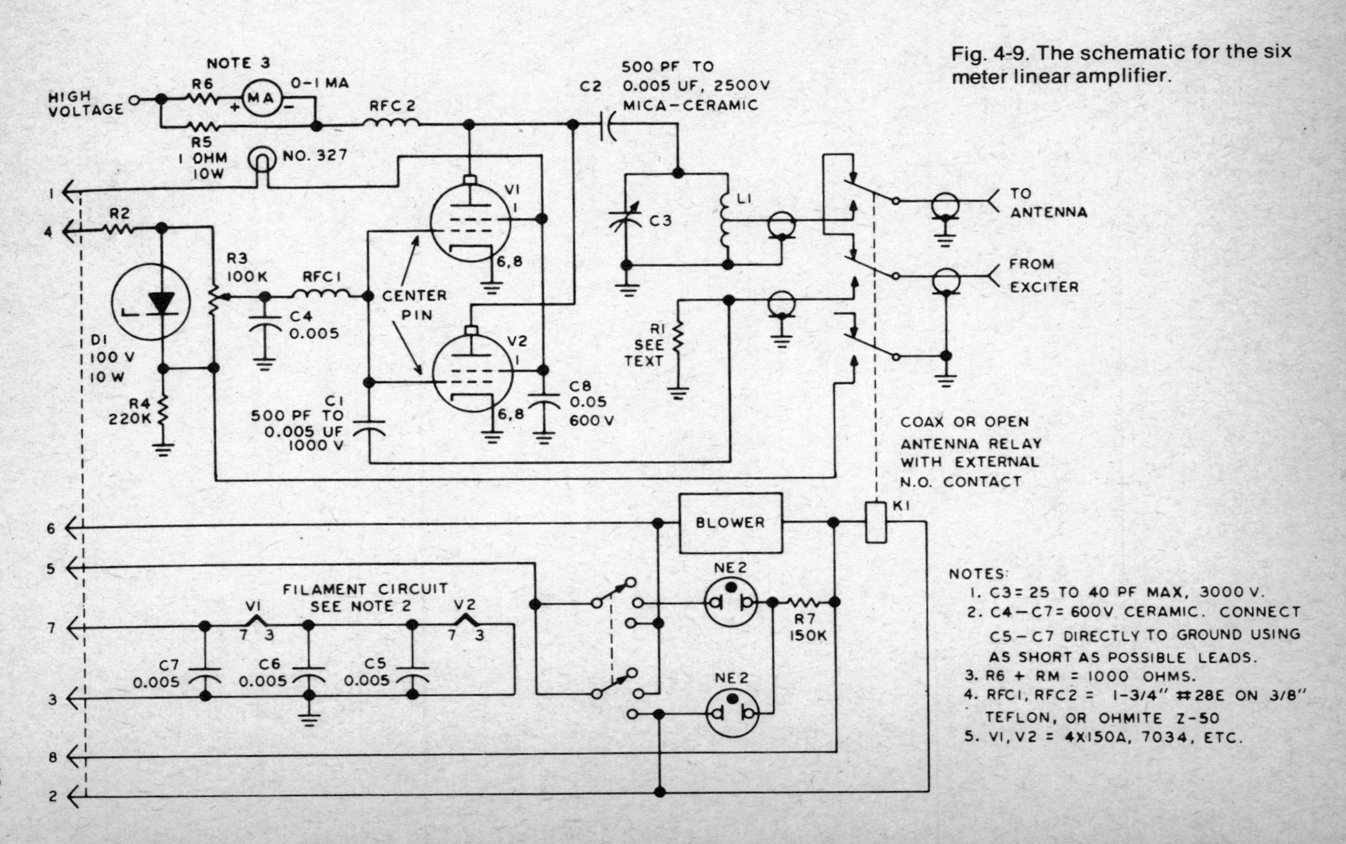

Fig. 4-9. The schematic for the six meter linear amplifier.

series. Slightly over 11.3 volts is available to the filaments which are also wired in series. This arrangement runs the finals at approximately 5% below their nominal 6.0 volts, which is within tolerance.

The DC Supply

The high voltage transformer is rated at 850 to 1000 volts center tapped at 350 to 500 mA (Fig. 4-10). Several power supplies have been built with a commercial surplus transformer available locally at $5.95. This transformer is rated at 1000 volts CT at 400 mA and provides 2200 volts at 450 mA in SSB service. Several nationally known supply houses list usable transformers at

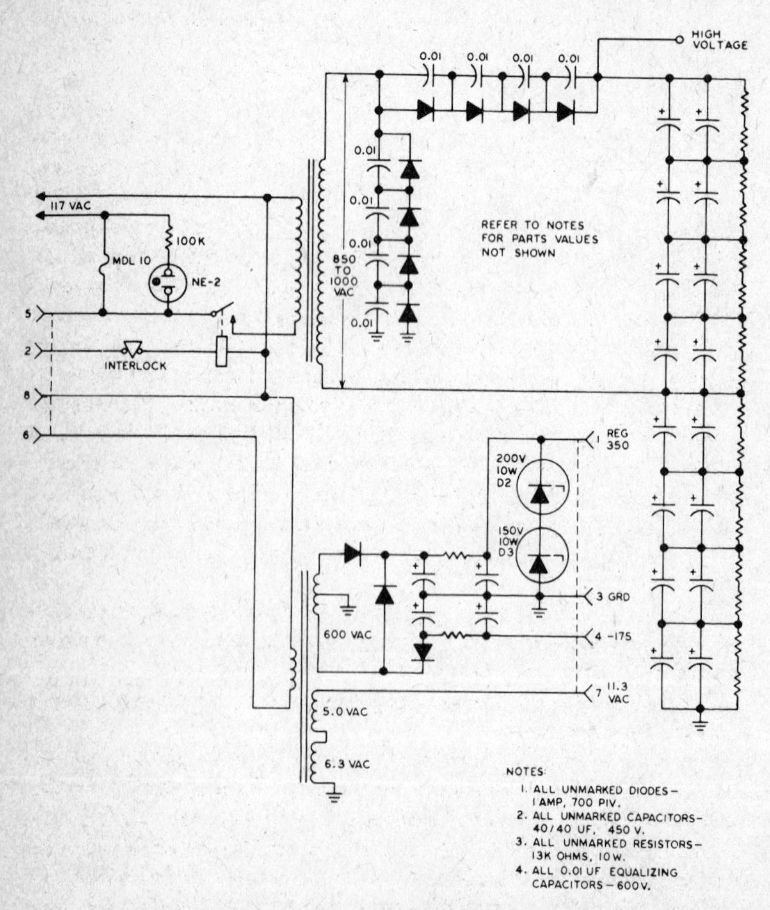

Fig. 4-10. Power supply diagram.

$2.00 to $5.00. In some cases it may be necessary to series or parallel two transformers in order to arrive at the proper voltage or current rating. A transformer of about 850 to 900 volts CT is recommended, otherwise the maximum voltage rating of the tube will be exceeded. Only the high voltage winding is used in order to increase the available current. Rectification is by eight 700 volt PIV silicon diodes in a voltage doubler and the filter consists of sixteen 40-40/450 volt capacitors in series-parallel. The total filter rating is 20 mF at 3600 volts. Twenty capacitors of the same type were purchased for $7.80 and used for all filtering purposes. The 10″ × 14″ × 3″ chassis on which the power supply is mounted is not crowded. All bleeders are 13K, 10W obtained at 20 for 59¢. Each bleeder section in the high voltage supply consists of two of these resistors in series. The remaining resistors are used in the bias and low voltage supplies.

Finishing Up

The controls are all on the amplifier and consist of a two pole, three position switch which turns on the 60 CFM blower and the filament, bias, and screen supplies in the "standby" primary relay in the "operate" position. This primary voltage is also fed to the antenna relay coil through the normally open contacts (spare) on the exciter PTT relay so that in "standby" it is possible to operate the exciter alone and in "operate" the linear is automatically switched into service. The only other panel control is the plate tuning capacitor. Neon lights indicate the position of the function switch. Inside controls adjust the bias in "operate" position and control the antenna loading. All of the control and low voltage wiring is contained within an 8 conductor cable; the high voltage lead is separate and consists of a length of RG-8U with PL-259 connectors which have been painted *red* to indicate their unusual function. Other connectors may be used if desired. Plate current metering consists of a 0-1 mA meter which with the proper multiplier indicates full scale when 1 ampere is drawn through a 1 ohm resistor which is in series with the high voltage lead.

The amplifier may be built in a Heath SB series speaker cabinet and fitted with Heath knobs in order to complement an SB-110. A 10″ × 5″ × 3″ chassis is supported within the case by the front and rear panels.

The panel is sprayed with Illbronze, "forest green" wrinkle paint, which is good for the Heath green.

Adjustment is simplicity itself. The exciter is tuned up per usual with the linear function switch in the "off" or "standby" position. The linear is then switched to "operate" and keyed momentarily while the bias control is adjusted to a static current reading of 200 mA. At this point a small amount of drive is applied and the plate tuning is peaked for maximum output as indicated by an SWR bridge in the forward mode or by using a monitor scope. Modulation is now applied until the plate current *peaks* at 500 mA. Some slight adjustment of the tank coil tap may be required at this time in order to load the amplifier properly.

The possibility of using commercial pull-out tubes should not be overlooked. These tubes are common in FM broadcast and aircraft ground station transmitters and as such are often changed on a time basis rather than on condition. They are available at hamfests and at surplus stores at nominal prices.

Amplifiers have been built with coax and open type antenna relays. The coax type has been found to be preferable; however, the open type will work quite well and will save the builder much cash.

Most of the parts are of a highly noncritical nature. The input and output coupling capacitors have been varied from 500 pF to .005 without degrading performance. Various tuning capacitors have been used and all have worked well. The plate coil in each case was adjusted to resonate at 50 MHz with the capacitor fully meshed. Blowers from 30 to 100 CFM have been tried, the only undesirable effect noted was increased air flow noise above about 60 CFM.

The most critical item is the tube socket. *Do not use sockets without the built-in bypass* and do not leave out the external screen bypass.

COMPACT 1200 WATT AMPLIFIER

The Circuit

The ceramic tetrodes are operated as low mμ triodes (Fig. 4-11). The control grids are tied to the cathodes. With conventional Class-B grounded grid operation the grids would be promptly destroyed. In this mode, however, grid current is nil, idling current is low (about 35 mA), and feedthrough power is high. It takes about 150-200 watts of drive, but this is an asset, since there is no necessity to swamp the input. Almost any transceiver drives it adequately.

A tuned-cathode input circuit is used to present a better load to the exciter. RF isolation in the filament circuit is afforded by a home-brew filament choke. Bias is developed through a 33K cathode resistor. When in the standby mode, plate current is virtually zero. During operation a relay shunts out this resistor. When not driven the tubes will idle as stated above.

The tank circuit consists of a roller coil and tuning capacitor from a BC-375 tuning unit. A Barker and Williamson bandswitching inductor could be substituted. The loading condenser could be aptly dubbed a free-loading capacitor. It came from a derelict broadcast set. A 500 pF mica capacitor in parallel with it provides plenty of capacity even for the 80 meter band.

The coupling capacitor, specified as .001 5 kV, is solved by using parallel "beer barrels" from discarded television chassis. A parasitic suppressor is included in each plate lead, but no adjustment is necessary, and no tendency toward self-oscillation is evident.

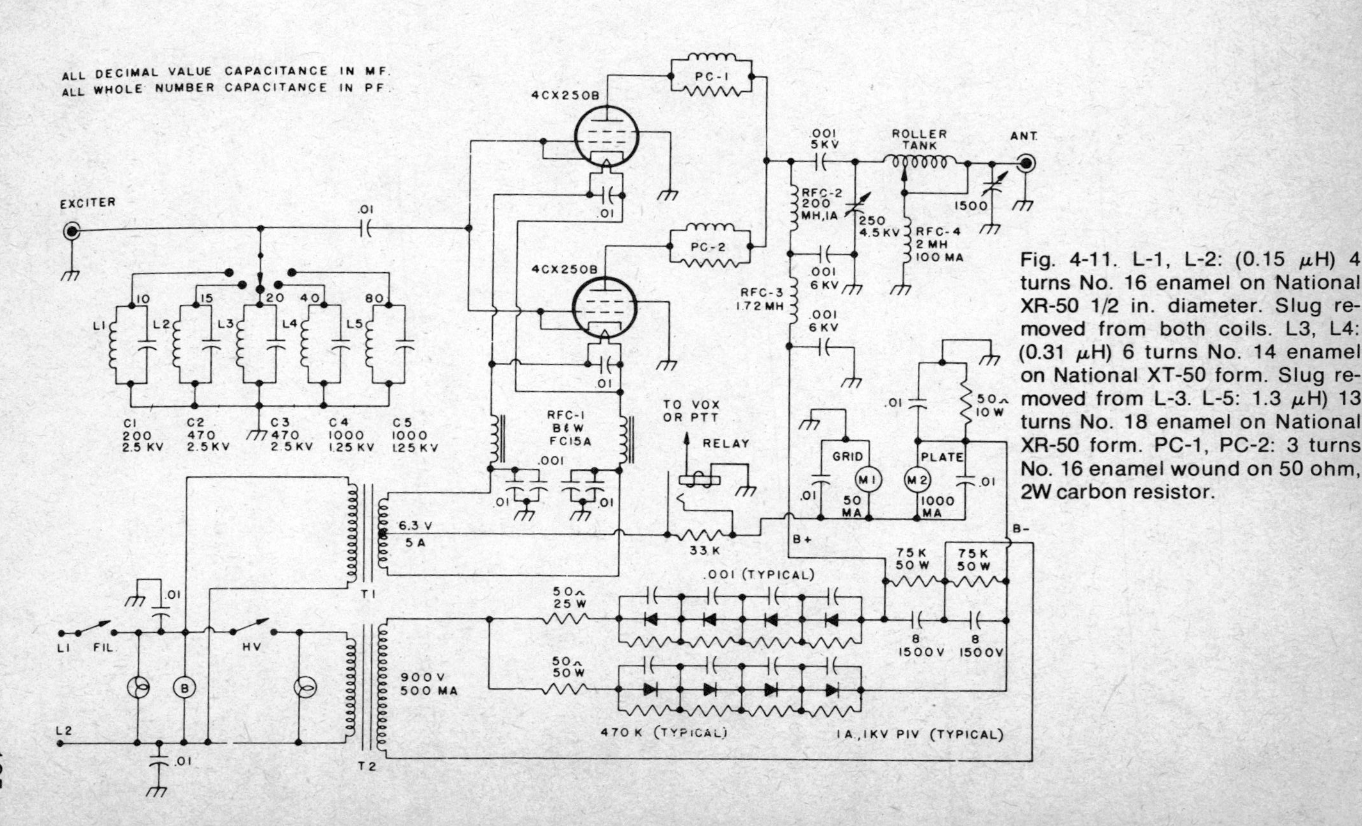

Fig. 4-11. L-1, L-2: (0.15 μH) 4 turns No. 16 enamel on National XR-50 1/2 in. diameter. Slug removed from both coils. L3, L4: (0.31 μH) 6 turns No. 14 enamel on National XT-50 form. Slug removed from L-3. L-5: 1.3 μH) 13 turns No. 18 enamel on National XR-50 form. PC-1, PC-2: 3 turns No. 16 enamel wound on 50 ohm, 2W carbon resistor.

The power supply is solid state using the conventional doubler configuration. Only 4 μF of filtering seems to do the job. By using electrolytics, much more capacity can be engineered into the allotted space.

Construction

The amplifier is built on two aluminum chassis which are fastened to the cabinet with self-tapping screws. The RF section, measuring 4 × 6 × 2 inches, contains the tubes and associated wiring. The shield enclosure, built on aluminum angle, stands 4 inches over the chassis for a total "height" of 6 inches. The RF assembly lies on its side and fastens to the back of the cabinet with a cork gasket as an air seal. A centrifugal blower delivers air through a hole in the cabinet into the bottom of the pressurized RF chassis.

The power supply is mounted on a 5 × 9 × 1 1/2 inch chassis. Silicon rectifiers, surge resistors, and associated components fit easily underneath. The oil capacitors mount topside with the power transformer sitting directly behind them. Two angle brackets secure it to the rear of the cabinet. Electrical connection between the transformer and the power chassis is made through a husky four prong plug. With no room left for the bleeders, they are secured *outboard* on the supply chassis.

All other components; tuning capacitor, coil, loading condenser, filament transformer and relay are bolted or screwed directly to the steel cabinet.

The layout of components is non-critical, but it takes a little planning to come out with space for everything.

Cooling

Entire articles have been written concerning the proper cooling of ceramic tubes. Needless to say, it is *not* sufficient to blow air *at* them. A modest investment will provide you with the chimney type sockets required to do the job. By pressurizing the bottom of the RF chassis and forcing the air to leave through the finned anodes of the tubes, you will find that correct cooling is accomplished. The builder is urged not to sample the temperature of external anode tubes because of the lethal voltage present! A high speed squirrel cage blower is strongly recommended. "Silent" blowers are more compact and less noisy, but that they won't work well against back pressure. Hot air leaving the tubes flows into the cabinet, but since this particular cabinet is itself air tight, provision must be made to release the pent-up air. Punch a number of holes in the cabinet for this purpose (just behind the filament choke). Additional holes located in the cabinet bottom near the bleeder resistors serve to carry excess heat away from these components in the process.

Power Supply and Filaments

The power supply is built around a husky 900 volt transformer which can be appropriated from an old television set. Silicon rectifiers in a voltage doubler develop 2400 volts DC. The bleeders are mounted over a series of holes where air, exiting from the solid cabinet, carries the excess heat away. Besides offering a margin of safety and equalizing the voltage drop across the series filter capacitors, the bleeders provide some degree of voltage regulation.

4CX250s require 6.0 volts on the filaments. Excessive voltage will seriously impair tube life. You have the choice of inserting a series dropping resistor in the primary winding of the filament transformer or rewinding the secondary of a 6.3 volt filament transformer. If you decide to do this, be sure to get the center tap electrically centered, or you will have a transformer that overheats. It is really not as difficult as it sounds and certainly worth the try if you have an old filament transformer available. Naturally you must keep your plate and filament supply separate so that the tubes can be preheated prior to application of B+. A time delay relay can be employed for this purpose.

The filament choke is a home made affair wound on a ferrite rod ala Lafayette radio. A search through old magazines will provide the details. A commercial choke can be substituted.

Metering

Plate current and grid current are monitored in the negative return leads as a safety precaution. This places B− above ground. A 50 ohm 10 watt resistor across the plate meter serves as a safety device should the meter coil open.

Tuned-Cathode Tank Circuit

Separate tuned circuits are provided for each amateur band. Coils are wound on slug tuned forms. The slugs are removed for 10-15 and 20 meters. A mica capacitor is soldered in parallel with each coil. Each coil and capacitor should be resonated, if possible, with a grid-dip meter. After all coils are mounted around the bandswitch the box is shielded (a Minibox will do) and secured to the amplifier front panel.

Tuning and Operating

The amplifier should be carefully checked for any sign of instability before any attempt is made to drive it. This is done by connecting a dummy load and applying plate voltage, running the coil, tuning capacitor and loading capacitor through their range while monitoring grid and plate current. If you are fortunate, you will see zero grid current, and the plate meter will sit on 35 mA like a rock. If a weak parasitic is found, it may help to compress the coils on the parasitic suppressors in the plate leads. If the amplifier is determined to be stable under all conditions it may be connected to the exciter. A relative output meter in the output of the linear is a *must*. Tuning is accomplished much as with modern transceivers; i.e., resonance should be found *quickly*. With a little practice all tuning and loading can be accomplished in a matter of several seconds. With the output meter in the line it is a simple matter to tune for *maximum output*. When properly tuned the grid current is very low; from plus two or three mA to minus one or two mA. Grid current should never be allowed to go beyond 10 milliamperes and the ideal condition is zero. This can be controlled with antenna loading and drive. Excessive antenna loading will produce negative grid current; excessive excitation creates high positive grid current. Under normal operating conditions grid current will flicker slightly (1-2 mA) with voice peaks. Plate current will peak approximately 1/3 to 1/2 of DC tune condition. Typically, plate current will talk up 100-150 mA depending on voice characteristics, etc. Shouting into the mike to produce higher readings will cause flat topping, distortion and citations. After a little experimentation the settings for tuning and loading capacitor and roller coil can be logged and will speed up the tuning process. The amplifier is quite rugged and is easily operated with some experience.

It should be noted that the amplifier presents a different load to the exciter than does the antenna. This means that some retuning of the transceiver is required to go from *barefoot* to *shoes* operation. This is not inconvenient, however, and takes only seconds.

A SIMPLE BIAS REGULATOR FOR LINEAR AMPLIFIERS

With the exception of those circuits using zero bias tubes, class AB2 and class B operation of linear amplifiers appears to have been largely neglected by the radio amateur. The reason for this is not hard to find. The little literature which is available on the subject invariably shows bulky and elaborate regulated bias supplies, often containing five or six tubes, being used to maintain the bias voltage at a steady value while the grid current surges to 100 mA or more with each transmitted syllable. Some form of regulation is, of course, essential, for the bias voltage would otherwise swing wildly with each surge of grid current. But this regulation can be provided so simply using semiconductors that it need no longer influence the choice of amplifier type.

The regulator described in this article possesses the following features: (1) It contains only five small components, costing approximately $3, and requiring about two cubic inches of space. (2) It is a two-terminal device which is connected to the amplifier at the point where regulation is required. No separate power is needed. (3) The bias voltage may be adjusted by the turn of a knob to any value between 40 and 100 volts (other ranges are available by changing the component values). (4) It can be used for any class of amplifier, radio or audio frequency, whether grid current flows or not, but as described above, it is of particular value for class AB2 or class B amplifiers. (5) It will hold the bias voltage within 1.0V of the selected value during grid current excursions up to 100 mA, or within about 4V for a current up to 400 mA. (Closer regulation could be obtained with additional components but is not necessary.)

(6) It can be used in conjunction with almost any bias supply of suitable voltage, and only a small current is taken from the supply, regardless of the value of grid current to be handled.

Bias regulation differs from the more common use of regulators in power supplies in that grid current flows backwards into the supply instead of drawing current from the supply. The shunt type of regulator is ideal for this application because it can be designed to draw only a very small constant current from the bias supply and yet accommodate a heavy flow of grid current from the linear amplifier while maintaining the bias voltage constant. The more commonly used series regulator does not possess this property, but must be set up for a power supply drain greater than the peak grid current to be catered for.

The design of this circuit is based on the use of shunt regulation, and evolves the simplest possible circuit that will give adequate regulation plus an adjustable voltage.

Resistors R1, R2, R3 form a voltage divider which applies a fraction of the total voltage to the base of the transistor, while the emitter voltage is fixed by means of the zener diode. The device regulates in such a manner that the voltage between base and the negative terminal just exceeds the zener voltage. Any further increase in voltage is then opposed by the heavy increase in transistor current which results. The voltage at which the device regulates is changed simply by altering the ratio of the base dividing resistors. The base could be connected to the moving arm of a potentiometer, with fixed resistors from either side of it to the plus and minus terminals, but if this were done, the bleeder current would increase needlessly as the regulated voltage was increased. With the circuit shown, the bleeder current (through R1, R2, R3) remains constant as R3 is varied. To allow for different bias supply units, the series resistor R (Fig. 4-12) should be chosen so that a current of between 15 and 20 mA flows into the regulator when it is set for maximum voltage (i.e., R3 at maximum). As the regulated voltage is reduced, the voltage drop across R

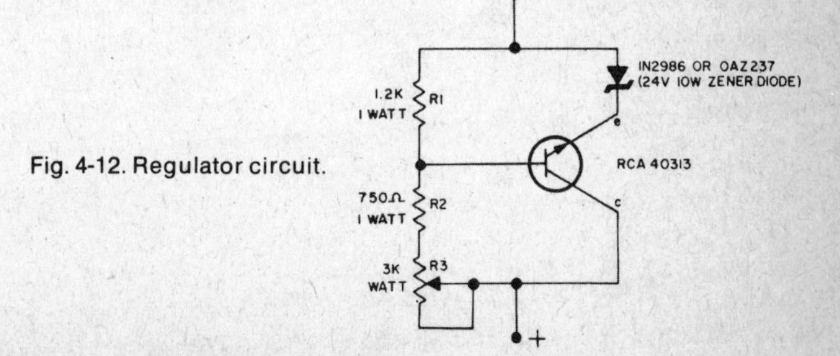

Fig. 4-12. Regulator circuit.

will increase and cause an increase in the current through it and through the regulator. This is of little consequence unless the bias supply is rated at less than 30 mA, in which case R could be made adjustable and coupled to R3; or, alternatively, R could be replaced with a constant-current device. These refinements, however, will not generally be necessary or desirable. The bias power supply unit itself need not possess good regulation; in fact, a poorly regulated supply giving a drooping voltage-versus-current curve is desirable to minimize the current changes described above.

If it is desired to substitute other components or alter the range of the regulator, watch the following points: (1) The higher the zener voltage the better the regulation, but it must be less than the required minimum regulated voltage. (2) The transistor must have an adequate voltage rating. (The RCA 40313, with a 300V rating, is good for the highest bias voltages likely to be required.) (3) Watch the dissipation—transistor and zener both carry the full grid current. (4) Determine the maximum base current (maximum grid current divided by beta). (5) Select the base divider resistors to pass between 10 times the base current.

Fig. 4-13 shows the regulator in a practical circuit for a linear radio frequency amplifier.

With the VOX relay open, maximum bias voltage is applied to the amplifier to bias it off. When the VOX relay operates, the bias falls to the value set by the regulator, and the antenna relay changes over. The diode shunting the relay coil is to avoid damage to the regulator, due to the "kickback" of the relay winding. Should the "slugging" effect of this diode cause the relay to release too slowly, the diode could be placed across the regulator unit instead. The positioning as shown has the advantage, however, of preventing sparking at the VOX relay contact due to the relay inductance.

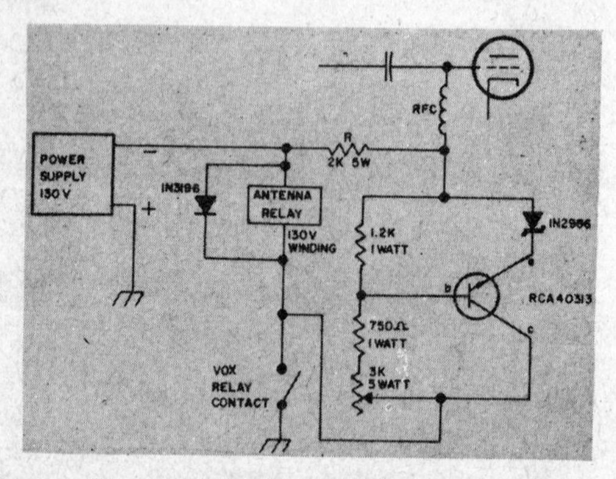

Fig. 4-13. Biasing the linear amplifier.

If the antenna relay arrangement is not required, simply omit relay and diode from the circuit. If the provision for biasing back the amplifier during standby is not required, the positive end of the regulator should be grounded directly and the transistor case may in fact be bolted to the transmitter chassis without any insulating washer.

The diagram shows a 130V power supply, but this could be considerably higher (or lower if only a reduced voltage is required). In either case, choose the series resistor R as described earlier.

The circuit of Fig. 4-13 has been designed to handle a considerable amount of grid current, but the same circuit without any change whatever may be used for class AB1 or other amplifiers which do not draw grid current.

POWER PERK

Here's an amplifier (Fig. 4-14) capable of running a full kilowatt on 432 MHz, without straining either the budget or the final tubes. The big difference between this amplifier and others that have been described is the use of vapor phase cooling. This allows a relatively small tube to dissipate a lot of power.

Other features of this amplifier are:

1. Simple construction requiring no machine tools.
2. Single ended operation with no balancing problems.
3. Large overload capacity.
4. No noisy blowers.

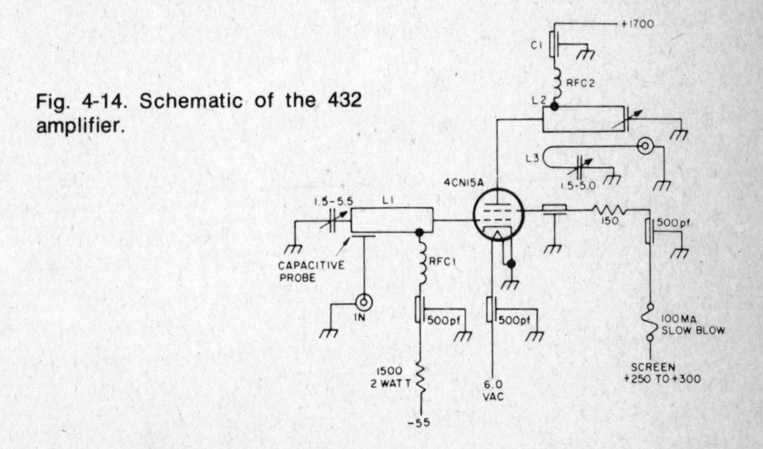

Fig. 4-14. Schematic of the 432 amplifier.

The tube used is a 4CN15A. Electrically, it is the same as the 4CX300A, but mechanically different in not having any air cooling fins. Fins are not required when vapor phase cooling is used.

Dissipation of plate power is accomplished by boiling water inside the plate trough (Fig. 4-15). This holds the plate temperature at about boiling temperature (100°C). As more power is dissipated the water boils harder but never gets any hotter. The tube anode is actually somewhat hotter than the water because it is not entirely immersed. Being inside the plate trough, the water is separated from RF and DC fields, and therefore causes no electrical problems.

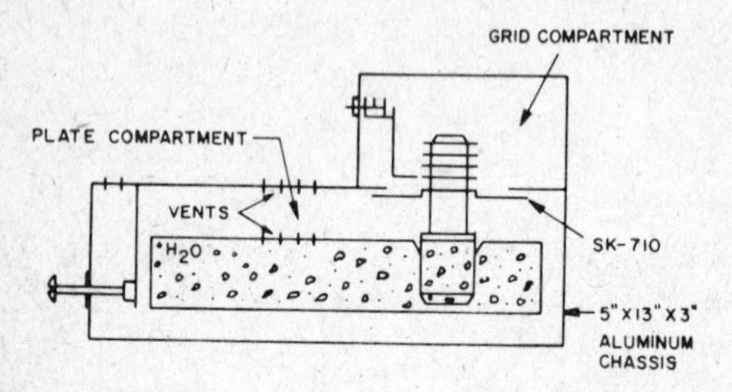

Fig. 4-15. Water inside plate trough boils to keep the tube plate temperature from rising above 100°C. Steam escapes from ventilation holes in the plate trough and plate compartment.

Construction

The construction of the amplifier is clear from the drawings (Figs. 4-16 and 4-17). Half wave, capacitively loaded lines are used in the grid and plate circuits. The grid line is made from flashing copper and tuned by a 5 pF

Fig. 4-16. Dimensions of the plate line box.

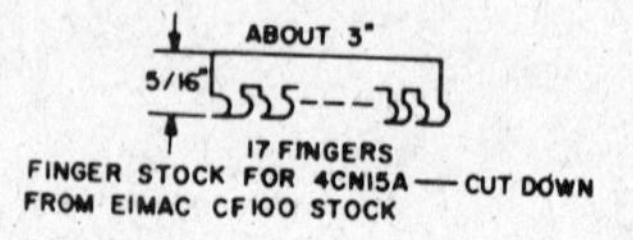

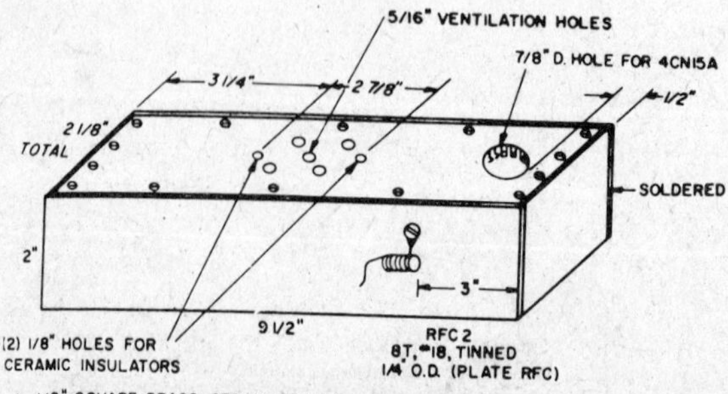

variable capacitor. Coupling into the grid circuit is by a capacitive probe also built from flashing copper. This is adjusted by bending to maximize the drive. A small fan is mounted over the grid circuit to cool the tube pins and prevent steam from coming up through the tube socket. A slight modulation can be heard locally from the fan blades passing near the grid circuit. A wire mesh shield could be used to prevent this effect.

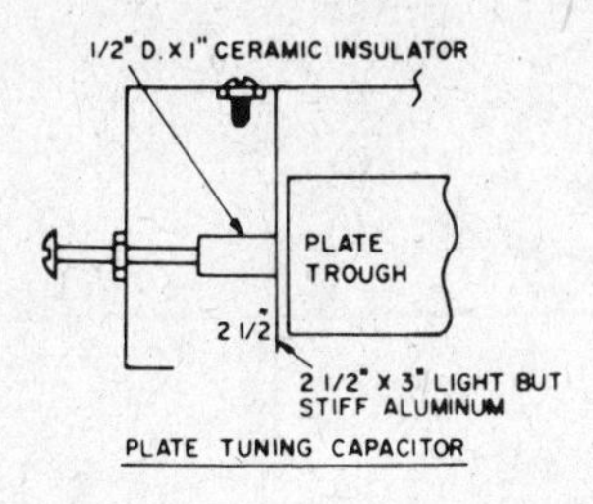

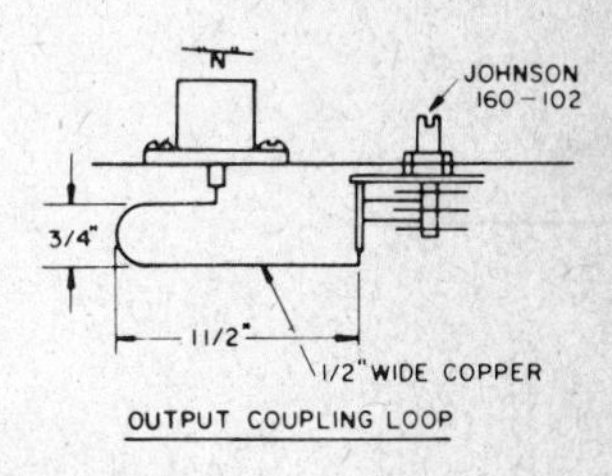

Fig. 4-17. Details of construction.

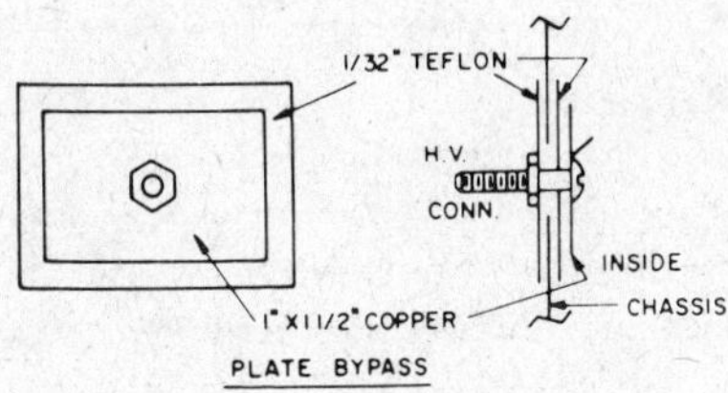

Brass stock 2″ × 1/16″ is used to make all sides of the plate line. Since, mechanically, the line is a water trough, it is made leak-proof by careful soldering of all joints. Acid flux is used and, after soldering, thoroughly scrubbed out. The top of the trough is made removable for cleaning.

Electrical contact to the 4CN15A is made from cut down Eimac CF100 finger stock.

The outside of the plate compartment is a 5 × 13 × 3 inch chassis. Ceramic standoff insulators are used to support the plate trough. In order for the tube anode to protrude as far as possible inside the plate trough (Fig. 4-18), the clearance left is insufficient for standoffs inside the plate compartment. For this reason braces are mounted over the main chassis. The height of the plate trough is adjusted so the anode seal is flush with the top of the trough. The plate tuning capacitor is a sheet of "springy" aluminum, operated by a 10-32 screw with a ceramic insulator between the screw and the aluminum. When the amplifier was first assembled, the insulator was not used and considerable arcing took place at the metal-to-metal contact.

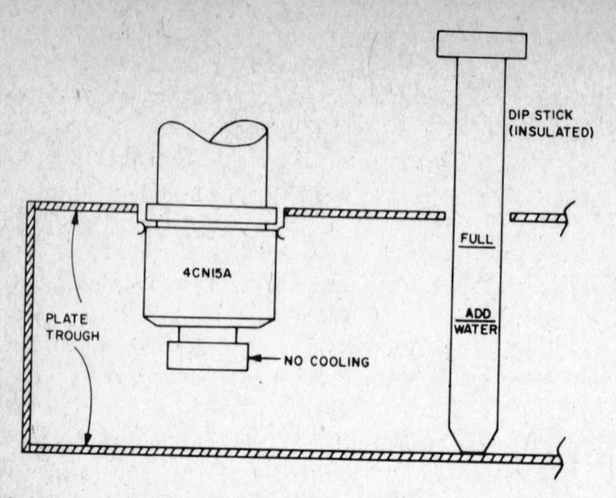

Fig. 4-18. View of water levels inside plate trough. Water must cover part of main body of the tube. Small portion on end of tube is a seal cover and cannot conduct any quantity of heat.

Operation

The maximum power capacity of the amplifier is difficult to determine. The maximum allowable temperature for the ceramic-to-metal seal is 250° C. Measurements of the seal temperature were made using Tempilac indicator with no RF applied. It was found that the tube can dissipate well over a kilowatt in a good rolling boil (see Fig. 4-21). About 200 watts is dissipated with no boiling occurring. Thus, in general, dissipation of heat is not a limiting factor for amateur power levels.

Manufacturers' specifications for CW operation limit the plate voltage to 2000 volts and the current to 250 mA. This gives an input of only 500 watts.

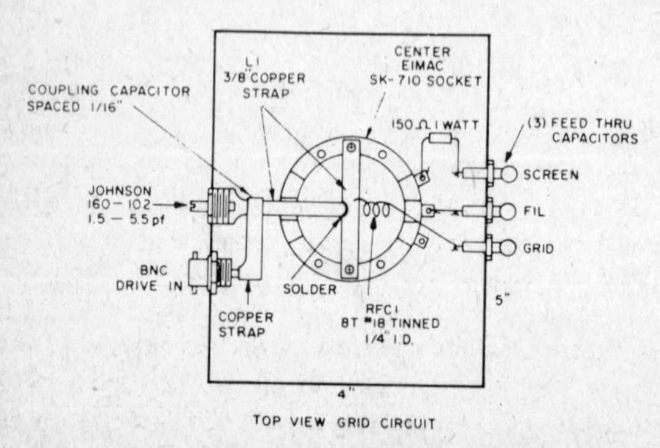

Fig. 4-19. Grid compartment details.

However, this amplifier has been operated for many hours at a kilowatt (2500 volts 400 mA) with no sign of tube deterioration and at 500 watts for general operation and at a kilowatt for schedules and band openings for the past three years using the same tube with no drop in output.

About 25 watts of grid drive are required for full output (Fig. 4-19). This gives 25 mA of grid current at about 90 volts bias. The screen voltage is adjusted to give the desired input. This requires about 250 volts for 500 watts input and 300 volts for a kilowatt. Screen current is usually not more than 10 to 15 mA and, under some conditions, will be 5 or 10 mA negative. The plate current dips at resonance but the best tuning device is a power output indicator. Another method of tuning at high power levels is for minimum boiling sound (this requires a reasonably quiet room!). Efficiency generally runs about 50%.

Screen modulation has been used for AM operation. Here the screen is dropped to about 100 volts for a carrier power of about 100 watts. About 300 peak-to-peak volts of modulation is applied from a modulator with 4000 ohms output impedance. This was a push-pull 6146 modulator with 8000 ohms output impedance, swamped by an 8000 ohm resistor.

The trough is normally filled to within 1/4" of the top and allowed to boil down to half the tube length before refilling (Fig. 4-20). An *insulated* dip stick is used to measure the water level with *power off* and the high voltage grounded. Remember, full high voltage exists on the plate trough. Filling is done by siphoning water from a bottle with a length of 1/4" plastic aquarium air hose. The hose is placed through a ventilation hole.

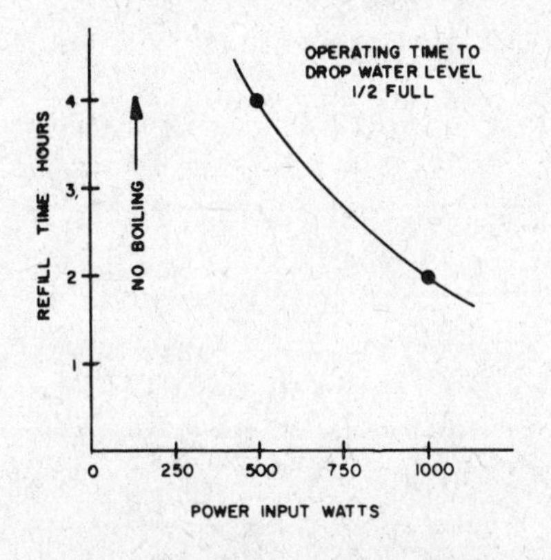

Fig. 4-20. Power input watts.

Distilled water is not required, since no voltages are impressed across the water. However, if tap water is used, deposits will build up inside the trough necessitating frequent cleaning and making distilled water a worthwhile convenience.

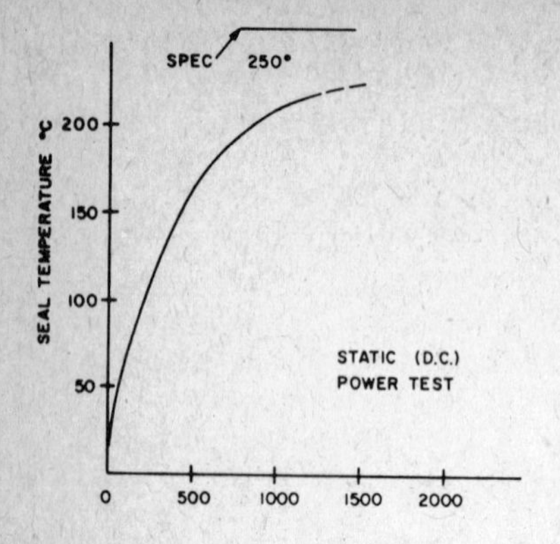

Fig. 4-21. Dissipation watts.

Modifications

Although finless tubes such as the 4CN15A are mechanically ideal for amplifiers with "poor man's" vapor phase cooling, almost any external anode tube with ceramic insulation can be used. Check the temperature rating on the seals—if they are 250°C or higher there should be no problems. Glass-to-metal seals are generally rated at 150° C which is not far enough above the 100° C boiling point to be safe.

A version of the 2C39 series, the 3CPN-10A5, has no cooling fins and can be easily used in this type of amplifier. Additionally, many of the 2C39As have their fins held on by a set screw and can be easily removed.

Chapter 5

50 Megahertz and Above

6 METER FET CONVERTER

The FET combines some of the best features of the vacuum tube and transistor and is rapidly being used in many new electronic circuits. FETs are divided into two main groups: the junction FET and the insulated gate FET. The junction FET was selected for this converter to simplify the construction. This is possible as the Source and Drain are interchangeable in the JFET. The determination of which element is the Source or Drain depends upon the applied voltage.

Three Texas Instrument TIS-34 FET transistors are used. T1 is the RF amplifier, T2 is the mixer and T3 is the crystal oscillator. Note the similarity to a vacuum tube circuit with the Gate (grid) and Source (cathode) resistors and the Drain (plate) connected to the tuned circuit. No fancy biasing circuits are required. Although a standard transistor can be used in the oscillator circuit, the simplicity of the FET oscillator is unique. The 50 MHz incoming signal is mixed with the 36 MHz signal from the crystal oscillator resulting in a 14 MHz output signal. In the crystal controlled type converter the receiver acts as a variable IF. If used with the Collins "S" line, for example, you will be able to cover 50 to 50.4 MHz of the 6 meter band by tuning the receiver from 14 to 14.4 MHz. If you have a general coverage receiver you can tune the entire 6 meter band.

The schematic of the converter is shown in Fig. 5-1. The converter is constructed on a 3 × 5 3/4 inch printed circuit board. Small inexpensive kits for etching copper circuit boards are available at most radio stores and mail order electronics firms. The actual process is not too difficult for the average ham. It is best to draw the layout on the copper side of the board in pencil and then fill in those portions to be retained with the resist paint. Actually, any type

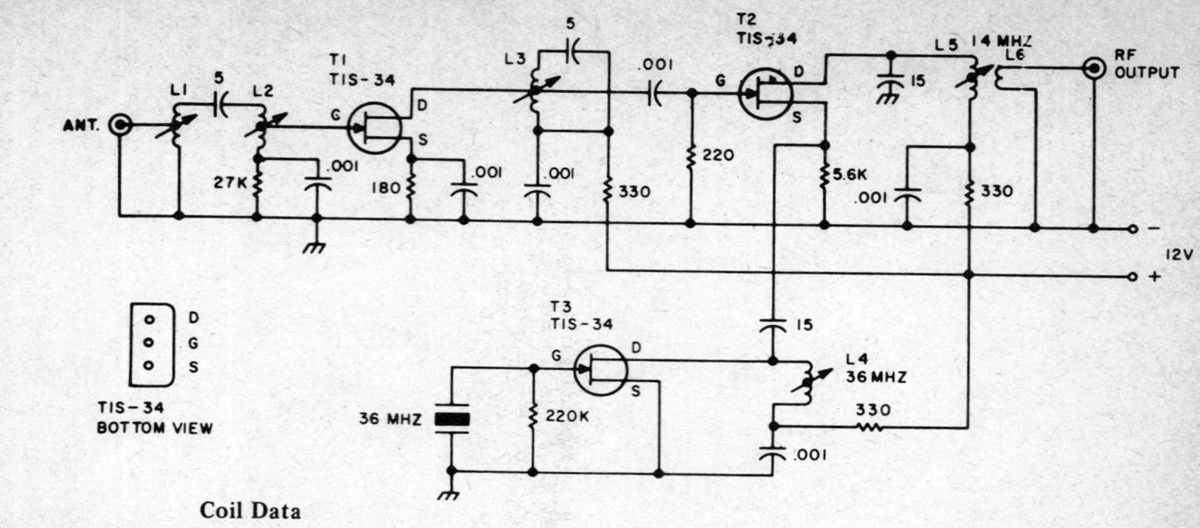

Coil Data

L1 - 10T #2BEC CW ¼'' tuned form tap at 2T.
L2 - Same as L1 tap at 4T.
L3 - 8T #28 CW-tap at 4T ¼'' form.
L4 - 16T #28 CW ¼'' form.
L5 - 40T #28 CW 5/16'' form.
L6 - 2T small hookup wire.

Fig. 5-1. Schematic of the 6 meter FET converter.

of model airplane paint works quite satisfactorily. The PC board is next placed in a small plastic or glass container and covered with the etching solution. (Not used in all kits.) The etching process takes 20 to 30 minutes, during which time the solution should be agitated by rocking the plastic container back and forth. When the etching process is completed the board is washed with water.

Following the etching process the resist paint may be removed with lacquer thinner or carefully scraped from the board. After the paint is removed, the board should be cleaned with steel wool for fine sandpaper. The resistors and condensers can now be mounted along with the transistors, crystal socket and coils as indicated in Fig. 5-2. As mentioned previously, the Drain and Source are interchangeable so you don't have to worry how the transistors are installed as long as the Gate is connected to the proper terminal. The transistors require a supply of 12 volts DC. This can be obtained from a small battery pack or a standard 12 volt power supply. The unit draws 18 mA. If intermittent use is contemplated the battery pack will be satisfactory, but if you plan to operate for long periods of time, a standard power supply is suggested. The unit is mounted in a small chassis box 6 1/4 × 3 1/2 × 1 1/8 inches.

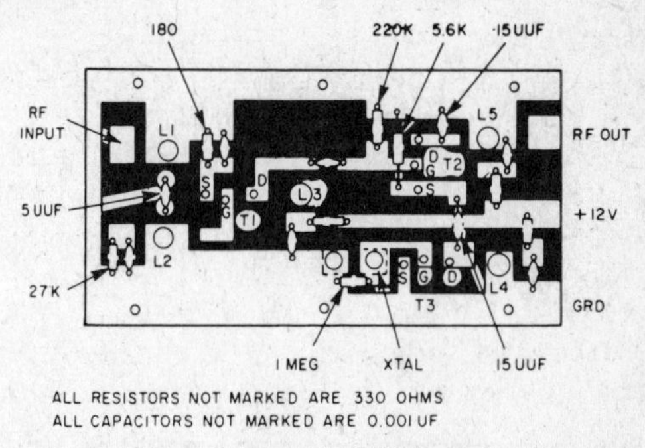

Fig. 5-2. Layout of PC board.

With the converter connected to the receiver, antenna and power supply, the coils can be adjusted for maximum noise in the receiver. This should allow the reception of signals and the coils can be peaked on the receiver "S" meter. In some areas channel 2 may cause some interference. If this is a problem, a tuned circuit consisting of a 45 pF trimmer and 5 turn coil 1/4 inch in diameter can be installed between the RF input and the antenna jack to trap out the channel 2 signal.

A VARIABLE RESISTANCE VFO FOR 6 OR 2

A considerable effort is made in VFO designs to provide mechanical rigidity in the tunable element, whether coil or capacitor. Yet, the variable resistance VFO is a completely rigid device which has seen little application.

Two models of the VFO are discussed. One features two miniature potentiometers of coarse and fine tuning. The other has only one standard, a lower cost pot for the economy minded. The circuits are identical except for the added pot. The schematic in Fig. 5-3 shows the standard series tuned Colpitts oscillator. If you prefer the parallel tuned version, simply modify it according to Fig. 5-4. The series tuned version seems to give slightly better stability with supply voltage variation.

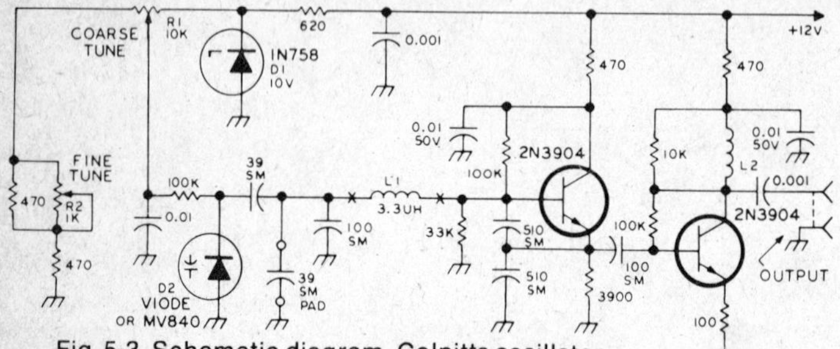

Fig. 5-3. Schematic diagram, Colpitts oscillator.

The voltage to the diode is regulated by the zener diode. The .001 capacitor is for filtering any AC or noise. It could be almost any value. The 100K resistor is for isolation of the diode. An RF choke could be used, but would introduce more reactive components into the critical tuned circuit. The output coil, L_2, is chosen to resonate with the length of coax to the next stage. If no coax is used, 15 μH is the correct value. If maximum output is desired, a smaller coil can be used and a trimmer capacitor added to tune the output rather than cut and try on the coax length. The 10K resistor across the output just loads it and broadens the response. The ouput should be at least 15 volts peak to peak into 10K ohms of good clean sinewave.

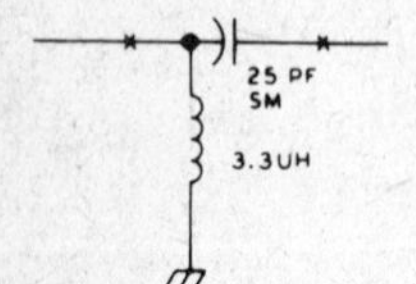

Fig. 5-4. Parallel tuned version.

The values given are for a frequency of 8.3 to 8.5 MHz. Changing the padder quickly puts it in the two meter range from 8.0 to 8.2 MHz. The unit is quite stable with voltage variation from 10.4 to 15 volts or more. You can shut the oscillator off in several ways such as: grounding the base, opening the emitter, or simply switching off the supply voltage. Normally the latter is not too satisfactory, but no turn-on drift could be detected. Everything is rigidly mounted to the PC (see Figs. 5-5 and 5-6) board. The components should be mounted on the back for easier soldering.

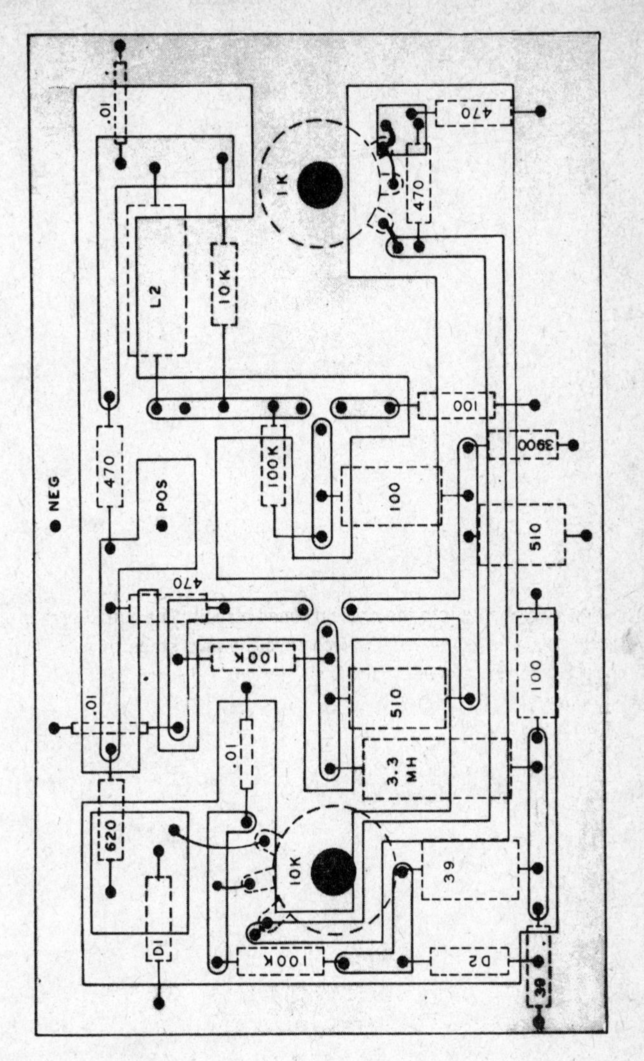

Fig. 5-5. Printed circuit board.

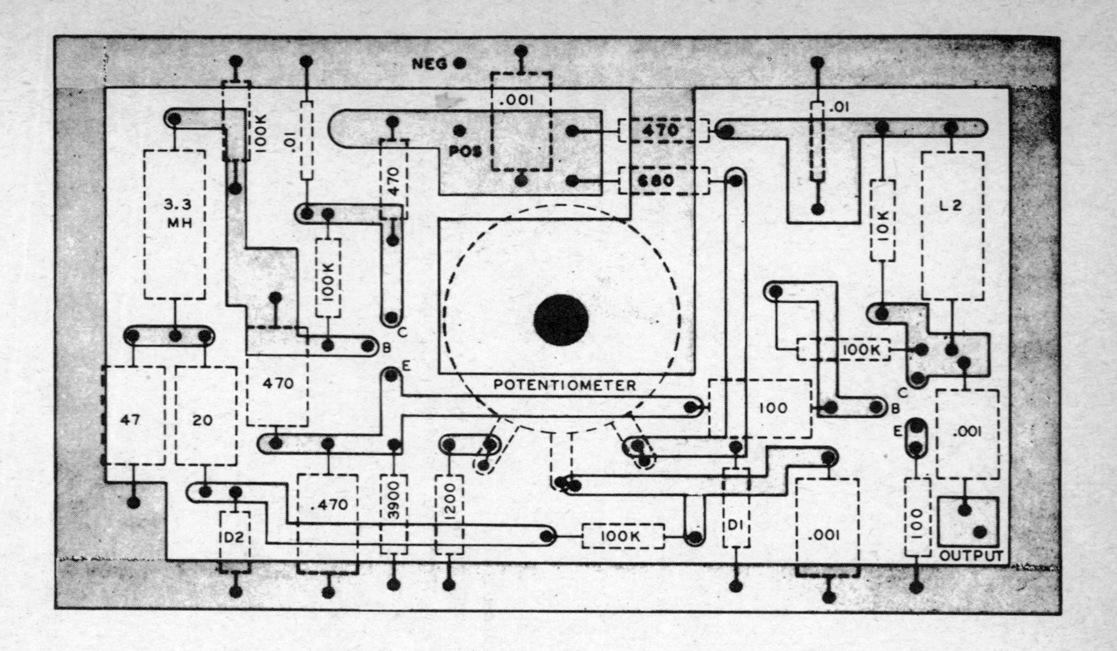

Fig. 5-6. Parts placement on other side.

The two pot model is recommended for SSB since it is easier to zero beat, or the one pot version can be used with a vernier.

Most verniers will restrict you to 180 degrees rotation, while the pot will provide 270 degrees. The single pot is quite adequate for AM work without the vernier, and is considerably cheaper.

6 METER JUNK BOX RIG

This transmitter could be called a "Glorified Sizer", running about 18 watts input into a 6V6. The schematic is shown in Fig. 5-7. This tube was chosen as a final because most old car radios from the junk yards happen to have two of them in the line-up. The 6U8 oscillator and doubler came from a TV set, as did the 12AX7, the slug-tuned coil forms, sockets, resistors, condensers, and the modulation transformer.

Lay out parts so that all leads can be kept as short as possible. Another point: isolate the modulator section from the oscillator and final. The rule is "shield until it hurts." This complete rig, less power supply, is built in a box 5 × 7 × 6 inches high, starting on a 2 × 5 × 7 inch chassis. The modulator is kept away on one end of the chassis, and the underside has an aluminum shield 2 x 5 inches separating the modulator wiring and sockets from the oscillator and final. If good construction practices are observed no difficulties should be encountered.

It is a good idea to check all the slug-tuned coils with a grid-dip meter to get them somewhere near frequency before they are installed in the chassis. Give them another check and adjustment after the wiring is completed and tubes are in the sockets, with the *power off*.

The receiver is a 12 volt auto radio, converted to 110 AC operation, with a converter in front of it.

The 6 meter converter (Fig. 5-8) is an accumulation of ideas, not necessarily original nor the best available, but "cheap and it works." As is the case of the transmitter, most of the parts, including tubes, can come from a junked TV set. Plenty of shielding is used between stages.

Receiver coils are also tuned somewhere near frequency before they are installed in the chassis, and touched up after the final wiring and tubes are in place.

With the IF frequency shown, the car radio will tune from 50 to 51 megacycles, which is the most popular part of the band.

If meters are a problem, just about any sort of a meter movement can be converted to 0–100 mA tuning meter for the transmitter with a proper shunt. All you're looking for is an indication, not a measurement, for loading the transmitter, and you can pick up a movement from an old tube tester, or something similar for next to nothing.

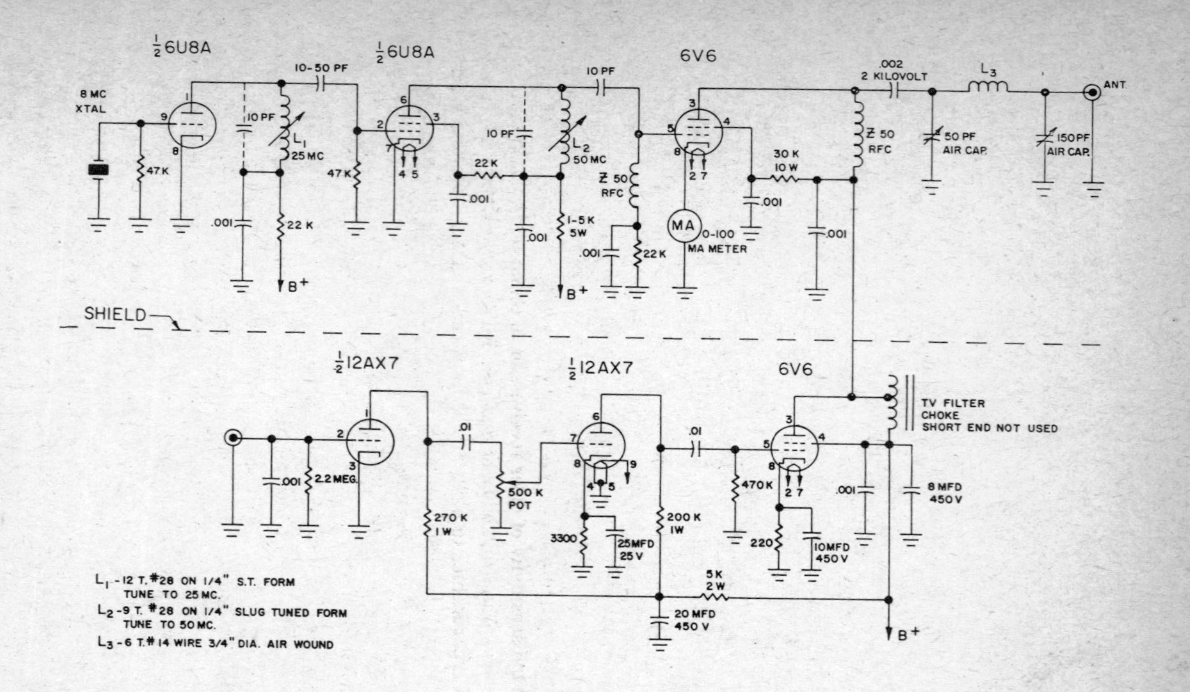

Fig. 5-7. Junk box transmitter schematic.

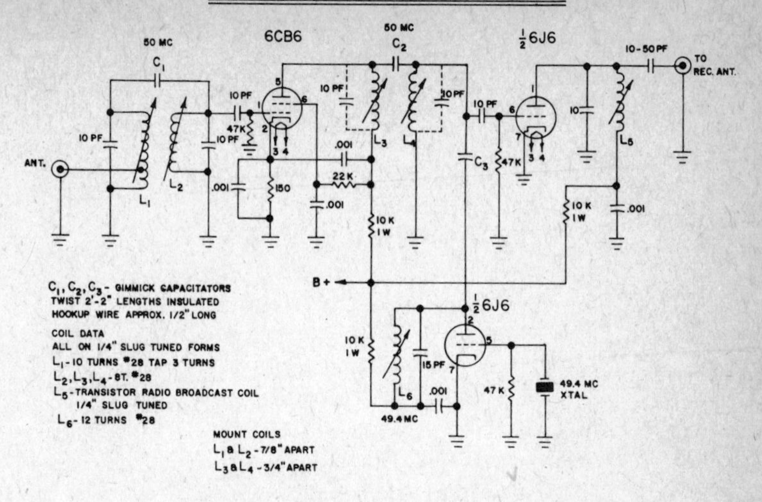

Fig. 5-8. Junk box converter schematic.

A DUAL-GATE FET PREAMP FOR 2 METERS

An RF preamplifier is a handy thing to have when there is a desire to improve the sensitivity of older receivers or to complete the construction of modern ones. The availability of dual-gate FETs makes it possible for the average experimenter to build a good front end for his receiver. This FET amplifier can provide about 20 dB of voltage gain to improve the sensitivity of fair or poor receivers. The preamplifier can also improve the signal-to-noise ratio.

The Circuit

Figure 5-9 shows the schematic of the amplifier. A common source unneutralized Motorola MFE 3007 is used. The input signal is applied to one gate, and the other gate is used for biasing. An RF choke and feed through bypass capacitor are used to improve isolation from the power supply. Experimentally tapped coils are used for input and ouput coupling.

Construction

The amplifier is built in a small aluminum box. The top is cut out and replaced with a copper plate, and a copper partition is soldered in the center as a shield. Both L1 and L2 are 3 1/2 turns of 18-gauge wire stretched to about 3/4 in. long. The coils are 1/2 in. diameter. MC 603 trimmer capacitors are used for C1 and C2. They are 1–28 pF, but a variety of small variable types could be used. The transistor is mounted on a clip in the center shield. L1 and

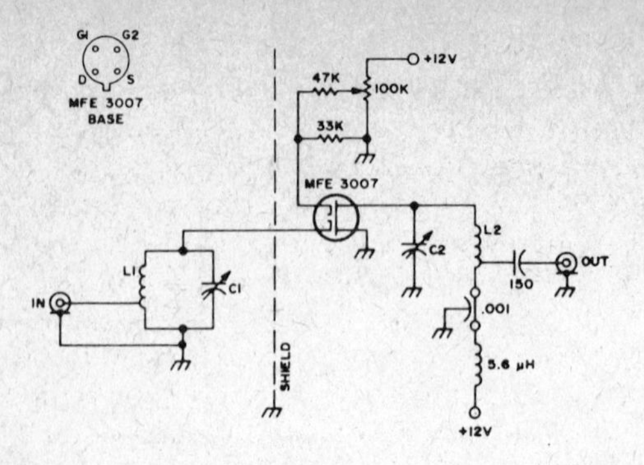

Fig. 5-9. 146 MHz RF amplifier.

L2 are mounted perpendicular to each other to minimize electromagnetic coupling. A 100 kΩ variable is mounted in the hole on the output side of the box. Venerable UHF connectors are used for both input and output terminals. Since the MFE 3007 is susceptible to gate breakdown, all leads of the transistor are shorted together with a short piece of wire. No problems are encountered with transistor damage, and when the wire is removed the amplifier functions normally.

Adjustments and Operation

The first step in tuning is to set the tuned circuits approximately on frequency with a grid dip meter, then see that they tune above and below the desired frequency. An AM receiver is used with the AVC turned off, the audio gain set at maximum, and the RF gain control on the receiver used to control the output level. A modulated signal generator is used to obtain an AC voltage at the speaker terminals. A VTVM with its dB scale can be used to measure the change in output signal. This setup makes it possible to estimate the gain of the preamplifier (assuming that the receiver is linear). With the gate bias set at zero, and a weak signal applied, all receiver RF circuits are tuned for maximum output.

After the tuned circuits are set up with zero bias, the 100 kΩ pot is turned slightly to increase the bias. This generally causes a decrease in output, but retuning the tuned circuits, mostly C2, should bring the output up to a point higher than it was previously. This process of increasing bias slightly and retuning is continued until the amplifier breaks into oscillation.

The bias is then reduced to provide stability and all RF circuits in the receiver are tuned to assure stability and proper tuning.

This preamplifer provides a good increase in signal gain that, in some cases, may mean the difference between Q5 copy on a weak signal, and just barely being able to hear it.

EASY PREAMP FOR 450 MHz

With the wide availability and low cost of used 450 MHz commercial transceivers and also with the new FM transceivers on the market, the 450 band is really enjoying popularity. Many 450 repeaters are also being combined with VHF repeaters. Good 450 coverage is essential to a remote base system if it is to be used from a mobile control point. The mobile receiver is often the weakest link since it is affected by local noise that does not bother the mobile transmitter or remote receiver. Since most 450 tube radios have trouble receiving signals weaker than .8 μV or so, addition of an RF preamplifier will make a significant improvement in receiver sensitivity and quieting.

This is a simple 450 preamplifier that can be built easily, using a minimum of parts. It will provide up to 10 dB of gain with a noise figure low enough to provide a major improvement when used with older tube type receivers.

Circuit

Figure 5-10 shows a single stage common base amplifier. The simplicity of this single stage amplifier allows for the use of a minimum number of parts. Addition of a second state would not be justified unless the receiver to be used were really poor. The 800 pF DC blocking capacitors allow the use of trough-line inductors to simplify construction. 1N914 diodes may be added in parallel at the input jack to protect the transistor from burn-out by a nearby transmitter. Other types of silicon NPN UHF transistors such as the 2N3839 may be used in place of the 2N2857.

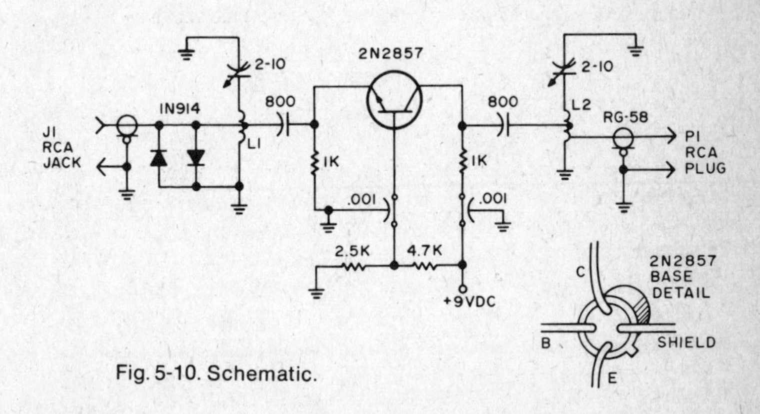

Fig. 5-10. Schematic.

Construction

The first step in construction is to cut out the six parts of the chassis as indicated in Fig. 5-11. The use of epoxy PC board results in construction that is much stronger or lighter, and is easier to work with, than in the case where copper or brass sheet is used. The PC board may be either single or double clad. It should be good quality material such as epoxy to prevent peeling when heated.

Figure 5-11 shows two views of the center partition, one from each side to show circuit construction details. One view is shown of both the side pieces, which are different. One hole only is drilled in each piece. The size of the hole for the antenna jack will depend on the type of connector used. RCA jacks are used almost universally in all receiver strips. The output connection from the preamplifier is a short length of RG 58 with an RCA plug on the end.

The pieces of board should be cleaned to facilitate soldering after they are cut out. Steel wool will do a good job here.

The center partition should be soldered to the bottom board first. Spot solder it together with two small blobs of solder so that its position may be adjusted if it is placed incorrectly the first time. Using a large soldering iron or gun, run a smooth bead of solder along the joint. The joint will be permanent after the entire length is soldered, and will be very difficult to remove. The HOT and GND ends of the chassis are added next in the same manner. The structure now should be quite sturdy, and at this point the internal circuitry may be added.

Mount feedthrough bypass capacitors in the holes in the bottom plate. Bend the leads of the transistor out axially from the bottom of the case, and solder the shield lead to the center partition so that the collector lead extends through the hole in the partition. The base lead will connect to the feed-through capacitor on the bottom plate. Next add the 1K resistors on opposite sides of the partition to the two free transistor leads as shown in Fig. 5-11. Keep the leads as short as possible.

The tuned circuits are added next. L1 and L2 are made from 1/4 in. diameter copper tubing and should be 8.6 cm long, including the length of the capacitor. Many types of capacitors may be used in this circuit, but they should have maximum of about two picofarads when at minimum capacitance. Ten picofarad capacitors peak near the minimum capacitance end of their range. If a piston type capacitor is used, the capacitor is mounted first and the tubing is cut to fit from the capacitor to the end of the chassis. The length of the coil is not extremely important since the capacitance may be varied. If another type capacitor is used, the coil tubing is soldered to the end of the chassis first. This is strong enough to support the capacitor while it is soldered from the end of the coil to ground by its leads. The hole in the end of the chassis is now used for access to the capacitor for tuning. Be sure to put the rotor lead of the capacitor to ground so that the tuning tool will not detune the circuit when it is touched. L1 is tapped at 5 cm from the ground end, and L2 is tapped at 5 cm for the collector, and one half inch for the output coax lead.

With the coils in place the 800 pF capacitors can be added. These are small disc ceramics stripped from an old TV receiver chassis. The critical point with these capacitors is size—the smaller the better.

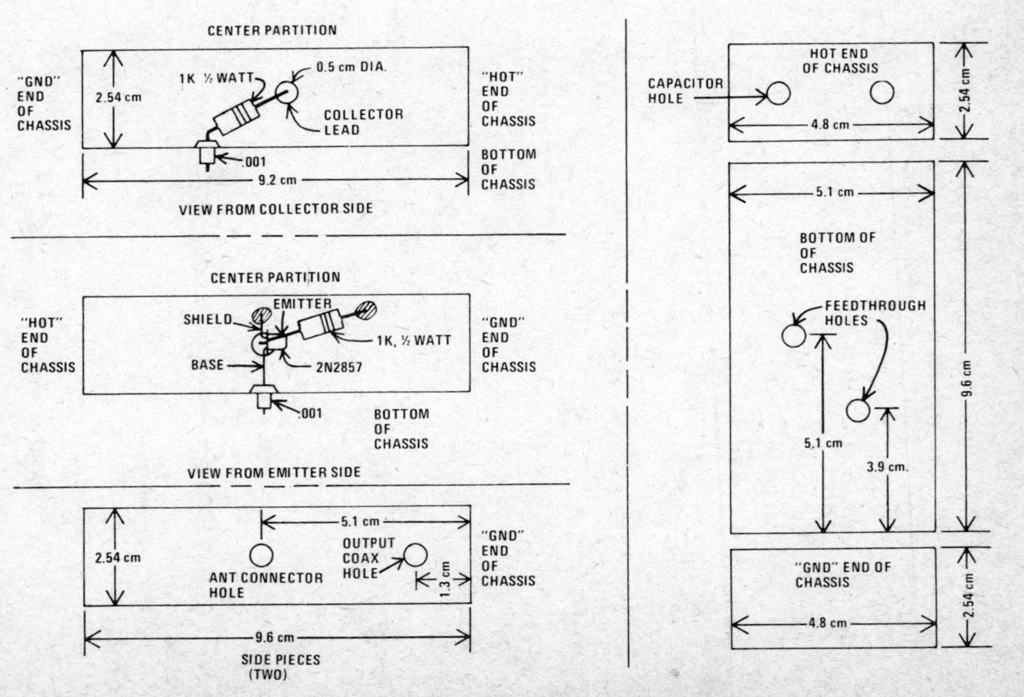

Fig. 5-11. Construction details and dimensions.

The two sides can be added now that the center circuitry is completed. It is easier to mount the input jack after the side panel is mounted to make soldering the long joint easier. On the output side it may be easier to solder the coax braid on first, since this may be difficult. The coax is passed through the hole with about 0.6 cm of braid exposed. The braid is spread out axially and soldered to the chassis. The insulation is stripped before mounting the panel and soldered after the side panel is secured. Be sure to run a smooth bead of solder around all corner joints. The internal wiring is now completed.

The 2.5K and 4.7K resistors are mounted externally under the bottom plate. It is necessary to drill a hole through the bottom to ground the 2.5K resistor. A top cover can be made by soldering bolts (with the heads cut off) around the edges and center of the top of the chassis. Poke the extended bolts through a piece of paper to make a template for drilling holes in the cover. As a final step the inside of the chassis can be cleaned with a solvent such as toluene, acetone, or lacquer thinner, and a stiff acid brush. Do not use ordinary rubbing alcohol; it will leave a residue. The cleaning will remove the rosin and any solder balls or metal filings that may be sticking to the rosin.

Operation

Just plug it in and tune it up. The amplifier should pull about 1.5 mA at 9V. This voltage may be obtained from a dropping resistor and a 9V zener diode, or 6V could be obtained from a 6V tube filament in a series connected 12V filament receiver system. If the preamplifier is mounted externally to the receiver it may be powered by a 6V lantern battery left permanently connected.

It may be necessary to reduce the value of the 2.5K resistor if the amplifier pulls more than about 1.5 mA. As the bias is increased, the current is increased; this will increase gain, but will also increase noise generated by the transistor. As bias is increased, instability may result and bias will have to be reduced by making the 2.5K resistor smaller. Instability is discovered by the appearance of spurious signals, or the capacitors may tune for a peak at more than one point.

The diodes may now be added and the preamplifier rechecked to determine whether performance has been affected. The entire front end of the receiver should now be retuned. The input circuit of the preamplifier should tune broadly and the collector circuit should tune rather sharply. This tuning should be smooth and within the range of the capacitors.

The preamplifier may not work well in duplex operation since it is susceptible to cross modulation problems typical to bipolar transistors, and the single input tuned circuit does not provide much selectivity.

2 METER FM TRANSMITTER

For the amateur who has had some experience in solid state VHF circuitry, this transmitter can provide a convenient and inexpensive way to

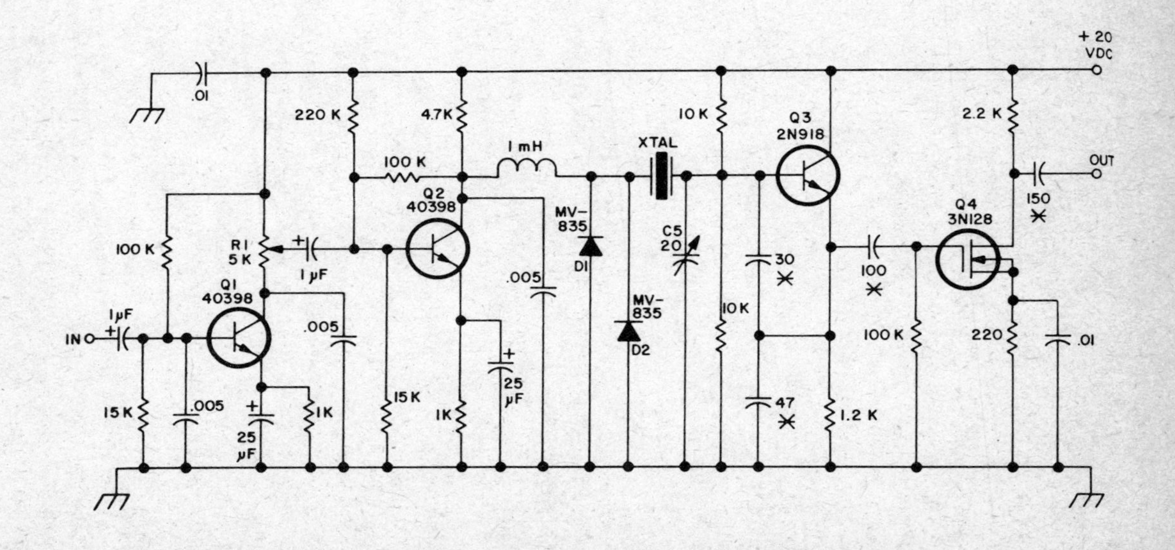

Fig. 5-12. VCXO board. Capacitors marked with + are electrolytic. Capacitors marked with * are mica. Other capacitors are disc ceramic. C5 is a small ceramic trimmer.

join the activity on the local repeater. This circuit could also be used as part of a handy-talkie or mobile installation.

The Circuit

The transmitter is built in three semimodules; the voltage controlled crystal oscillator (VCXO), the amplifier-doubler (Amp-Doubler), and the power supply. Figure 5-12 shows the VCXO. Two 40398s amplify a ceramic or crystal microphone input to drive a pair of MV-835 varicap diodes. Bypass and coupling capacitors were selected to give a one dB 300 to 3000 Hz bandpass. A small amount of negative feedback was used on the second 40398 (100 K resistor) to reduce excessive gain, while slightly reducing distortion in that stage. It may also help to improve bias stability, thus improving the frequency stability of the oscillator since the varicaps are biased directly by the collector of the 40398. Using the direct coupling eliminates the need for a coupling capacitor and separate bias for the varicap. The varicap would be biased at the same voltage as the collector quiescent voltage anyway. The only drawback is the fact that the frequency stability is dependent on the bias stability of the 40398. Using two varicaps in parallel provides more dynamic capacitance change, even though it doubles the total capacitance. Direct FM is used on the 18 MHz fundamental cut crystal. The crystal was cut for a 20 pF load capacitance. The 2N918 oscillator uses an old and familiar circuit to drive a 3N128 MOSFET buffer. The common source buffer provides about 1.3 rms at the input of the Amp-Doubler. The use of a MOSFET buffer reduces the problem of frequency "pulling" caused by tuning or load changes in the Amp-Doubler.

A 24V rms 20 volt-amp power transformer, a bridge rectifier, and a large filter (Fig. 5-14) capacitor provide 28 volts to drive the last three stages of the Amp-Doubler. A dropping resistor and zener diode with another large capacitor give clean regulated voltage for the VCXO and first doubler.

The Amp-Doubler (Fig. 5-13) is the most interesting and most difficult part of the transmitter. Many basic considerations go into VHF transmitter circuitry. As frequency decreases there is a six dB per octave increase in transistor gain. This can give a stage much higher gain at lower frequencies, and cause low frequency oscillation. This type of oscillation can usually be traced to poor power supply bypassing, component self-resonance, or RF choke resonances with other circuit capacitances. Each state must be bypassed so as to be effective at lower frequencies as well as at VHF. This is the reason for dual bypass capacitors. Any RF chokes should be low Q. Wirewound resistors and ferrite beads have good success, but ordinary carbon resistors can be used with a Q of almost zero. Collector RF chokes can be eliminated by using a coil that is part of the tuned circuit to supply B+ current. Interstage coupling was accomplished by experimentally tapping coils to obtain an approximate match. Efficiency could probably be much

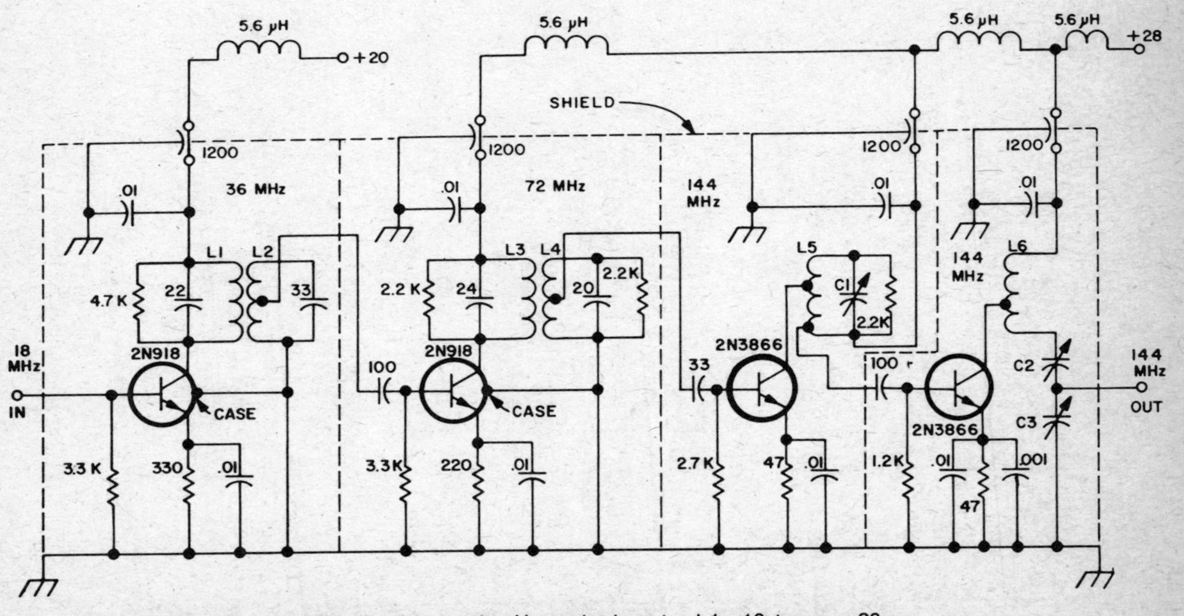

Fig. 5-13. Amplifier-Doubler strip. Unmarked parts: L1—10 turns 26; L2—10 turns 26 center tapped; L3—8 turns 24; L4—8 turns 24 center tapped; L5—5 turns 24, 3/8″ long, tapped at 1.5 turns and 3 1/4 turns from the "cold end"; C1—2.5 to 11 pF; C2—5.5 to 18 pF; C3—3 to 15 pF. (L1 through L5 wound on 10/32 Glastork forms. L1, L2—iron slug, L3, L4—brass slug.)

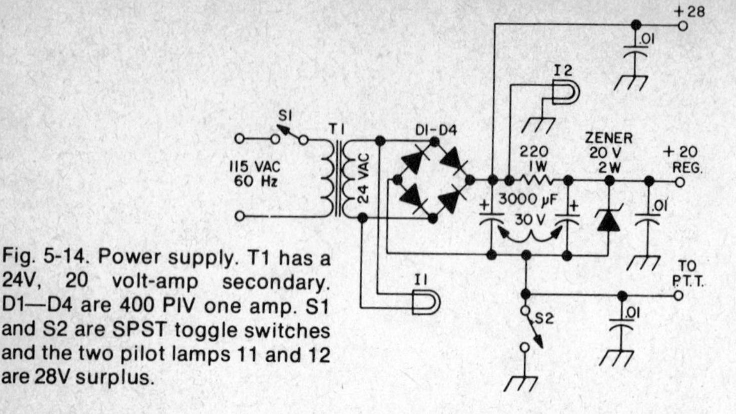

Fig. 5-14. Power supply. T1 has a 24V, 20 volt-amp secondary. D1—D4 are 400 PIV one amp. S1 and S2 are SPST toggle switches and the two pilot lamps 11 and 12 are 28V surplus.

better, but each stage could be easily driven to meet or exceed its power rating. Resistors are used across tuned circuits where necessary to reduce any tendency toward parasitic oscillation. Since the first two stages use double tuned circuits, the chances of unwanted harmonics from the oscillator getting through are greatly reduced. This makes two tuned circuits at each multiplied frequency.

Construction

A popular small size LMB cabinet (LMB No. CO-3) was used for this project. The first step was to secure parts for the power supply and assemble it on the chassis provided with the cabinet. The VCXO was assembled quite easily with the PC board (Fig. 5-15) layout provided for it. All parts on the VCXO were mounted by soldering directly to the board (Fig. 5-16), including the HC 18/U crystal. The board was a double clad epoxy type with enough of the foil on top of the board etched away for the components to pass through. The top foil is then grounded to the outer bottom foil. Since it was difficult to find a small 5K audio taper potentiometer in a small size at low cost, a standard transistor radio control was used. The lugs for the switch on the pot were simply cut off and not used. The frequency trimmer capacitor with about 20 pF could have been used. Tubular components such as resistors and electrolytic capacitors were mounted perpendicular to the board to conserve board space. After completeing the assembly, the shorting wire was removed from the leads of the 3N128. The leads were shorted together to prevent damage to the insulated gate junction. No trouble was encountered with the VCXO, and both units that were built worked the first time. It was possible to hear the output of the VXCO in a nearby FM receiver on the two meter frequency. As a matter of fact, with about a foot of wire on the VXCO, it could be used as a transmitter with a range of about fifty feet.

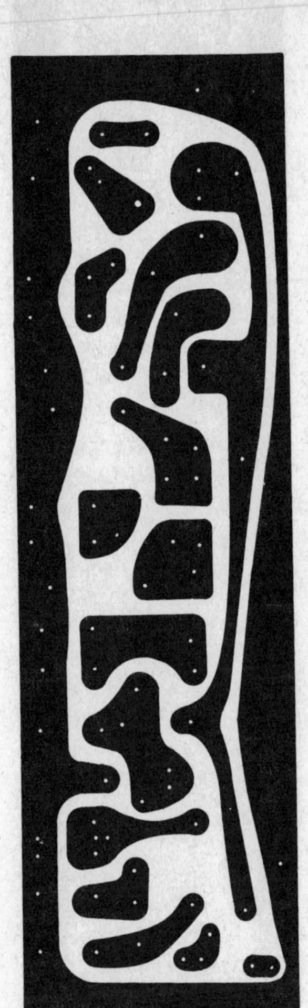

Fig. 5-15. Full size printed circuit VCXO board layout.

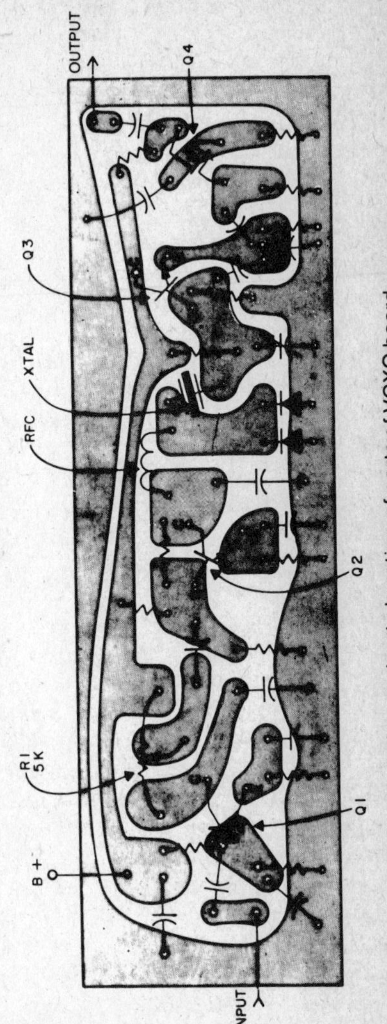

Fig. 5-16. Approximate location of parts of VCXO board.

As previously mentioned, the Amp-Doubler is the most important part of the transmitter. Generally, circuit layout and construction are most important in VHF transistor amplifier design. The chassis should be highly conductive and copper is generally used because of its availability in PC board stock. Since the copper foil is thin, it is light and easy to heat for soldering, and the epoxy board provides excellent strength. Epoxy board provides excellent strength under soldering and unsoldering of experimentation. Short and direct wiring is standard procedure for RF circuitry since any conductor has a resistive and inductive component that may be significant compared to other circuit impedances. All grounds in a stage should be as close together as possible because the chassis may have significant inductance between ground points. Good emitter bypassing is important for good gain in common emitter stages. Unwanted interstage coupling causing instability can be noticed by strange tuning characteristics as well as outright oscillation. Care should be taken to see that each tuned circuit tunes smoothly to a relatively broad peak, and drops off symmetrically on both sides of the peak. Of course, the circuit should be checked to see that it tunes both above and below the desired frequency so that the peak is at actual resonance. Sharp peaks indicate instability caused by regeneration. This is similar to the effect of a Q multiplier, set just before actual oscillation. If sharp peaks or strange tuning characteristics appear, they can usually be eliminated by putting resistors across tuned circuits. Resistors between 470 and 4700 ohms usually will give desired results. Using resistors in this way can be thought of as a cheap and dirty cure for poor components and layout, but it works. In any case, complete stability should be attained before putting the transmitter on the air, and efficiency is a secondary consideration. The common practice of tuning and experimenting with a minimum of shielding, and closing the shielding after the circuit is stable provides a safety factor on stability. Heat dissipation is a major limiting factor in solid state circuitry. Care must be taken to assure that transistors are not destroyed during prolonged periods of tuning since mismatch conditions produce highest heat dissipation. In normal operating, transistors in the Amp-Doubler strip are running near or exceeding their continuous duty dissipation. The transistors will become quite hot if the transmitter is keyed continuously, but normal FM transmissions are seldom more than a few seconds long on a busy repeater. If adequate heat sinks were provided, the transistors would be within their ratings, however.

Specifically, the Amp-Doubler housing was built of double clad printed circuit board. Partitions between stages stiffen the structure, provide shielding, and holes in the shields allow coupling. Components were assembled by soldering their leads together as closely as possible. The coils were wound on Glastork 10/32 coil forms. L1 and L2 were wound on the same form in the first section of the Amp-Doubler. After final tweaking, the windings were about 1/4″ apart on the form. When the first doubler is completed, it can be checked. Drive and power was determined by measuring the voltage drop across the emitter resistor. 1.5V was obtained across the 330Ω resistor, giving 45 mA. This makes about 80 mW, but transistor and other component characteristics will vary, making different transmitter versions different. After it was established that the first stage was being driven, the output was tuned with the aid of a grid dip meter used as a wavemeter. No instability was encountered until the following stage was assembled. At this time, resistors were added to stop the instability. Drive to the following stage was read by measuring its emitter voltage and tuning the previous stage for a maximum. The slugs in L3 and L4 were made by cutting off brass bolts and sawing a notch in them to allow tuning. The other stages are capacitor tuned. Coil forms are mounted vertically in the housing through holes in the chassis, and secured with epoxy. L5 and L6 were mounted perpendicular to each other to reduce spurious coupling. After drive and output were obtained in all stages, final adjustments were made to obtain desired power levels. Coupling taps on coils, and coupling capacitors were adjusted for drive. Emitter resistors were adjusted, making them smaller to increase transistor voltage and power, or larger to provide greater safety. This final tweaking is mostly a matter of personal feel, looking for smooth tuning, good efficiency, and good power. After everything was set, the coils were doped, cold solder joints cleaned up, and components epoxied down securely. A hair dryer was found to be very helpful to make the epoxy flow smoothly and cure fast. Some care was taken to prevent epoxy from connecting RF points to ground because it does have some conductance. Finally, the last portions of the shield were soldered in place as well as possible, considering the lack of space while reaching down inside the shields. A last clean-up can be made with a solvent such as toluene or lacquer thinner to remover solder rosin and other small foreign particles.

Conclusion

The one watt power proved completely adequate for use in a repeater system, and increasing power would be very expensive compared to the cost of a one watt stage.

AUDIO BOOST FOR MOBILE TRANSCEIVERS

Small portable VHF transceivers as well as some mobile units suffer from a lack of sufficient audio power output. This is particularly true of portable equipment when it is used in a mobile installation. This simple audio amplifier/loudspeaker unit was specifically developed to boost the audio output level of a portable transceiver. However, the circuit has far wider application in solid-state receiver and accessory units. The circuit can be used as the complete audio section (preamplifier and power output stages) in a receiver and is also ideal for use as the audio section in a multitude of receiver accessory units such as outboard product detector adapters, audio selectivity units, etc.

Basic Amplifier Unit

The heart of the accessory unit, and indeed almost the entire unit, is a microcircuit unit developed by Bendix Semiconductor, Homdel, N.J. Designated the BHA-0004 and available inexpensively. It is a complete audio preamplifier/power amplifier unit in a plastic case measuring about 1 x 2 x 3 inches. Certainly other integrated circuit AF amplifiers are available, but what makes this unit unique is that it will deliver 5W continuous (not peak) power output using 12-14V DC, requires no heat sink of any sort and requires only a few external components. The high impedance input (20 KΩ) and low impedance output (3-8Ω) make for easy interface with a detector stage output and a loudspeaker.

Figure 5-17A shows the internal circuitry of the BHA-0004. Basically it consists of a preamplifier stage and a class B complementary audio power stage. The idle current and center voltage are preset for correct operation over a wide range of load conditions. A 20 mV maximum input will produce 5W

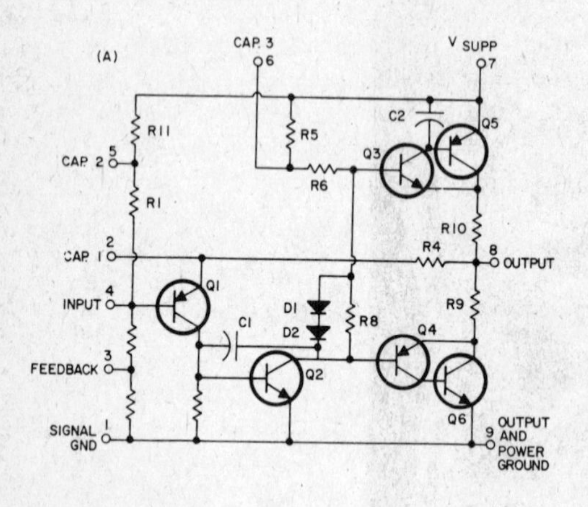

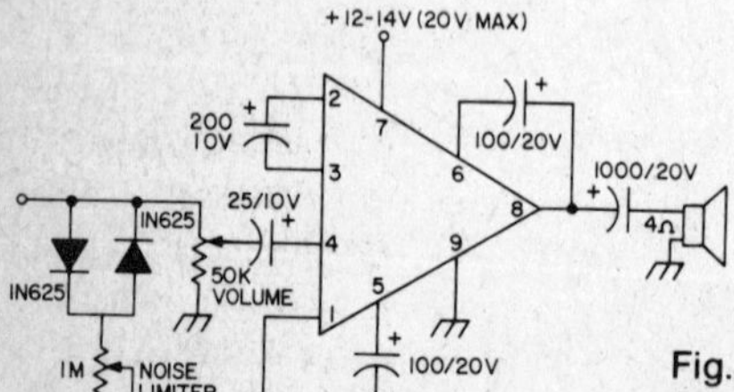

Fig. 5-17. Internal schematic of the BHA-0004 amplifier unit (A) and schematic of the amplifier as used for a audio boost accessory unit (B).

output into 3Ω with a supply voltage of 12–14V DC. The distortion over the 300–3,000 cycle range is less than 1%.

Figure 5-17B shows the BHA-0004 connected up with its external components for use as an audio boost accessory. A very simple noise limiter circuit is shown being used ahead of the amplifier. It may, of course, be eliminated for use of the unit as a straight audio amplifier. It was added because of the specific use for which the accessory unit was intended—operation of a portable tranceiver in a mobile situation—since most portable transceivers do not contain noise limiter circuits. The capacitor values shown are not critical in an application where only speech response is required and similar units of perhaps as little as half the capacitance values shown may be used.

Construction

The BHA-0004 is simply fastened to the back of a loudspeaker face by epoxy cement. This method of mounting appears simple but is actually extremely secure. In fact, it may be undesirable in some cases since the unit cannot be removed without breaking it, other than by the use of special epoxy cement solvent. The capacitors are wired directly to the amplifier unit and between the unit and the potentiometers mounted on the loudspeaker enclosure wall. There are no adjustments that need be made to the completed unit.

CHEAP AND SIMPLE FOR SIX

This six-meter converter has been designed for maximum simplicity and low cost. It uses three VHF bipolar silicon transistors in the common RF amplifier-mixer-oscillator arrangement. Surplus switching transistors were used to provide low cost, but if parts are bought new, transistors designed for use in FM broadcast receivers can be used. If the parts are bought new, it should be possible to build this converter for $12 or $13. Since the crystal makes up almost half of this price, a surplus crystal can cut costs considerably.

The Circuit

50 MHz signals from J1 are coupled to the base of Q1 by L1 in Fig. 5-18. D1 and D2 help to prevent burnout of Q1 when the voltage across L1 exceeds about 0.2 volts. Q1 is a neutralized common emitter amplifier. The 10 pF capacitor from L2 to the base of Q1 and the 560 pF bypass on L2 provide neutralization. These values are determined by experimentation. L2 couples the collector of Q1 to the base of W2, the mixer. Q3 is a Pierce overtone oscillator which is also coupled to the base of Q2 through Cx. The 50 MHz input signal and the 49 MHz oscillator signal are mixed in the mixer to produce the difference frequency of 1 MHz in the output at J2. The RF choke in the collector of Q2 is used to avoid using another tuned circuit here. The gain is ample without any effort to match

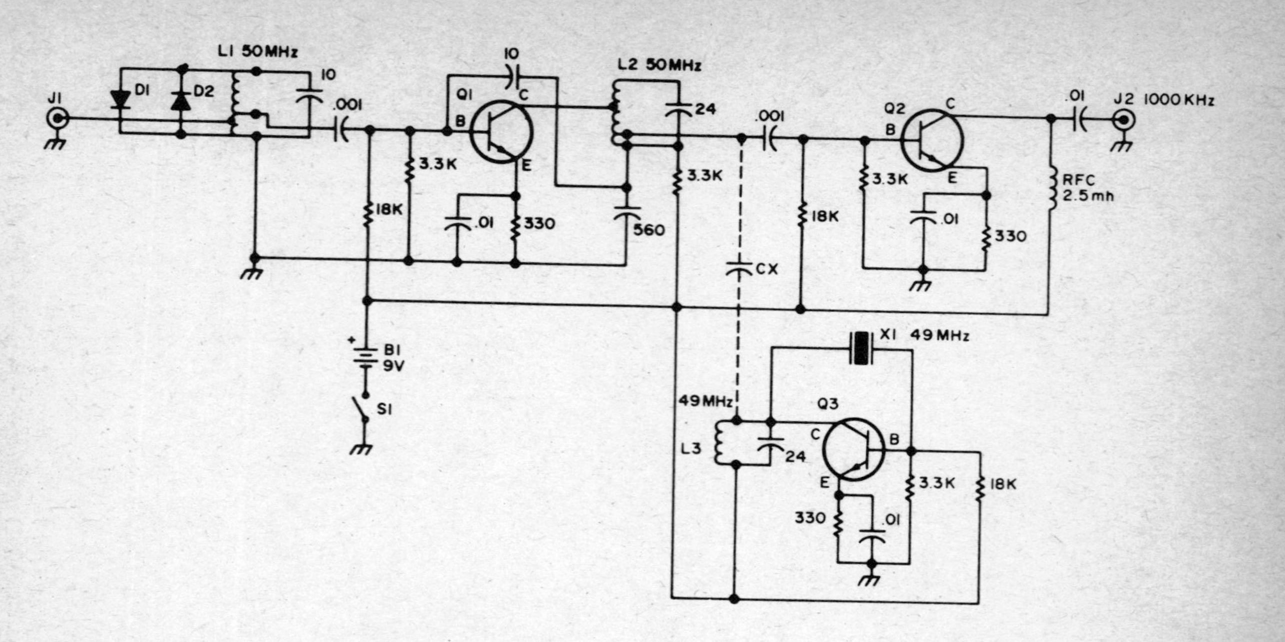

Fig. 5-18. Six meter converter schematic. Refer to the text for unmarked components.

impedances. If a crystal other than 49 MHz is used, only L3 need be changed to provide for the different output frequency.

Construction

Figure 5-19 shows the printed circuit board layout and parts placement. If you have done a lot of PC board work, you will probably want to use a printed circuit board for your converter. Once the layout is determined, making the board and soldering the components is easy. However, the initial cost of a good printed circuit board kit is high, and would probably not be desired for just one project. Perforated board is available if you decide not to use the printed circuit variety.

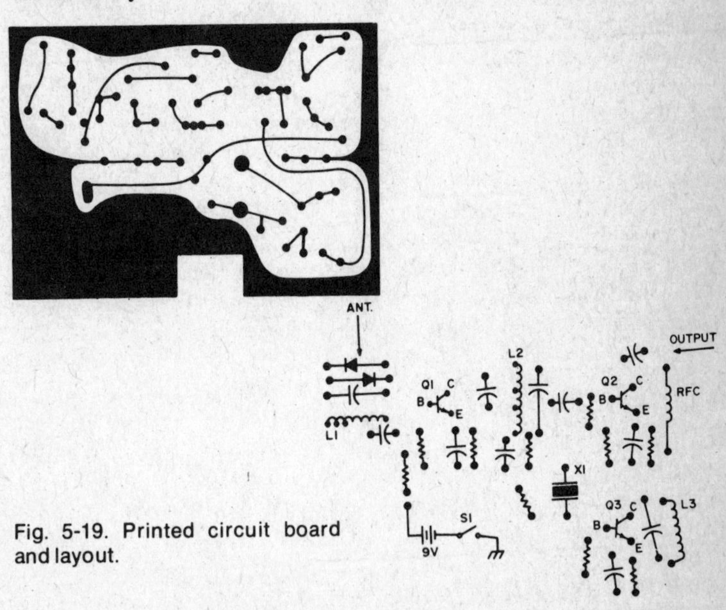

Fig. 5-19. Printed circuit board and layout.

The next thing to worry about is winding the coils. The wire is enamel covered and about No. 21 in size. The coils are wound on a pencil which is removed after winding, L1 is thirteen turns tapped at three and four turns from the bottom. L2 is ten turns tapped at three and four turns from the bottom. The fourth turn goes to the collector of Q1. L3 is nine turns.

Diodes D1 and D2 are 1N34As. Some type of high speed switching diode would probably be better, but the 1N34A works nicely. In the original version a 2N708 was used for Q1, and 2N917s were used for A2 and Q3. They were randomly selected from the junk box, but if transistors are bought new, the 40242, 40243, 40244 series are suggested. The 2N3478, 2N4259, 2N706A, or almost any other VHF silicon NPN transistor could also be used. The stable bias circuit will allow for changes in transistor characteristics.

X1 is a 49 MHz overtone crystal. Other crystals in the 40 to 50 MHz range were tried with equal results. If the 49 MHz frequency is used, output from the converter will be in the AM broadcast band, making it usable as a mobile receiver with a car radio. Cx is the coupling capacitor from the crystal to the mixer. It consists of two pieces of hookup wire twisted together to form a gimmick capacitor, and it is soldered on under the printed circuit board. The wires can be about one inch long.

The board is mounted in a 4 by 3 by 2 inch aluminum box. A shielded box is recommended to help prevent IF signals from going through the converter directly to receiver. The board is mounted with solder lugs soldered to the grounded outer foil of the PC board. The other end of the solder lug is bolted to the side of the box. A total of four solder lugs is used for this.

S1 is a d. p. d. t. slide switch. Any small switch will work here so long as it will fit into the box. If the switch were mounted differently, or a different switch were used, the cut out in the circuit board might not have been necessary. The battery mounting problem is the next segment of the construction. There almost isn't enough room for the battery. The battery is of the standard nine volt transistor radio type, and was mounted with a small dab of glue on the inside of the aluminum box. Battery holders are available, but are expensive and take up space. A connection between the battery and the circuit board is still necessary, however. For about 13 cents you can buy a connector which will fit the battery. Be sure you check the polarity of the leads coming from the plug; the color of the leads is sometimes confusing.

The last things to be mounted after the board is in the box are the input and output connectors. A phono jack for the output and a BNC connector for the input were used. These items happened to be available and they fit well in the small space. After the connectors are mounted it may be difficult to get the board in and out.

Alignment

A grid dip meter is required for tuning the converter. First check for output from the oscillator coil, L3 at the crystal frequency. If output is not detected, the coils can be tuned by a process called "knifing." The knife is made of a non-metallic tuning wand or other kind of insulated shaft. A small brass slug is attached to one end and an iron ferrite slug is attached to the other end. The brass slug can be obtained from a brass bolt, and the iron slug can be removed from a slug tuned coil form. When the brass end is inserted in a coil the inductance of the coil will decrease and the resonant frequency will be higher. The iron slug will lower the resonant frequency of the tuned circuit. By using this method and watching the output, you can determine whether the coil should be compressed to lower the frequency, or expanded to raise the frequency. Be sure to use a weak signal for final tuning. The tuned circuits should be close enough to the proper frequency for the converter to work before any tuning is done. The tuning should be rather broad and noncritical.

Results

When used with a good receiver the sensitivity of the converter is very good, and the leakage from a signal generator with the output at zero almost pins the S meter of the receiver. The main disadvantage of using bipolar transistors is cross modulation. A nearby FM broadcast station can be heard at spots on the dial. There is also some feedthrough into the converter IF frequency. In this case, however, six meter signals, ignition noise, and power line noise have usually been stronger than the spurious signals. Changing the IF output frequency by changing the crystal frequency might relieve the problem of BC feedthrough, but it is mainly a problem of shielding the leads from the receiver to the converter. Using an FET for Q2 would improve the cross modulation characteristics of the converter, but would increase the cost.

2N5188 TWO-METER EXCITER

The RCA 2N5188 power transistor (Fig. 5-20) is a type rated at 4 watts maximum dissipation and is available inexpensively. It was designed for core-driver and line driver as well as class C RF service. Its F_t of 325 MHz means it will operate well up thru 50 MHz in class C operation, and fairly well up thru the 144 MHz or two meter band. Some 2N5188 transistors are better at 144 MHz than others, so the "hotter" ones should be used in the last two stages of the exciter illustrated here. These transistors were used throughout the exciter.

The 2N5188 has a maximum rating of 25 volts from collector to emitter, and 5 volts from base to emitter. The maximum power rating of 4 watts is for a case temperature of 25°C. Allowing for some temperature rise (at a derating of 23 mW per ° C) and the use of small heat radiators still means that about 1 to 2 watts input can be run into these transistors. In this exciter, the output amplifier runs about 1 to 1 1/4 watt input with from .4 to .5 watt output on CW, with a 12 volt battery power supply. An 18 volt supply and large heat radiator should result in more than one watt output. CW keying probably could be done in the base bias resistor return connection to ground of any stage of two stages.

The RF exciter was built on a piece of copper plated Bakelite or epoxy board 4 x 8 inches in size. This fits into a 4 x 8 x 2 inch chassis as a bottom cover and shield can. The copper plating makes a good ground surface for mounting small variable condensers and can be soldered to with a 25 or 50 watt soldering iron. Bypass condenser and emitter leads should be very short to help maintain stability.

Two types of heat radiators were used on the transistors. Both types are snug fitting on the TO-39 transistor case. The "fin" type is less expensive unless one happens to find the smaller "ribbed" type in the surplus market.

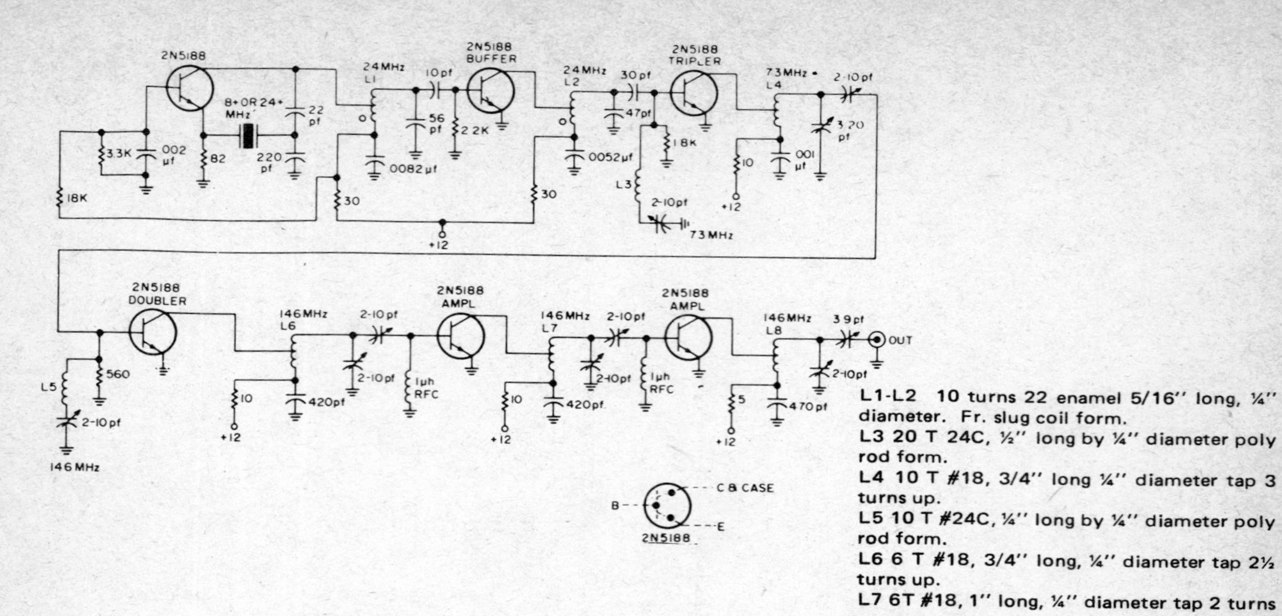

L1-L2 10 turns 22 enamel 5/16″ long, ¼″ diameter. Fr. slug coil form.
L3 20 T 24C, ½″ long by ¼″ diameter poly rod form.
L4 10 T #18, 3/4″ long ¼″ diameter tap 3 turns up.
L5 10 T #24C, ¼″ long by ¼″ diameter poly rod form.
L6 6 T #18, 3/4″ long, ¼″ diameter tap 2½ turns up.
L7 6T #18, 1″ long, ¼″ diameter tap 2 turns up.
L8 5T #14, 5/8″ long, 5/16″ diameter tap 1 turn up.

Fig. 5-20. Schematic for the 2N5188 exciter.

The first two transistors do not require a cooler since their collector current runs between 15 and 30 mA, at 12 volts, but the following tripler, doubler and amplifiers do warm up and need one type or the other. The collectors are connected to the TO-39 case, so a large cooler or radiator does increase the 8 pF output capacitance somewhat. This effect is taken care of in the circuit design.

The 2N5188 transistors were found to be quite stable at two meters in grounded emitter circuits, without any form of neutralization. Tapping the collector leads into the lower end of each coil can be used for impedance mismatching to eliminate the need for neutralization, and also to provide a fairly high Q tuned circuit. The output load resistance of the 2N5188 varies over a range of from about 600 ohms down to about 100 ohms in this exciter, depending upon the DC collector current. The base to emitter impedance is lower, ranging from about 50 ohms down to 25 ohms. These low values can be matched to the tuned circuit values of 1000 to 2000 ohms by either using adjustable link coupling or by small coupling capacitors. The latter method was used in this exciter. The total C and L values in each circuit were chosen to have a Z of from 15 to 25 in order to reduce unwanted harmonics with only a single tuned circuit between each stage. Many circuits shown in transistor handbooks and in some magazine articles are so heavily loaded by the transistors that the operating Q may be closer to 3 than to the 15 or 20 needed for harmonic suppression. A tripler should be used as a tripler not as a combined doubler, tripler and quadrupler, since only the tripled frequency is useful. The same reasoning holds true for doubler and amplifier stages.

The crystal oscillator is a type that will function with either 8 or 24 MHz crystals for output at 24 MHz. The slug tuning adjustment is more critical with 8 MHz crystals than with 24 MHz overtone crystals. The buffer stage was lightly coupled to the oscillator to insure good oscillation starting ability with 8 MHz crystals. Both types of crystal were tested in the exciter with good results in either case.

A buffer stage at 24 MHz was needed in order to drive the tripler stage to 72 MHz, or higher. The tripler stage requires a higher value of base bias resistance than for a doubler. The rule used here is to try different values in each stage so as to produce enough RF drive to the next stage, but not run into the problem of too high a peak voltage across each base to emitter. Remember, the absolute maximum base voltage on these transistors is 5 volts, so often low values of resistors, or none, (only an RF choke) is required in each stage.

The doubler and tripler stages function by waveform distortion across the base to emitter bias resistor. This means that there are double and triple frequencies in the input side of these transistors and a low impedance path is needed, just as in a parametric doubler using varactors. Small high Q series tuned LC circuits were connected from base to ground in each frequency multiplier stage. The output power increases greatly as these circuits are series tuned to resonance with the transistor output circuit also tuned to resonance. This effect is particularly noticeable with low power supply voltage and low driving RF power in each stage. The 2N5188 is a double diffused epitaxial planar transistor of the silicon NPN type, not too far different from a varactor diode. Probably some parametric frequency doubling action takes place since a straight doubler or tripler generally has very low RF output at low supply voltages.

Small German-made ceramic (low cost) trimmer condensers were used for these series tuned circuits as well as for coupling and parallel tuned two meter collector circuits. These condensers have high enough Q and mount easily, or they can be suspended by heavy leads but do not tune smoothly when there is much RF current flowing through them. Small Erie or CRL adjustable flat circular types might be used. Other piston type trimmer condensers are available, but are more expensive. The German mass-produced units may not be available in radio stores, so substitution may be required in constructing this type of exciter.

Standard sized red coded ferrite slug coil forms were used in the oscillator and buffer stages. Number 18, 16, or 14 wire was wound over a 1/4″ and a 5/16″ diameter steel drill as a temporary winding form for the higher frequency circuits.

The coil data is listed in a separate part with the proper tap points for 12 volt operation. A 15 or 18 volt supply may require a little change in coil tap position and in the values of bias resistors throughout.

The output circuit is matched to a 50 ohm load with a small 3.9 pF fixed condenser but another 2 to 10 pF trimmer (or 2.5 to 7 pF) would be advisable in this position in order to couple the output into an antenna coax line or into 75 or 93 ohm coax lines to a larger amplifier stage. A tube, or a larger RF power transistor, might be driven with the 1/2 watt output available from this exciter.

The parts layout starts with the crystal socket and progresses along one side of the 8 × 4″ board and back again along the other 8″ side, leaving 1/2″ clearance on all sides. This permits the unit to be fastened to the lips of an 8 × 4 × 2 inch chassis.

Tune up is not difficult if a 0 to 300 or 500 mA meter is connected into either the − or + leads to the battery. All stages will draw very little current, except for a few mA in the oscillator state since it has some + bias on the base to make the stage start oscillating easily. As soon as the crystal oscillator coil is tuned correctly the DC meter reading will increase from about 12 or 15 mA to 30 or 40 mA even if all other stages are detuned from resonance. Then as each stage in turn is tuned to resonance, the DC meter reading will increase. The total DC reading will be over 200 mA when all circuits are correctly aligned for maximum output reading into a one or two watt RF wattmeter. A 50 ohm 1 watt resistor termination across the output jack and a diode RF voltmeter (5 volt range) across the resistor may be also used for reading RF output. Even a nearby two meter receiver S meter may be used to tune up the exciter into some form of dummy load. Nearly all circuits have to be adjusted carefully and the variable stage coupling condenser readjusted for getting maximum output. Too much capacity in these coupling condensers will lower the Q of the tuned circuits too much for good spurious signal suppression. Too little will mean less output. An RF meter in the output of the exciter will measure spurious as well as desired frequencies. The two meter receiver S meter is a selective device reading only the desired frequency. If the various adjustments do not result in agreement on the two RF metering systems, one can suspect spurious oscillation or excessive harmonic power in the output system. The adjustments for maximum radio receiver and RF wattmeter or diode voltmeter readings should be the same. If not, start reducing coupling capacity values and retune each circuit for resonance and maximum output.

Some 2N5188 transistors have better gain at two meters so these can be used in the two meter amplifier stages. The lower frequency stages are much less critical so the weaker transistors can be plugged into these parts of the circuit.

SOLID STATE EXCITER FOR 450 MHz

The UHF bands of 432 and 1296 MHz appear due for a large change and advance from now on as concerns portability, size, and usage. The availability of plastic transistors that work up in the 1296 MHz region, tenth watt resistors, small coils, crystals, and terminals, make construction of complete crystal controlled exciters possible in a 5 x 10 cm Minibox with room left over for two 9V batteries!

This exciter uses doubling in all the multiplier stages. A comparison is also made with a similar unit which used tripling instead of doubling. Be sure to check on the final results as outlined later here. It may save you lots of time on such units.

Tools and Accessories

For the construction of a crystal exciter of this small size, there are certain things you should collect before you start. Without them you will lose time improvising and will not have such a compact unit.

A low wattage soldering pencil with a stock of small copper tips and several files to shape the tips are needed. You'll see why when you tackle the small terminals.

Other small tools: two pairs of tweezers, one flat, one pointed; the smallest sidecutters and needlenosed pliers you can get; a steel scriber; a set of No. 60 to 80 drills, with maybe some extra ones around No. 75 and 76; "coffee stirrers," which are just flat pieces of wood 7 mm wide by 10 cm long by 2 mm thick, with a pointed lump of wax (high-Q high temperature coil wax, that is) on the end for holding windings on the coils (use it like cement); several solid insulating rods of Lucite and/or Bakelite, 3 and 6 mm for filing into insulated screwdriver blades and insulated picks; small 10¢ screwdrivers that you file down for scraping around pins, on copper, etc.; emery cloth and crocus cloth for polishing brass plate capacitor sufaces; Exacto knife for cutting holes in fiberglass sheet; dentists' tin shears for cutting small brass or copper pieces; a jeweler's saw and plenty of blades, umpteen teeth to the inch; and all the rest of the usual tools you may have around.

Another handy tool is a coffee-stick with a 3 mm square 1000 pF bypass capacitor cemented on the end with 3 mm leads. This is the last word for testing working bypass capacitors. Take any bypassed terminal or brass plate used for bypassing and connect this test capacitor across it, just by pressure. You don't have to solder it. Do it with all power on and watch the RF power output meter. If it makes any difference when you add it to the existing bypass, that bypass is not right. Remember, it isn't only the *amount* of pF at 432 MHz, it is also the length of the leads, if any, and the shape and position of the components being bypassed.

Components, Including Terminals

Resistors are easy, but you've got to put in a stock of 1/8th or 1/10th watters if you want to do the best job on small units like this exciter. Not all of the HEP56 transistors are exactly alike, so you may have to trim up the emitter resistors in the final tests. Have some of each of the following, in ohms: 22, 33, 47, 100, 220, 330, 470, and in thousand ohms, 1, 2.2, 3.3, and 4.7.

Most of the capacitors are easy, but you should have a good stock of the small high K ceramic bypass capacitors like .001 or .005. These should be the real small ones, say 3 cm square. You'll also use a lot of the small dipped mica silvered type DM capacitors for coupling, in values of 1, 3, 5, 10, 15, and 30 pF. They seem to do the job very well for VHF/UHF. Some multiplier stages like to have a variable coupling capacitor, and then you should use a small circular, or mica, trimmer. So far, in the UHF region, the mica compression trimmers seem the best for tuning; they are thin, long and narrow, and do a good job. It is easy to tell which is maximum and which is minimum, unlike some rotary ceramics. Values in pF of 1 to 12, and 2 to 25, are good.

Winding wire should be on hand, and phenolic forms of 3, 5, and 7 cm OD. Wire sizes can be 22, 24, 28, 30, and 34.

Various kinds of insulation materials are useful, such as small strips of linen-base Bakelite 2 or 3 mm thick for terminal strip pin mounting, putting between the coils and the copper-clad baseboard, etc.

Terminals. Instead of making up small individual little planks with three pins each, make the terminal strips as in Figs. 5-21A and 5-21B. It worked out excellently, and from now on that's it for me. Notice in Fig. 5-25 how everything goes together on those pins, three for each stage.

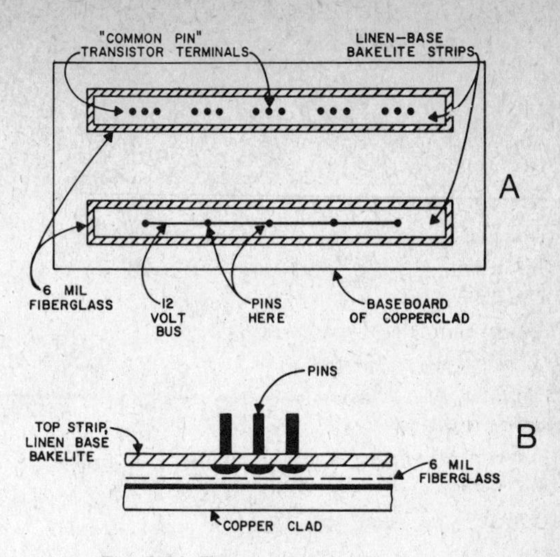

Fig. 5-21. Terminal pin layout.

Note the 6 mil fiberglass under the strip to avoid shorting the pin terminals to ground. The terminals are just plain pins of nickel plated brass.

Figure 5-22 shows a handy bench type power supply using two lantern type batteries with 6W DC capacity, rated at .5A. This is a maximum, by the way, but will give you plenty of sock from a hill or mountain top with a good beam of 432 MHz.

Check the grounded terminals in your car so you can plug this unit in for mobile work if you wish. It is possible to arrange the baseboard to be isolated from the chassis (minibox) if you have to. The only difficult part is the bypassing of the output connector.

The meter is not too complicated. As the oscillator stage and each doubler stage are built, connect in a meter as in Fig. 5-22. When this stage is tuned up and the current adjusted to what you want, solder in an emitter resistor of the proper value, and remove the meter and pot.

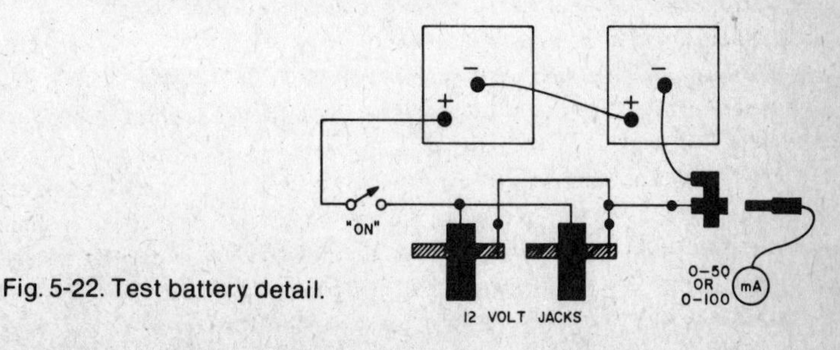

Fig. 5-22. Test battery detail.

Fig. 5-23. Emitter resistor check.

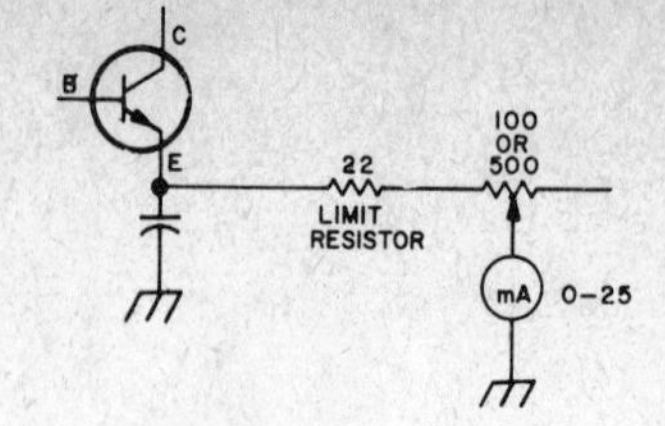

The Circuit

The final schematic, Fig. 5-24 and the parts layout, Fig. 5-25, are given now for clarity of reference in the following details. This circuit, using doubling multiplier stages only, was made up as a direct comparison to a three-transistor tripling exciter. The tripling exciter is a nice unit and it turns out to be excellent for a signal source, as a local oscillator, for calibration purposes, etc., but the fact remains that this doubling unit puts out more than four times the RF power at 432 MHz. This is somewhat to be expected as there are four transistors working on the job against only three in the tripling unit, but it was worthwhile to check them both in actual operation. It might be possible to put out more power with more expensive transistors, but this section describes a means of getting crystal control at 432 MHz which is *not* expensive. It fits in a little 5 × 10 cm Minibox.

The total current drain runs at only 25 to 30 mils at 12V, or less than .4W DC power, leaving another 5W battery capacity in those lantern jobs for amplifiers.

Construction

The stage-by-stage method is used here in order to get the most out of each transistor and at the same time check the possible variable parameters involved and to avoid marginal operation. The oscillator should start every time; the load on it should not be too large; the multipliers should tune nicely above and below the desired doubling frequency, etc. The emitter circuit current meter and the RF output connector can be shifted as each stage is finished, and checked for frequency and power output.

The oscillator is assembled and wired as in Fig. 5-26 and connected for tests. There is a small Bakelite plank between L1 and the baseboard, and another one under the crystal socket, to keep the terminals from shorting out to ground. Take care in soldering those little pins.

The base of each pin is surrounded by insulation which keeps the wires and components from shorting to ground. The thin copper baseboard solders with a touch, and there is plenty of room for the few resistors and capacitors needed, and they all fit nicely in the layout as shown in Fig. 5-25. There is even room for another stage, which might be an RF power amplifier. There is only one wire

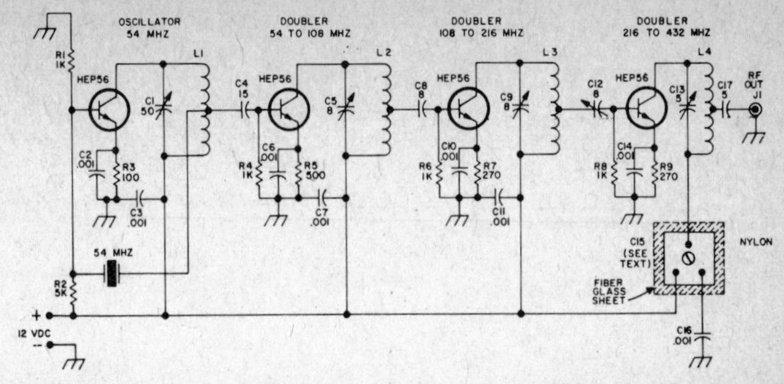

Fig. 5-24. Schematic. Transistors—All Motorola, HEP56, L1, 20 turns, centertapped, No. 26 DSC, on 5 mm OD phenolic, air core; L2, 10 turns on 5 mm form, No. 26; L3, 5 turns on 5 mm form, No. 28, about 1 cm long; L4, 3 turns, tinned bus wire, No. 18, about 1 cm long, tapped 1 turn from grounded (RF) end, 5 mm diameter.

on each of the three coupling capacitors that does not have a pin support, which is the centertap on the multiplier coils, and this holds up one end of a silvered mica which is too light to shake loose. Purists for mechanical rigidity can put three pins on the planks under the coils if they wish.

Use an external pot of 100 or 500Ω for the emitter resistor while tuning up, then when the desired value is found—which depends somewhat on the drive required for the next stage—R3 is soldered in place. This test setup is shown in Fig. 5-23.

L1 and C1 should resonate well above and well below 54 MHz. Do *not* rely solely on a grid dipper for this. Check it out as shown in Fig. 5-26.

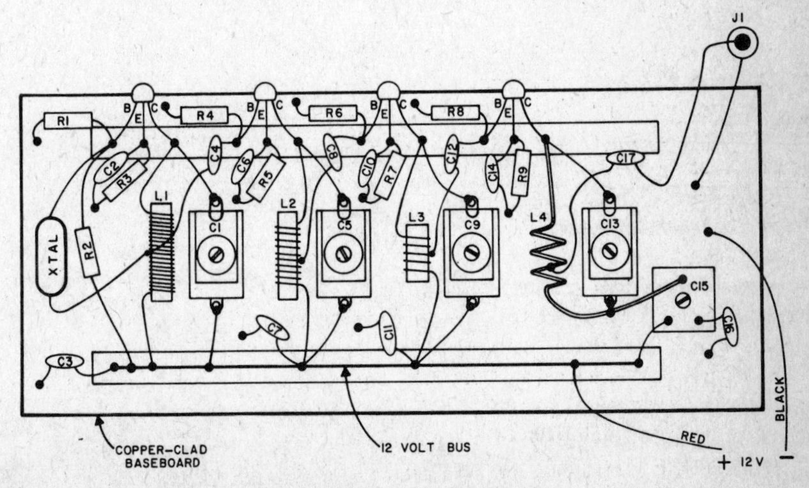

Fig. 5-25. Pictorial diagram and parts layout. Board size is 4×9 cm.

You should also listen to the carrier on a communications type receiver, not in the CW position, but with plenty of AF gain so you can hear spurious signals, and other assorted squeaks and groans. These can easily be eliminated by proper tuning and bias, if you know they're present.

As C1 is reduced toward resonance, the RF should increase to a maximum and then drop off with a snap. Back off slightly from resonance and find the position where it comes on every time and still has near maximum power output.

Trouble

One problem was caused by connecting the output cable to the centertap of L1. The oscillator works perfectly with the alternate output connection shown in Fig. 5-26, and it also works very well when the centertap goes to the next base through C4, as in the final schematic shown in Fig. 5-24.

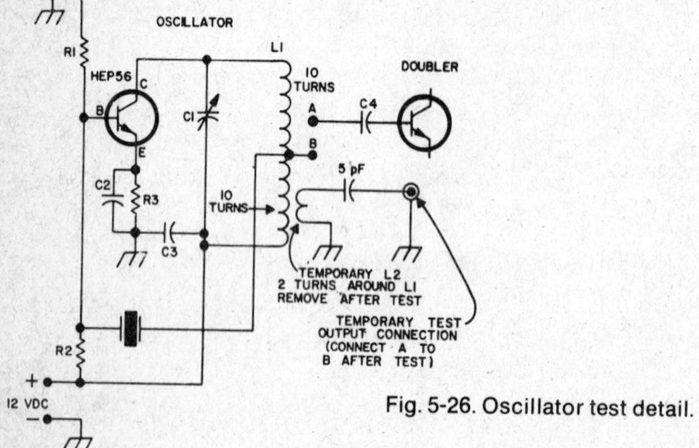

Fig. 5-26. Oscillator test detail.

This oscillator shows good, stable RF power out of about 50 mW at 18 mA of collector current. Listening to it on a receiver, .5 to 54 MHz, it stayed in the passband (of the receiver) nicely, no matter how the oscillator was tuned.

Note that the base of Q1 is *in phase* with the collector except on the crystal frequency. The crystal reverses the phase and it then oscillates, but only on the crystal frequency. Anywhere else it is degenerative—that is, it has negative feedback.

Doubler, 54 to 108 MHz

The emitter resistor check of Fig. 5-23 was also used here. The final value for best doubling to 108 MHz was 500Ω. Don't forget that different transistors, even of the same number and manufacturer, may require slightly different bias voltages, especially in harmonic multiplier service.

The tuning capacitor across L2, which is C5, was checked for a return to ground or a return to the low end of L2 and no difference was seen.

A choke coil was also substituted for the base resistor R4 and again no difference was noted, so the resistor was left in. An RFC here has caused trouble in the past with spurious oscillation under certain conditions.

Second Doubler, 108 to 216 MHz

This one went together like a charm with the only difference being C8, the coupling capacitor from the previous stage. This showed a preference value of pF for maximum power out when a variable capacitor was used. A fixed one of 8 pF was installed as the best value. Not really critical, but it is good to have the best value.

The best emitter resistor value was found to be 270Ω for this stage. Again, a two turn coil was wrapped around L3 for an output check on 216 MHz, and was later removed. Again, 5V DC was found at the output of the diode monitor on 216 MHz.

Doubler, 216 to 432 MHz

The best emitter resistor value checked out at 270Ω the same as the previous stage, with a collector current about 5 mA.

C15 is a brass plate bypass put in for security at 432 MHz. C16 was also added for a small improvement.

Various output taps on L4 were tried, with one turn from ground showing up as best. And there you are, on 432 MHz.

This complete crystal controlled exciter, fitting into a 5 × 10 cm Minibox, using doublers, gives RF power output which is at least four times greater than the same type of unit using tripler stages.

A 2 METER CONVERTER FOR AN AM-FM BROADCAST RECEIVER

Most two meter converters have had their output on either 10 or 20 meters. For the beginner, this alone could be a drawback in that he may not have a general coverage receiver. This project (schematic shown in Fig. 5-27) uses an ordinary FM broadcast band tuner or receiver. Using the tuner has several marked advantages over the conventional AM/SSB receiver. First, since it's FM to begin with, you'll get the added quieting and noise suppression that's lost in an AM/SSB receiver. Second, depending on the complexity of the tuner, it may have muting and AFC provisions. The muting can be used as a squelch and the AFC keeps those off-frequency stations on your dial without retuning. Third, maybe you're a newcomer to the two meter band and don't know the more popular frequencies that are in use in your area. With this converter you can cover the entire two meter band, MARS frequencies and some of the commerical band from 144 to 164 MHz.

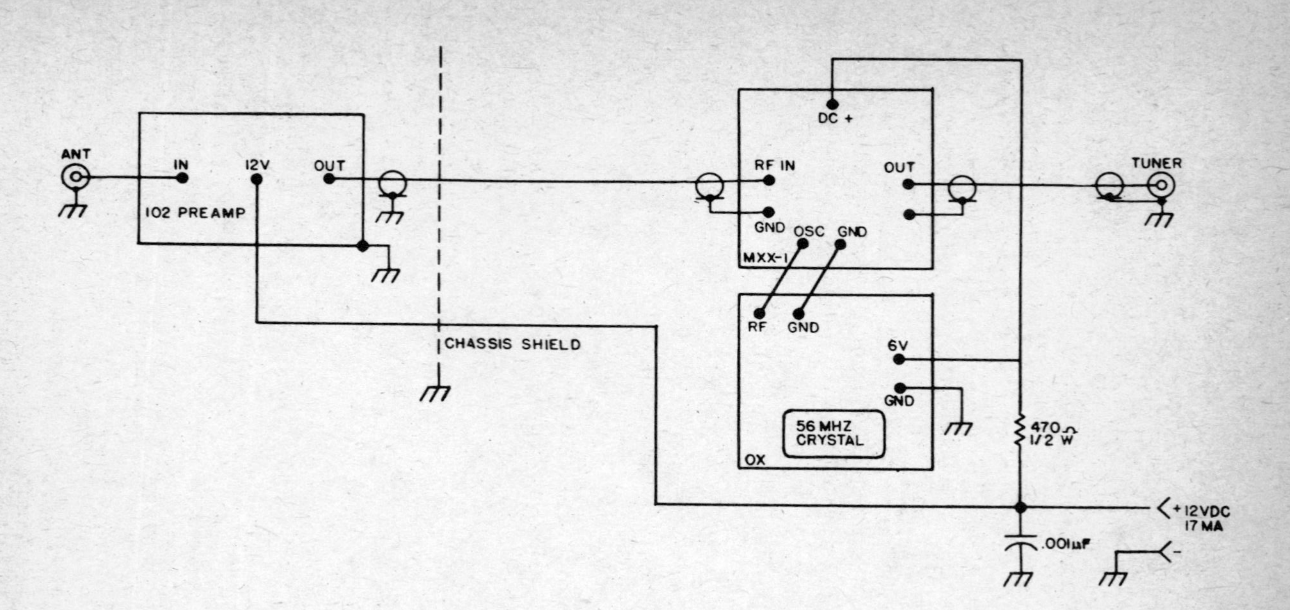

Fig. 5-27. Block diagram of the converter showing the method of connecting the modules.

Modules are used to their fullest extent to simplify construction and to eliminate those hard-to-find parts. The unit is designed into three subassemblies which are placed on a small chassis. An RF amp, mixer and oscillator make up the three modules. The RF amplifier used is manufactured by Vanguard Labs (196-23 Jamaica Ave., Hollis, NY 11423). Their 102 preamp has less than a 2 dB noise figure with a power gain of 24 dB at 150 MHz. The preamp comes from the factory tuned to the frequency of your choice with a bandwidth of 2–4 MHz. This particular model uses a neutralized J-FET. From the preamp, the signal is passed to a mixer stage where a 56 MHz signal is injected from the local oscillator. After converting the signal down to the 88–108 MHz band, the signal appears at the tuner output jack. For example, if the input signal is at 144 MHz, then the converter's output will be 88 MHz. The lower part of the FM dial was chosen because of the general lack of stations there. The stations that are present are mostly lower power and educational.

The mixer and oscillator kits are from International Crystal (10 North Lee, Oklahoma City, OK 73102). The MXX-1 and OX are the mixer and oscillator kits, respectively. Both are the HI kits with a 56.0 MHz EX crystal used in the oscillator.

Construction is simple, straightforward and noncritical. The case used is a small cowl Minibox which measured 5 × 10 × 7.5 cm. A shield is mounted across the preamp chassis to suppress any unwanted FM broadcast signals. The module layout is not particularly critical, although the relative positions should be followed for best results. A SO-239 is used for the antenna input, and a phono connector is used for the converter's output.

The kits are easy to assemble and go together without any problems. Entire construction time for the OX and MXX-1 takes about thirty-five minutes. On the OX board use the coil with the red dot. The MXX-1 coil and capacitor are the coil with the green dot and the 4.7 pF capacitor. After completing the boards, double check your wiring, especially transistor placement. Make sure you have soldered the terminals that are factory staked into the boards. The oscillator and mixer operate on 6V DC. A quick check of oscillator output can be performed by using a field strength meter or by measuring current flow in the collector.

Next mount the boards and preamp using the screws and stand-offs supplied. Connect the antenna to the preamp input. If the distance is over 2 cm use shielded cable. Now connect the oscillator RF output to the mixer's OSC input with a short piece of wire. The mixer output uses RG-58 which goes to the TUNER jack. Preamp "OUT" to "RF" mixer input uses miniature coax cable. The only thing left is for power, which is primarily 12V DC for the preamp and 6V for the mixer and oscillator. The preamp is connected directly to the 12V source while the mixer and oscillator receive their power through a 470Ω resistor.

After all connections are made connect an antenna. Use coax for the converter's output to the tuner (Fig. 5-28).

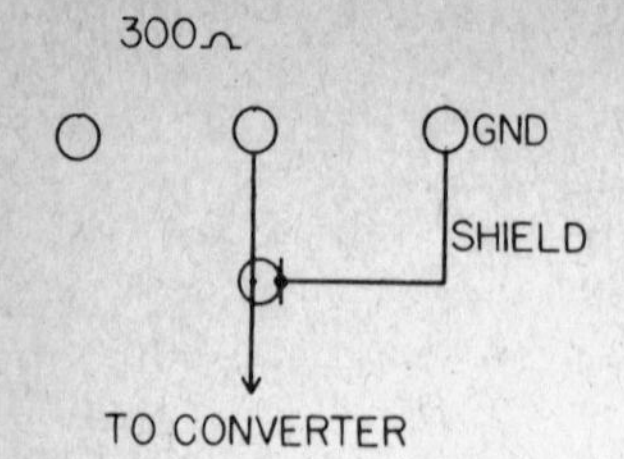

Fig. 5-28. To use coax with a 300Ω input, connect the shield to chassis ground and the center conductor to one of the 300Ω inputs.

With a signal present on the tuner, adjust the mixer coil for maximum signal. If the tuner is not equipped with a meter, adjust for maximum quieting. Since the output is untuned, some TV stations may be heard along with the normal two meter activity. This can be eliminated by using an 88–108 MHz bandpass filter. These are available through most parts jobbers.

Oscillator stability is quite good. A 1/4λ whip brings in most of the stations that many commercial units receive. Coverage of MARS and the commercial band are an added benefit. Mobile telephone, fire and police departments, and the government weather broadcasts, are all received with full quieting signal levels.

The only problem you may encounter is low sensitivity. This happens on some of the older tuners.If your tuner has an input sensitivity of around 2μV,no added amplification should be necessary. If not, adding another International crystal module, the SAX-1 RF amplifier, between the converter and tuner, should bring the sensitivity up to a respectable level.

VHF DUMMY LOAD WATTMETER

The Unit

This unit (Fig. 5-29) is similar to some 60 watt models which may be found around many commerical two-way radio shops. This dummy load has provision for connection to an external relative ouput meter. This external output meter may become an accurate wattmeter if the following criteria are met:

1. Frequency bandwidth of ±10% of calibration frequency.
2. RF output kept within power dissipation of dummy load.
3. Accurate initial calibration.

These criteria may be easily met in amateur VHF operation if only one band is considered for each set of calibration data. Since most VHF amateurs operate on 50 MHz, 144 MHz or 432 MHz, the ±10% frequency limitations may be easily met. This limitation gives a 10 MHz bandwidth at 50 MHz, 29 bandwidth at 144 MHz, and 86 MHz at 432 MHz. The limitation to the power ratings of the dummy load is only common sense, for if a resistive network is overloaded, the impedance may be drastically increased, caused by damage to the load resistors. The calibration limitation may be overcome if a standard, previously calibrated unit, or commercial unit is used.

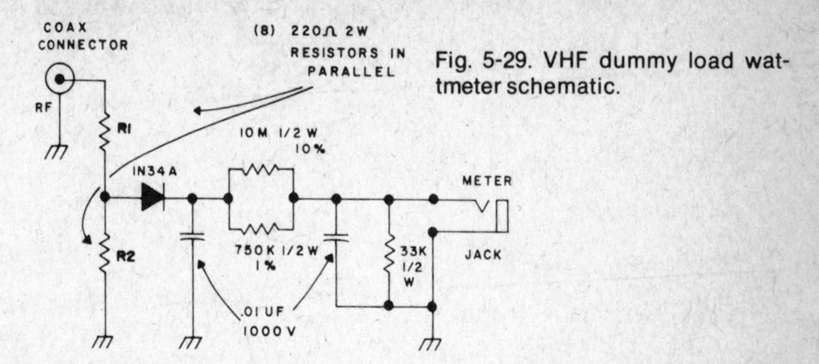

Fig. 5-29. VHF dummy load wattmeter schematic.

The unit may be built for less than six dollars (less meter movement) if all new parts are purchased. It consists basically of sixteen 220 ohm resistors in a series parallel arrangement. The metering circuit consists of a germanium diode pickup with necessary RF filtering. The meter movement is generally a VOM, but any 50 μA meter movement should suffice. Exact physical layout is not extremely critical, but it is suggested that the layout be made similar to the unit shown in the accompanying photographs. This unit is acceptable for 60 watt output transmitters without modification. The power capability may be increased to about 200 watts if the resistive network is suspended in 1 quart of oil. If this is done, care must be taken to keep the metering circuit out of the oil. The lead from the diode to the resistive network must, of course, be partly submerged, but keep the diode itself out of the oil.

Calibration

Calibration is best accomplished by using a Bird "Thruline" or similar commercial VHF inline wattmeter. Second choice is a Bird "Termaline" or similar dummy load-wattmeter. In both cases, a graph should be created by plotting meter divisions on the horizontal axis, and power on the vertical axis. The meter shunt should be placed at minimum resistance and increased to give maximum reading at the desired power level or, if a VOM or VTVM is being used the range switch should be placed on a high voltage setting and reduced a setting at a time until the desired reading is obtained. The transmitter should be adjusted for various power levels on the standard wattmeter and the voltage or current reading on the new meter recorded on the graph. In the case where the standard meter is of the dummy load-wattmeter type, it will be necessary to switch the coax from one unit to the other. Do not retune the transmitter, for each unit will present almost the same load to the transmitter (50 ohms). Take the reading and record as with an inline type of meter. The points on the graph should now be connected with a smooth curve.

If multi-band use is expected, the graphs must be made for each band. Use of the wattmeter now requires only the connection to the transmitter, setting of range switch to the proper level, and reading the graph.

The uses of this dummy load-wattmeter are as varied as the amateur mind can devise. One very important use is determining the losses of 50 ohm coax. Measure the output of the transmitter at the transmitter. Then measure the output at the end of the length of coax. The losses in the line become apparent. The loss of dB may be calculated by the standard power ratio formula, $10 \log_{10}$ Power out of coax/Power into coax.

Another use is the determination of efficiency of final amplifier stages. This efficiency may be calculated by Power out (measured by dummy load-wattmeter)/Power in (measured by plate current/plate voltage meter) × 100%. A third use is determining once and for all which amateur really has the most output. This list may be expanded by the builder to suit his own tastes.

Conclusion

This dummy load-wattmeter is not a Bird "Termaline" nor should it be regarded as a substitute for any other laboratory equipment. However, with a little care in calibration, (assuming a 5% accuracy standard is used for initial calibration) the accuracy should be within 10%, and this, is not bad for a wattmeter costing less than $10.

ONE-TUBE $10 2-METER TRANSCEIVER

Even in this age of semiconductor sophistication, LSI and varactor tuning, the true dyed-in-the-wool VHFer sometimes enjoys a little weekend fun working solely from the junkbox. True, the results are not always earth-shattering technological breakthroughs, but they often produce the most-used pieces of equipment in the area.

Simplicity is the byword (See Fig. 5-30 and 5-31). A carbon microphone was employed simply because it was handy, not necessitating a trip to the local parts supplier. The earphones are the vintage 1–2 kΩ magnetic type, used for the same reason. A multitude of possibilities exist, though, for the reader not desiring to make a straight carbon copy. An additional audio stage could easily produce room-filling volume. Conventional crystal oscillators could be worked in to produce rock stability. The tube could be replaced entirely by a VHF transistor. But then it wouldn't be a junkbox project.

In the circuit shown, the right-hand triode section of the 3A5 functions as a 144 MHz oscillator on transmit and as a two-meter superregenerative detector in the "receive" position. The left triode section operates as a headphone audio

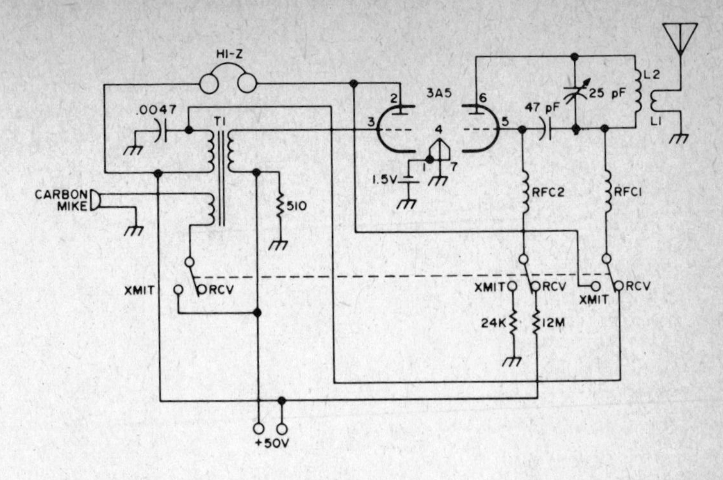

Fig. 5-30. Circuit diagram of what may be the world's most inexpensive 144 MHz transceiver.

amplifier during "receive" and as a microphone amplifier/modulator during "transmit." Switch S1 acts as the manual T/R.

DC power, as evidenced in the diagram, is such that full battery operation could be supplied for portability if desired. Although RF power input could be appreciably higher if more voltage is applied (up to 80–90V DC), resultant radiation when in the receive position might prove objectionable. A great deal, of course, depends upon the antenna configuration being employed and the degree of 144 MHz population in your area.

Performance

Construction is left pretty much to the builder. As with any VHF project, short to-the-point leads are the vogue. Main tuning is conducted by adjusting

Parts List

C1 – 25 pF var.
C2 – 47 pF
C3 – .0047 μF
L1 – 1 turn link of #16 e., ½-in. dia.
L2 – 4-3/4-turns of #16 e., ½-in. dia., ½-in. long
R1 – 510
R2 – 24K
R3 – 12 meg
RFC1,2 – 1.8 μHIRC-CLA
T1 – Triad A-21X
V1 – 3A5

Fig. 5-31. Parts list for 144 MHz transceiver.

the setting of C1. Incidentally, hand capacitance is a factor here—and one that must be reckoned with. For this reason it is advisable to completely house the rig in a small Minibox or the like and insure that C1 is physically secure. Likewise, L1/L2 are critical and should be mounted to sustain vibration and other environmental considerations without causing a change in operational frequency to take place. Both the variable capacitor and inductors, of course, should be well away from any slight heat that might be generated by the 3A5.

Even if a carbon microphone is employed, it is suggested that this *not* be constructed in conventional hand-held tranceiver form. The reason for this is that a wall-mounted mike button forces the user to induce severe vibration both through hand-action and speech—which can affect transmitting frequency. Again, the T/R switch can play a role. You may be listening in one frequency, through the switch, and find yourself transmitting 250 kHz up the band—all caused by the affect of physical switching. So be judicious in where you position your components.

In operation, this is no tropospheric DX hound. Like any superregenerative design, the receiver is extremely sensitive to weak signals but is affected markedly by stronger signals on the band. As a result, operation in a highly congested metropolitan area can be disappointing. Not because you're not being heard, but because your receiver will tend to blanket out.

If fed into a highly directional antenna array, however, results can be extremely rewarding.

THE FM "AUTO-START"

The auto-start concept is simple. Suppose you want to send a message to a buddy of yours; after fiving him a few calls with no reply, you key an audio encoder on your transmitter, which turns on a tape recorder attached to your friend's receiver. The tape recorder remains on for about thirty seconds or so, allowing you to transmit a short message, which in turn is recorded. Not only is it a useful gadget, but it's a lot of fun, as well.

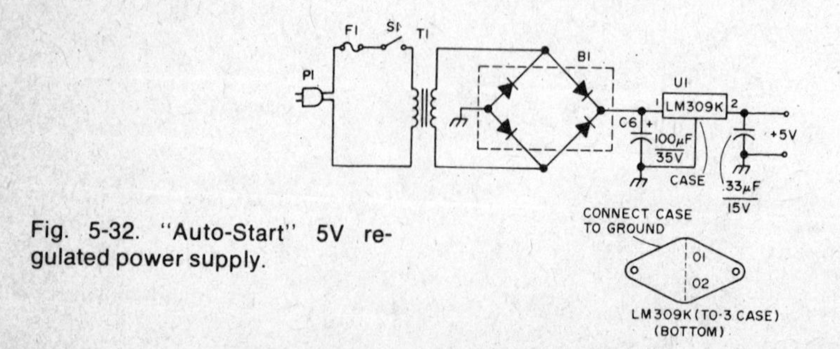

Fig. 5-32. "Auto-Start" 5V regulated power supply.

The Circuit

The auto-start is built around the Signetics 567 tone decoder, which has to be the greatest invention since stretch socks (Figs. 5-31, 5-32 and 5-33). The receiver audio output is attenuated and fed to pin 3. Whenever a tone of the proper frequency makes its way into the IC , pin 8 goes low. The decoder frequency is determined by R3, R4, and C2; in the configuration shown in the schematic, the frequency is equal to $1.1/(R_3+22,5000)C_2$ with R_3 in ohms and C_2 in farads. This formula assumes that the 5K trimmer is set mid-way. The values shown in the schematic are for a frequency of approximately 1 kHz. Should you decide to select another frequency, be sure to make (R_3+R_4) fall between 1kΩ and 20kΩ. Bandwidth is determined by C_3. The value shown gives a bandwidth of approximately 10 for an input frequency of 1 kHz. The bandwidth may be decreased by using a larger value of C_3, but the 10% bandwidth should be sufficiently selective to avoid false starting. For those more adventurous souls who wish to experiment, the formula for the bandwidth is

$$\%BW = 1070\sqrt{V/f.C_3} \text{ where } V = 200\text{ mV}$$

where C_3 is in μF, f is the decoder frequency in Hz., and V is the input voltage in millivolts.

R_6 and C_5, along with a 741 operational amplifier, form a timer. When pin 8 of the 567 goes low, C_5 discharges through R_5, thereby lowering the voltage on the 741's inverting input. As long as the inverting input's voltage remains below the noninverting input's voltage, the 741's output remains high, pulling Q_1 into conduction, which in turn pulls the relay. The normally open relay terminals close, turning on the tape recorder—just how they do this will be explained a little later on. Once C_5s voltage goes above 2.2 volts, the approximate voltage on the non-inverting input, the op amp output goes low, and the relay opens up, turning off the tape recorder. With the values of R_6 and C_5 shown, the timing period is about thirty seconds long. To increase or decrease the length of the timing period, raise or lower the value of R_6.

Little need be said about the power supply; if you use an alternate source, however, be sure it's well-regulated, as the 567 can get a bad case of the "funnies" with a poor power supply.

Construction

The circuit, as one might expect, is not critical and may be constructed to meet the aesthetics of the builder.

In the way of a chassis, Radio Shack puts out a line of "deluxe" cabinets at reasonable prices, in case you're tired of chassis that look like they're made out of surplus aluminum siding.

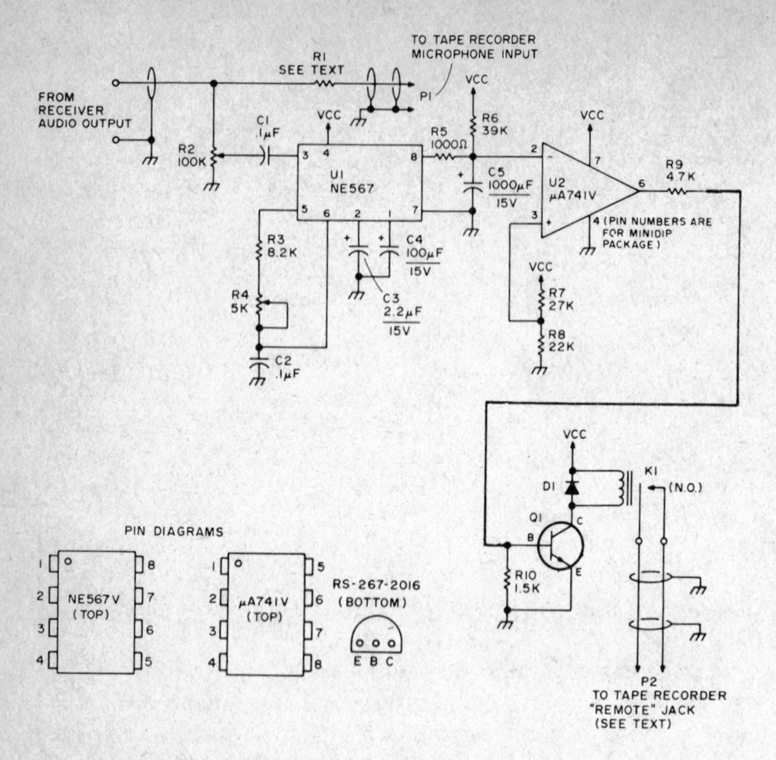

Fig. 5-33. The FM Auto-Start schematic.

The Tape Recorder

Most small recorders, both cassette and reel-to-reel, have a sub-miniature phone jack labeled "remote." In normal use, a small SPST switch on the microphone is connected to the jack, allowing the operator to turn the recorder on and off while holding the mike. When the input to the jack is a short, the tape recorder is turned on, and when the jack's input is open, the tape recorder quits. If your tape recorder has such a jack, you're in luck; just take P_2 and plug it into the remote jack. If your tape recorder has no such jack, the best bet is to sever one of the wires going to the capstan (the little shaft that rides against the rubber idler wheel) motor and attach each end to the normally open contacts of K1.

Before using the auto-start, you'll have to build audio encoders for all the people from whom you wish to receive messages; fortunately, this poses no problem.

After assembling the encoders, measure the frequency of the 567 by feeding the output of pin 5, a square wave, into an oscilloscope or frequency counter. Next, tweak the encoders until their frequencies are reasonably close to that of the 567.

Now the real test: have a buddy hook up an encoder to his transmitter, and attach the auto-start input across your receiver's speaker terminals (the speaker should be left connected). Have your friend key the encoder in cycles of three seconds on (about the time needed to totally discharge C_5) and ten seconds off. Adjust R_2 until the encoder just barely triggers the auto-start, and leave it at that position, or slightly higher. If the auto-start doesn't trigger, it may be necessary to (a) turn up the receiver's audio gain, or (b) adjust R_4 slightly.

After you have the auto-start working nicely, plug P_1 and P_2 into your tape recorder, and set it in the record position. Since P_1 feeds the audio output of the receiver directly into the mike input of the tape recorder, it will be necessary to attenuate the signal with R_1, starting with about 220kΩ and working *down*, so as not to blow the recorder input to bits on the first try.

A word of caution: Any formal message transmitted into the auto-start must be treated as any other piece of traffic; i.e., it must be kept on file.

Once you have your system working, the possibilities are endless. It is handy for recording another fellow's signal if his audio goes sour, so that he can get some idea of his problem's nature.

2M BEER CAN CAVITIES

The subject of resonant cavities seems to be surrounded by an air of mystery and confusion. They are things that are used by people who work UHF and microwaves and seem to have found little use in the VHF field. One reason for this is that few amateurs seem to fully understand what they are, exactly what they do, their construction, and their applications.

Cavities are basically an infinite number of shorted quarter-wave tuned stubs in parallel. They take the configuration of a cylinder with a rod in the center, with one end shorted and the other end tuned by a capacitor. If you will try to forget your mental picture of expensive-looking silver-plated plumbing for a moment, try to image a shorted quarter wave stub. Now connect another one in parallel. Now add more until you have a circle of them so that the last one added is touching the first. You will have a can with a rod in the center, which is precisely what a cavity is.

The cavity has many advantages over a conventional tuned circuit. It has a very high Q and low loss. By substituting the front end tuned circuits of a converter with a cavity, the selectivity can be improved to the point that no additional selectivity is needed to eliminate images and spurious responses that are less than a megacycle from the desired frequency. Two cavities are needed, not to get a narrow enough bandwidth, but to get a wide enough bandpass to be practical. Image rejection of better than 60 dB is easily obtainable for signals removed by 1 MC from a 2 MC bandpass at 2 meters. By adding one or two more cavities, rejection of better than 100 dB is possible. This means that it is not necessary to use double conversion or a high IF to get desired selectivity. With the use of cavities tuned for a 1 MC bandpass, it is possible to use the BC band IF with few problems. This should make construction of mobile converters easier.

Cavities have a very high Q and low loss. Aside from the advantage of image rejection, this can noticeably improve the effective sensitivity of many low noise converters. Because of the high voltage gain, the S/N ratio of an amplifier is better. There is a fixed amount of noise at the grid of a tube. The more voltage that is applied to that grid, the more the signal will be out of the noise. The Q of a cavity effectively raises the voltage of a signal. When used to feed an amplifier, this means that a signal is amplified without the addition of any noise, as the same signal is driving the grid of the tube with more voltage compared to the tube noise, in comparison with the voltage that the same signal would deliver with a conventional tuned circuit. On a marginal signal, this means the difference between Q5 copy and no copy.

A cavity is nothing more than a glorified can with a rod in the center. So why not use a can with a rod in the center? The tin plated soft drink cans (12 oz.) are just about the right size for a two meter cavity. They are easy to solder to and are easily available. The rod can be a length of copper tubing of about 1/8" or 1/4" diameter, available from auto supply stores. The rod is soldered to the bottom of the can in the center. Holes are drilled for the proper taps in the side of the can. The cavity is tuned with a ceramic trimmer of about 5-20 pF for the two meter version. A six meter cavity can be made by soldering two cans together after removing the bottom of one can, and tuning it with a 7-45 pF trimmer. The taps should be twice the distances given for the two meter cavity as impedance depends on distance relative to overall length. Some experimenting with the taps will produce an optimum point. Average distances are given as a starting point, and will work quite well, but each cavity should be adjusted for best S/N ratio. Fig 5-34 shows the basic construction.

As for practical applications, Fig. 5-35 shows a transistorized two meter converter that can be built using cavity input. Transistor Q1 is an emitter follower and is used to match the impedance to the grounded base amplifier, Q2. The emitter-coupled circuit has a small amount of loss because of slight impedance mismatch, but it is low enough to be negligible, and the direct coupling makes up for some of the losses found in coupling networks. The circuit shown in capable of a sensitivity of better than .1 microvolt for a readable signal. The IF output of 50 MC is chosen to work into a six meter converter for mobile use. With the use of cavities, the IF could be changed to any frequency from six meters down to the broadcast band by just changing the crystal, and L4, 5, and 6 to the correct frequency. Two diodes, wired back to back are used to protect the transistors from strong signals. Any small

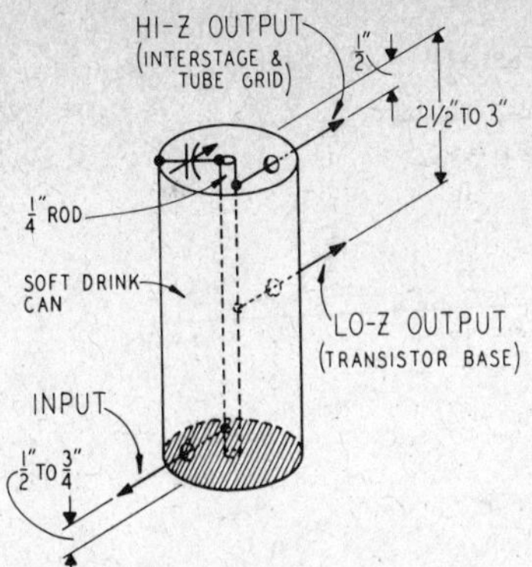

Fig. 5-34. Basic cavity.

general purpose diodes with a very low PIV will work. Don't underestimate the voltage gain of a cavity. The Q can run anywhere from 50 up into the thousands, depending upon the loading. With a Q of 1000, 10 millivolts input can give up to 20 volts output, or enough to destroy the base-emitter junction of a transistor. Think about that for a second. 100 millivolts isn't much when you consider the leakage of an antenna relay at this frequency. When the diode breaks down, it loads the cavity so that the Q and the voltage gain drop to a very low value. This should happen before the base-emitter junction breaks down as the diodes are at a higher impedance, therefore a higher voltage point.

The loss can be reduced more by using a glass piston trimmer instead of a ceramic type. Just drill a hole by the top of the cavity and mount it. A BNC connector can be mounted on the cavity for the input, if desired.

Figure 5-36 shows how the cavities are used for tube circuits. Figure 3-35 is for a grounded cathode or cascode amplifier. Figure 5-36 is for a grounded grid. The best advantage of the cavities is in the front end of a cascode, as most advantage can be taken of the voltage gain. The difference is slight, however, if care is taken to match impedances carefully and reduce loading.

The cavities for any of these circuits are bandpass tuned for the desired portion of the band. It is possible to stretch about 2 MC at two meters by careful adjustment. They are bandpassed by conventional tuning procedures. The tuning is quite critical, however, and must be done more carefully, as a small fraction of a degree of rotation can detune the cavity a considerable amount, so proceed slowly. The piston trimmers have the advantage of finer control.

The physical mounting of the cavities is easily accomplished by the use of small copper straps. The outside of the can is cold for RF and can be grounded at almost any point or points. The simplest way is to mount the can on its side and strap it to the chassis at both ends, making doubly sure to make good contact. Mounted this way, they take up little space.

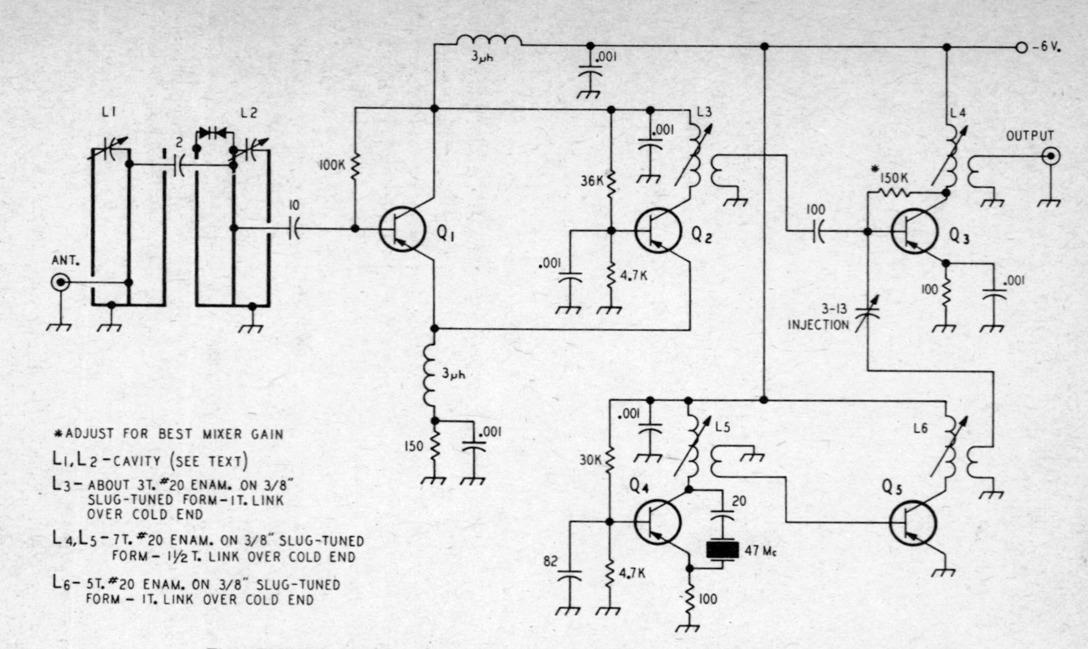

Fig. 5-35. Low-noise two meter transistor converter. Q1 and Q2 are 2N2360, 2N1742, 2N2495, etc. Q3, Q4, and Q5 are VHF transistors such as 2N2084, 2N1743, etc.

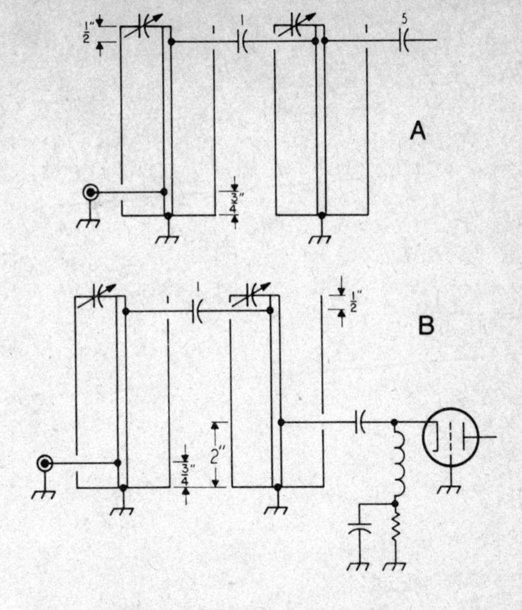

Fig. 5-36. Grounded grid input.

BUILD YOUR OWN 2M FM

All you have to do is look once at the price of commercial walkie-talkies to see how the effort of building your own can be justified. But there's a satisfaction that comes with operating something you've put together yourself, which goes well beyond the knowledge that you've beaten the manufacturers at their own game. Building up a portable FM transceiver is not a snap, but it really isn't particularly difficult either; and the time you spend will be amply rewarded by a good solid working knowledge of what's in those little radios you hold in your hand. Besides, it can be great fun!

The construction process is described here in its entirety—first, the receiver, then the transmitter. But since the project is to be a miniaturization job as well as simple construction, there are certain specifics involving components that must also be considered. In the main, these are dealt with individually.

Miniature Components

Capacitors. Bypass units can be the Lafayette thin units, where a 0.01 μF job can be found which is only 5/16 in. square by 5/64 in. thick. And the 1000 pF ones are only 11/64 in. square. For the lower values used in coupling and for fixed tuning capacitors, use the Elmenco dipped silver-mica jobs.

Resistors. Resistors can be the Ohmite 1/4 watters, but for the sake of miniaturization, you'll be better off if you get a selection of Allen Bradley 1/8 or 1/10 watt midgets. They're *really* small!

Crystals. The crystals should be the small plug-in kind, about 400 by 175 mils because repeater input channel frequencies do vary across the country. The most prevalent in the U.S. is 146.34, with 146.46 being Canada's prime choice.

Coils. The 8 and 24 MHz coils are the 9050 units from J.W. Miller, and are very handy for modifying to suit transistor input impedances, as well as having good, stable, mechanical tuning of the cores.

There isn't much else on the strip except thin copper-clad, 1/16 in. linen-base Bakelite or fiberglass strips, four Motorola HPE 55s, and two Motorola HEP 75s (2N2866).

Various colors of subminiature wire will help also.

Special Tools

Don't worry about particular tools; they're not too special, as you will see, but you should prepare a little, in order to do a real good job. You must have the usual set of good, small tools and it helps to thin down by grinding the already thin needle-nose pliers to get into those really narrow places you will find in back of the mounting strip. Use the same treatment on some small side-cutters also, because you will be cutting off a lot of small wires in even smaller places.

A collection of small low-cost screwdrivers will be handy, too—file them sharp and very small for special places. Sharp-pointed tweezers are handy as well.

For drills, go down to size 65 (35 mils). Depending on how lucky you are, your drill chuck may not take those little drills. Some of them don't. Then you have to lay out another $3 or $4 for a jeweler's chuck, which will take a No. 80 drill (13 1/2 mils).

You do not have to drill the component-lead holes exactly to size but the closer you do the more rigid the parts will be when mounted.

Various fiber TV tuning tools are useful for the trimmer capacitors, and several lengths of 1/4 in. Lucite and Bakelite rods make good insulated screwdrivers also.

A slightly unusual aid is a "coffee stick" with an arrowhead-shaped lump of coil wax stuck on the end. When you're winding small coils with small wire it is very handy to put a drop of wax on the coil and let it sink in and cool. You can do this with the tip of your small iron, and it sure helps hold all that tiny wire in place. All the common filter chokes use this method. It's good for a lot of receiver coils to come later, too, and for holding the extra turns wound on the Miller coils for base impedance matching.

Be sure to have plenty of subminiature clip leads with flexible wire of various lengths from 1 in. up to 1 ft.

Have a good selection of Arco midget trimmers on hand also, such as the 400 series, which are just 1/2 in. long.

No. 48 or 49 bulbs are good for checking RF as you go along the multiplier chain from stage to stage. You should always be able to light one of these, which glow red (dull) at 20 mW. Use a matching series trimmer, as little as 5 pF for 147 MHz and less than 1 pF on 450.

Have a roll of plain masking tape to hold down strips and things while working on them. A small drill vise or the "third hand" bench vise helps, too.

Dos and Don'ts. These hints apply especially to a multiplier chain including a straight-through amplifier used as a phase modulator, which is the circuit being described in this section. It has an 8 MHz crystal oscillator and ends up to 147 MHz, so you must be *sure* of the frequency of each stage as it operates. Do *not* rely on your receiver or on grid-dipping the inductors first.

These simple and inexpensive test accessories will help you in this work. Don't try rushing this multiplier along with using an absorption wavemeter (see Fig. 5-37) at every stage.

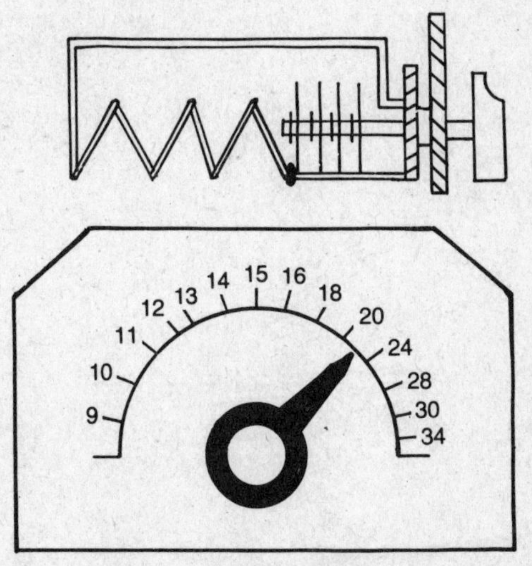

Fig. 5-37. Typical absorption wavemeter.

Getting right down to the point, here is a list of handy items to have on the bench while you're building crystal oscillators, phase modulators, multipliers, and amplifiers. Using the absorption wavemeter, any circuit under test can be checked for the real and exact frequency at which it is resonating or oscillating, by lighting a bulb on RF or using a diode detector with a meter.

When the absorption meter resonates with the RF in the collector circuit which is lighting the bulb or actuating the meter, a dip in the light or on the meter will show. This indicates the *real* frequency of the main body of the RF present. Some transistor collector circuits not tapped down on coils are especially notorious for this, and may exhibit two frequencies at the same time. For example, there may be energy at 72 and 96 MHz present. This is an indication of mistuning, or overloading, or both. Tap the collector down on the coil, don't load it so heavily to the next stage, check it carefully with the diode meter, and don't worry about a small remnant of off-frequency energy. After all, a multiplier is bound to have some of this present. Just get the *main* amount on frequency and be happy. And be sure the *next* stage also peaks on *its* desired frequency.

A grid-dipper in the *diode* position can also be used for this work. A one-turn link around the low end of the grid-dipper coil and a cable will get you into small places in a rig where you cannot insert the whole dipper.

The diode detector. Figure 5-38 shows the schematic of one of these useful pieces of equipment which allow you to listen to your transmitter multiplier stage as you build it, and check the actual frequency at the same time. With a good variable capacitor you can generally run over three to one in frequency range—up to the UHF region at least. From there on up things get a little more difficult.

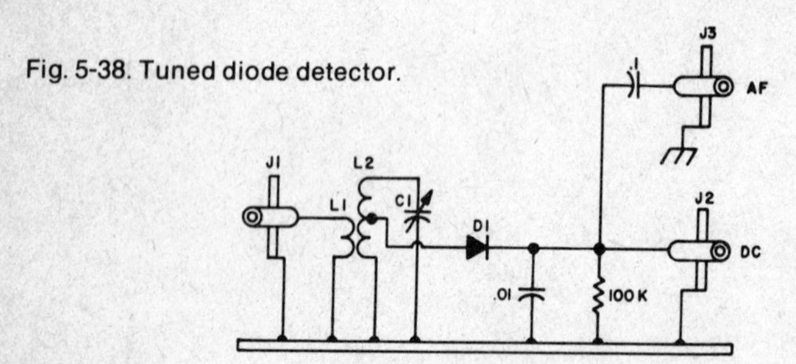

Fig. 5-38. Tuned diode detector.

These "receivers," because that's what they really are, although of low sensitivity, are especially helpful in transferring known frequencies on a signal generator to a homemade set of wavemeters.

The meter. This should be as sensitive as possible. Lafayette has good ones down to 50 μA. Use a tap switch to put resistors in series to bring the voltage range up to 10V or so for use with an active portion of the rig such as the 1W final circuit.

The AF amplifier. This item should not be neglected as it is at times a great aid to getting a trouble-free, *noiseless* carrier which you can then modulate and be proud of. The valuable RCA handbook, "Transistors, Thyristors, and Diode Manual," has a lot to say about "discontinuous jumps in amplitude or frequency as various levels of drive are encountered." These little termites can be seen on the meter or *heard* on the AF amplifier or can show up on both. Figure 5-39 shows a mounted version of the AF amplifier used here for this purpose. It is a worthwhile and handy little piece of equipment to have in a lot of situations, in both receiver tests and transmitter tuneup. Just plug it into J3 of the diode detector in Fig. 5-38 and hear those unwanted clicks, whistles, rushing noises, squeals, etc., coming from what you may have wishfully thought was good clean RF in your multiplier drive!

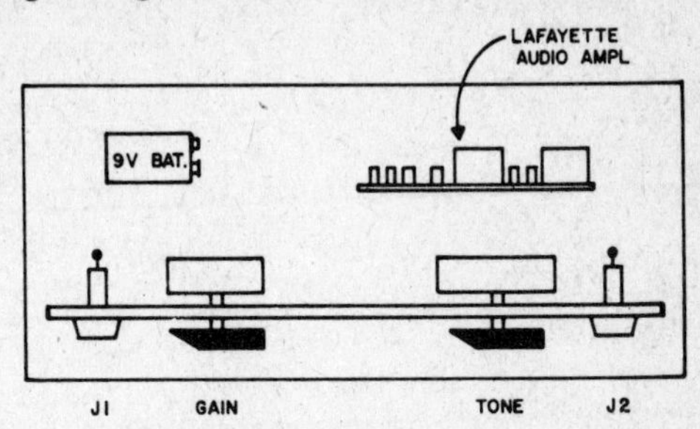

Fig. 5-39. Handy AF assembly, top view.

Important notice! Overdrive is especially to be avoided in multiplier chains with transistors. Superregeneration is one of the indications. Believe me, it can be a very nasty bug!

Diode detector cable probes. Have a collection of these on hand as in Fig. 5-40. You can also use them to feed RF into a pilot light, connect up to your lab receiver, etc.

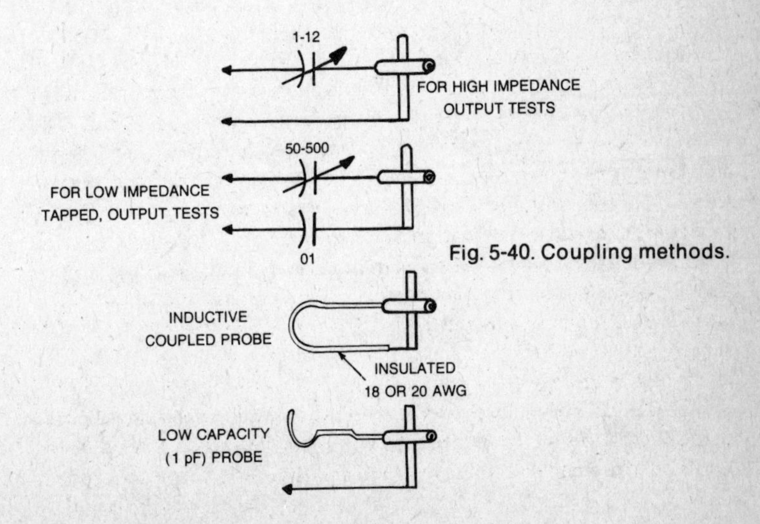

Fig. 5-40. Coupling methods.

Handy meter jacks. Figure 5-41 shows an elementary but flexible and useful metering method for checking total or only one stage current.

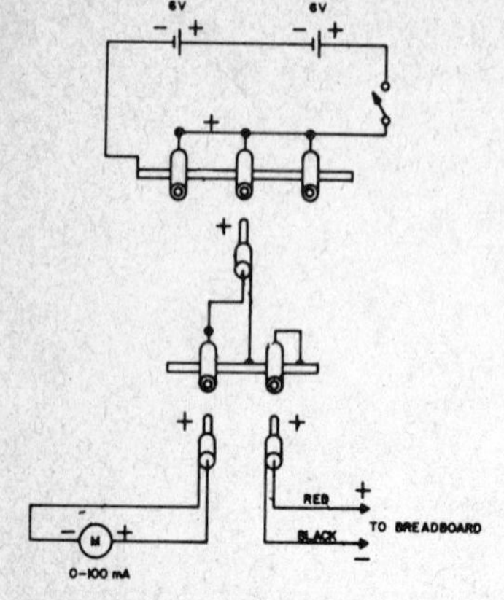

Fig. 5-41. Handy metering jacks and plugs.

Foundation Receiver

The basic design shown here is for a low-cost single-conversion utility receiver for 2 meter FM; particular attention is given to easy-to-build IF and discriminator modules for the 10.7 MHz section. The RF is tunable from 144 to 148 MHz, with a switch for AM use. This is a complete portable receiver, *not* tied down to a large AC communications receiver.

Discriminator action, with sample, is shown for easy understanding and homebrewing. Double conversion with crystal control can be added later.

The schematic of Fig. 5-42 shows how easy it can be. Remember, this is just a basic receiver which, without double-conversion, is a relatively broadband, easy-to-tune job, but it sure pulls in those interesting repeaters!

Front End. Simplicity is the word in this module. You can check different transistors for low noise, coils for RF, or add a low-noise stage; and the tunable oscillator is easy to change to a crystal oscillator for repeater operation. All three stages are tunable from 144 to 148 for coil, sensitivity, and selectivity experimentation, and to allow you to check the AM section of the band as well as repeater work in your neighborhood.

The oscillator tuning dial also relaxes preliminary oscillator crystal frequency requirements by allowing you to find out what crystals you will want later, and order them without rushing the deal. The link coupling at low impedance permits easy switching from tuned to crystal control, if you wish to retain the tunable feature.

The RF and mixer stages are tuned by small variable capacitors mounted on the baseboard with small brackets made from copper-clad. Small pointer knobs allow peaking of these circuits. The RF stage has a trimmer capacitor

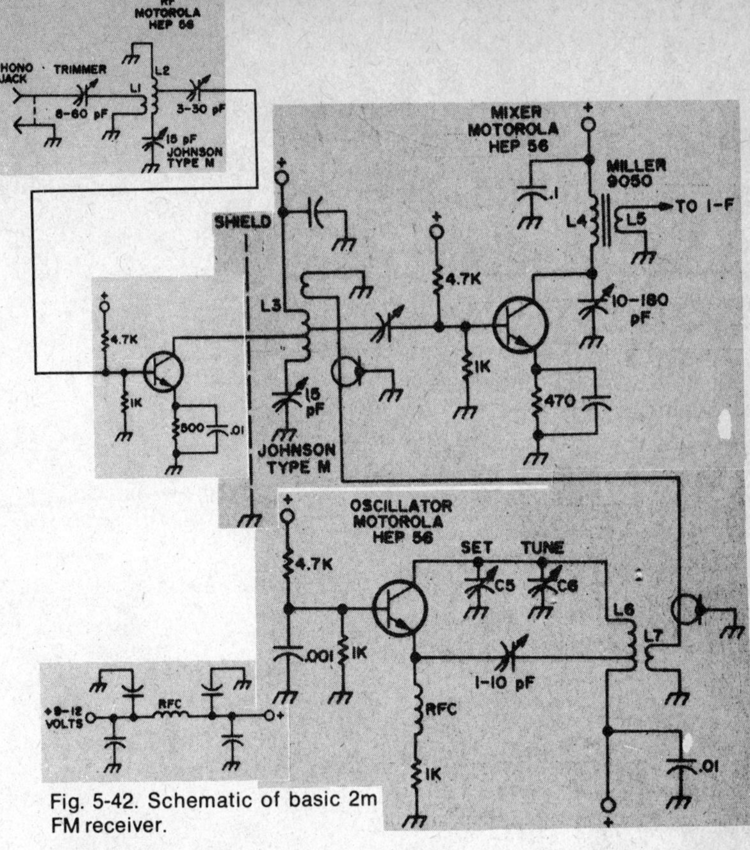

Fig. 5-42. Schematic of basic 2m FM receiver.

feeding the base which is quite useful, resulting in a welcome balance between gain and self-oscillation. The mixer also has a trimmer for its base input, which permits a selectivity adjustment for this circuit.

The tunable oscillator was mounted on the Miller slide-rule dial for mechanical stability as shown in Fig. 5-43, and works quite well—with the broadband IF of course.

The two-meter band can be spread out from 10 to 90 on the dial by trimming L7, increasing C5, and using a smaller C6.

Oscillator coupling can easily be adjusted for maximum conversion efficiency via L4 and L7, and the cable between them is a good place for the crystal-tunable switch mentioned above. To start up, adjust L7, C5, and C6 for the range 133 to 137.5 MHz as a local oscillator for the IF of 10.7 MHz to be used later. I tuned up the whole front end using the diode detector of Fig. 5-38 tuned to 10.7 for the IF section. When there is lots of 10.7 MHz energy out on L6, such as to deliver 5V DC out of the diode, you've got a good front end!

10.7 MHz I-F Stage. The reliable and sure-fire Motorola HEP 590 IC was used here—25 dB gain, no self-oscillation, what else would you use? Figure 5-44

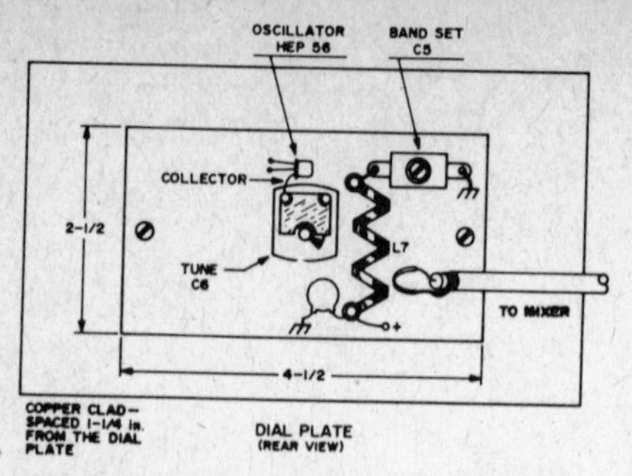

Fig. 5-43. Dial-mounted local oscillator.

shows the circuit, using Miller half-inch shielded coils on the input and output. Note that the 590 is simply turned leads-up and soldered onto a few resistor

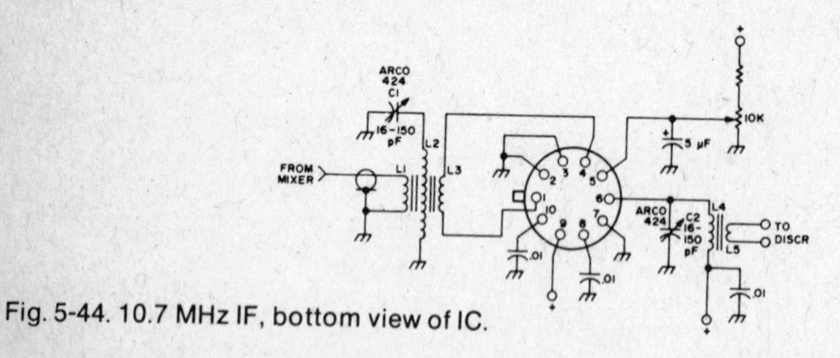

Fig. 5-44. 10.7 MHz IF, bottom view of IC.

supports, with a shield, as in Fig. 5-45. A gain control is used, which may or may not be kept in later as you wish. With the limiters that can be added, the gain control is *not* needed.

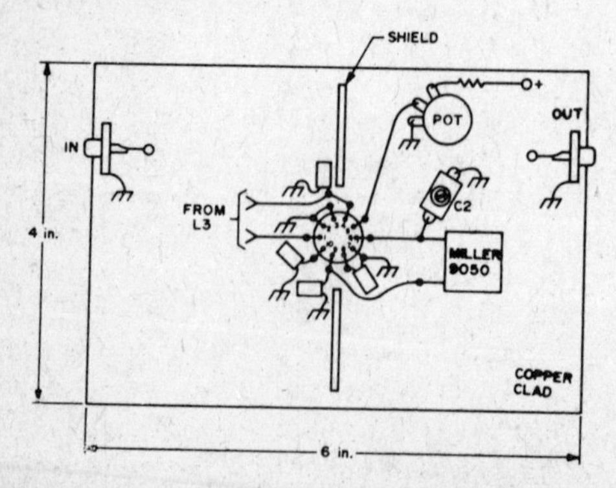

Fig. 5-45. IF shielding as seen from top side of board (bottom view of IC).

A B+ filter is included in each module, and a 100Ω resistor with a 10 μF capacitor may be needed also to cut out motorboating when more stages are added later, if you go to double conversion. Be sure *not* to return L3 to *ground* DC-wise, as the needed bias is supplied internally through pin 4. Pin 9 is the main B+, along with the cold end of L3. Gain control can be obtained through a pot in the pin 5 lead, where maximum gain is reached with pin 5 at *ground* potential.

For new readers, the internal and external circuit of the Motorola IC 590 is shown in Fig. 5-46. This IC, which is very useful for frequencies up to at least 6 meters, has extremely interesting features, among which can be noted the absence of internal feedback (even at 50 MHz), the high gain, and the excellence of the gain control at pin 5, either manual or automatic. For this receiver, mainly intended for experimental FM use, no AVC is used. Later, if you add double conversion, the limiter section module will eliminate the need for AVC.

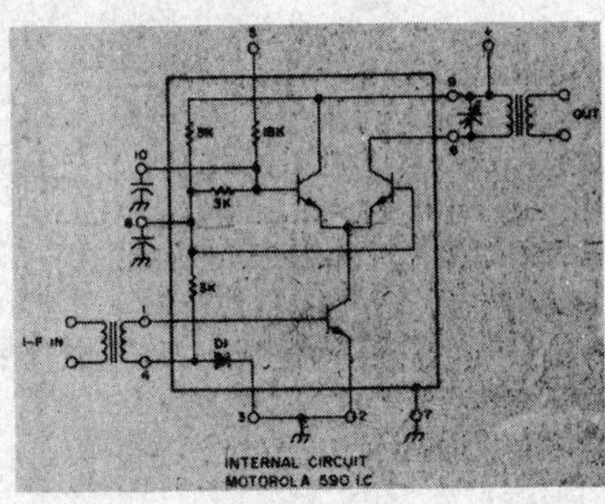

Fig. 5-46. Internal circuit of Motorola's HEP 590 IC.

Trimmers are shown for C1 and C2, but fixed capacitors of the proper value may be used to allow tuning of the IF coils at 10.7 MHz by the variable tuning slug cores in L2 and L3. Note that these Miller half-inch gems have very good electromagnetic as well as electrostatic shielding, due to the cup-core type of construction, and are available for use from 30 MHz down to 135 kHz. They can also be easily opened for the addition of primary or secondary low-impedance windings.

Discriminator. Figure 5-47 shows the circuit where L1 is a simple tuned coil in the collector circuit, with *no* coupling requirements other than a one- or two-turn link. When the primary of a discriminator transformer has to be coupled just right to the centertapped secondary it is not a job for the usual experimenter at his bench. With the link you can't go wrong. At least I haven't so far. Just tune L1 to 10.7 MHz, put a turn or two around it and another turn on L4 and away you go. Figure 5-48 shows the discriminator DC output curve, which handles about 25 kHz for 2-0-2V.

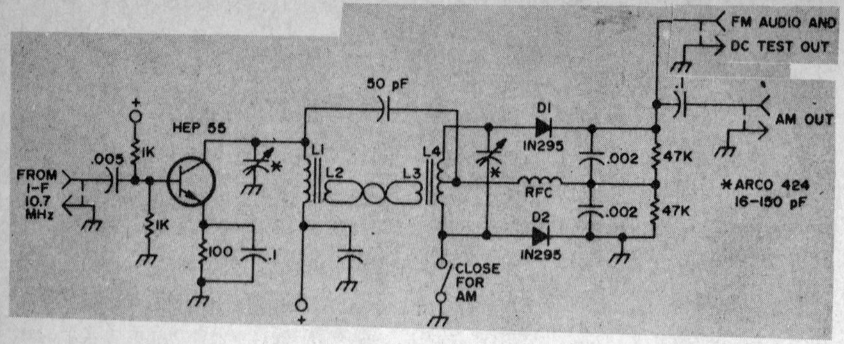

Fig. 5-47. Discriminator circuit.

FM Transmitter Strip

The transmitter section measures 1 in. wide by 8 in. long, and it puts out over a watt on 146-147 MHz, with low-cost components. This miniaturized transmitter is a logical step toward design and ultimate construction of a "shirt pocket" portable transceiver. The parts for that one jump up a little in cost, because it takes a lot more tools to make subminiatures, such as stereo microscopes, special materials and skills, jeweler's tools, and so on.

Shape Factor and Assembly Method. These are important features, as you will see, allowing the homebrewer to build a complete FM rig in a Minibox and still have room enough left over to change components for repairs or design improvements if needed. You can also substitute slightly different components if you have to.

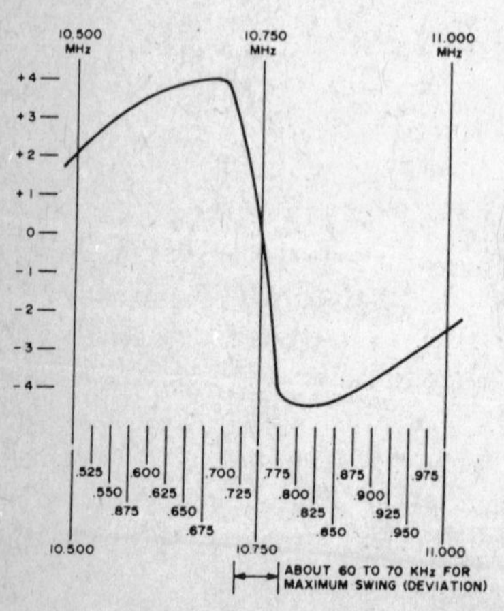

Fig. 5-48. Discriminator DC output curve.

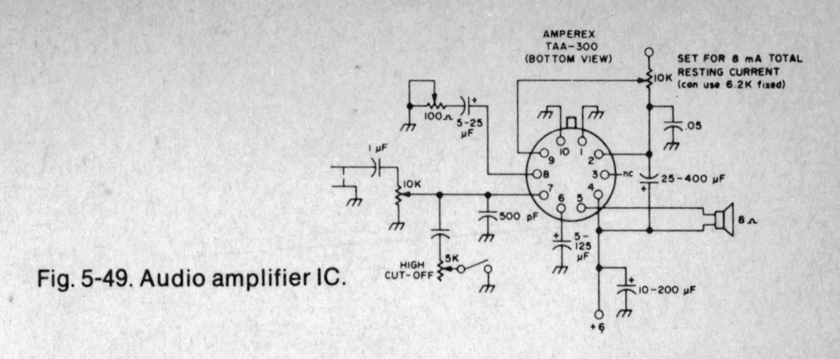

Fig. 5-49. Audio amplifier IC.

Figure 5-50 shows the method, using a copper-clad baseboard on which is mounted a drilled 1/2 in. high strip of insulating material holding all the components. Bypass capacitor leads to ground are no longer than 1/4 in., shielded coils are used, and all tuning is done from one side.

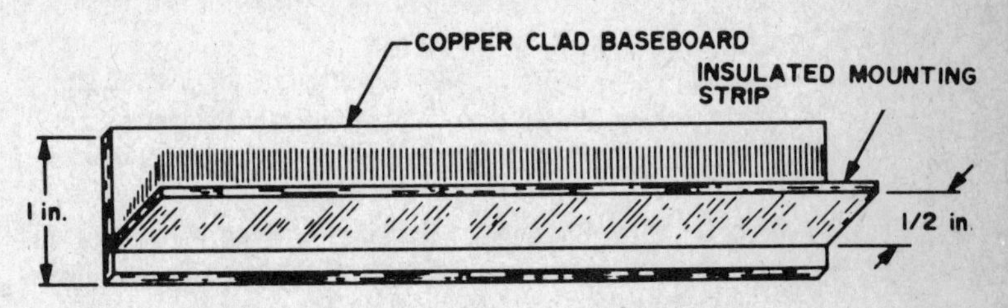

Fig. 5-50. Baseboard/mounting strip configuration.

Components are all on one side of the circuit board, and their leads and connections are on the other side. On the wiring side, every connection is spread out in front of you, with room between each one for good soldering; no resistor supports or other metal tie points are needed.

The B+ lead is red subminiature wire and goes from filter to filter along the strip. The RF lead is green and goes from the coil output tap of each stage to the next base coupling capacitor; the rest of the connections practically fall in place for soldering together. As you can see, there is still room left over!

The detailed planning of the holes to be drilled becomes a large portion of the work. Figure 5-51, component side and top views, shows how to start this off. The next step is to make a life-size drilling template using the components you have or intend to use. Most of these are not critical and you may substitute without trouble providing you keep thinking "little." Even here, you can go bigger with the components if you want to, but your overall package size may expand. You can also go smaller if you plan carefully and cram everything together a little tighter. The reason for this will be evident if you study the circuit, where you will see that no critical wires cross over each other, and that the power amplifier is well away from the oscillator.

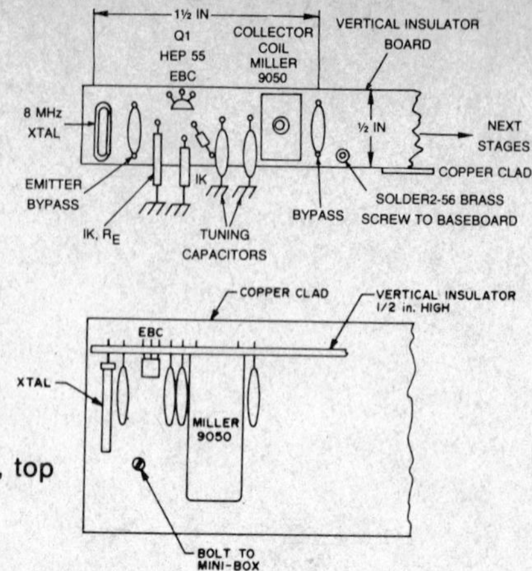

Fig. 5-51. Component layout, top and side.

Ultimate size is actually up to you, and you can judge for yourself after laying out the parts on hand. If you send for a selection of Lafayette Radio very thin and small capacitors, you will have an easier task to get it down in size.

Figure 5-52 shows two methods of preliminary fastening of the vertical strip to the copper-clad baseboard. This will start off the assembly, and after the wiring you could hardly tear them apart with your fingers.

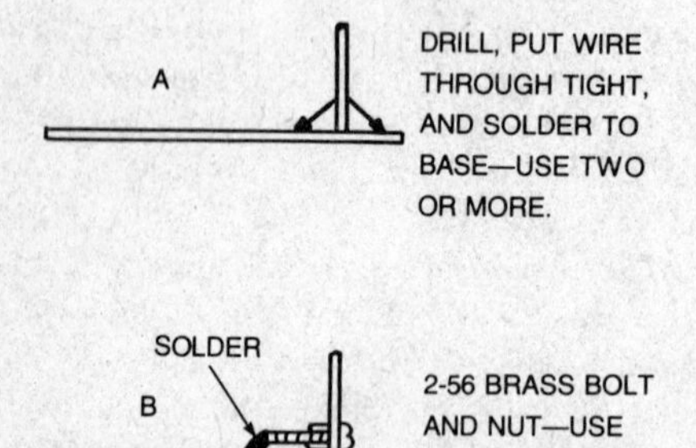

Fig. 5-52. Methods of fastening insulated strips to baseboard.

You can also make strip modules of any length you want such as modulator AF, receiver sections, etc., as shown in the receiver plans. This makes the task of repairs or improvement changes easier later on. These shorter strips can be fastened end on to each other and fastened down to the baseboard as shown.

Miniature Filters. Do *not* try to make up frequency multipliers *without* RF filters in the DC line to each stage, unless you care to experiment with RF phasing in battery leads—and that isn't good! You can make up "dime" filters without much difficulty if you follow the simple details below. Materials needed are tiny resistors (any value over 1k Ω), some 36 or 38 AWG wire (double silk covered), coil wax (you can use paraffin wax if you can't get coil wax), and small capacitors, such as the Lafayette types. Use 0.01 for HF and 0.001 (1000 pF) for VHF. Figure 5-53 shows the circuit. The main thing is to interpose an RF trap in the plus lead between each stage and any other.

The series method shown in Fig. 5-53 is best for high-gain amplifiers because it puts *several* filters between the high-power output stage and the sensitive first stage. However, if the filters are very good you can bring the battery leads from each stage to a common point, but this must be checked carefully if you have to do it.

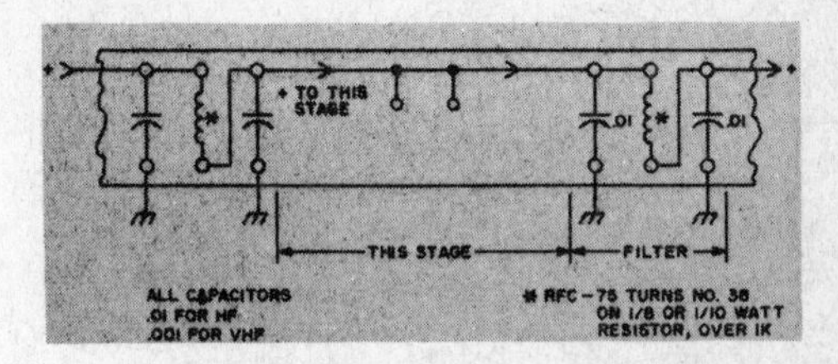

Fig. 5-53. Miniature filters, interstage.

Clean and tin carefully each resistor form lead close to the body, then melt a thin layer of wax onto the resistor to hold the wire from slipping when you wind it on. Solder one end of the wire onto one lead and then random wind 75 to 100 turns of 38-gauge wire onto it, and wrap the end around the other lead ready to solder. Put a drop of wax on the coil before soldering to hold the wire turns in place. The wax should penetrate the whole coil. Most types of insulation on 38-gauge wire will disappear as soon as solder and heat are applied, so you don't have to bare the wire first. Now you have an RF choke, and if you keep the capacitor leads real short to ground, the filter will do the job for you.

It works fine even up to 450 MHz if you use four capacitors, each to different points on the ground plane of copper-clad, as in Fig. 5-54.

Phase Modulator. Phase modulation results in a type of frequency modulation of the carrier at the RF output jack which the usual FM receiver cannot distinguish from true FM. Being crystal-controlled, it is used by practically all the FM mobile and base stations in the U.S. so it is 100% okay here. And of course with the crystals in there, you *will* be on the amateur FM channels, providing you buy them right. You have to pay around $7 for these but it seems well worth it.

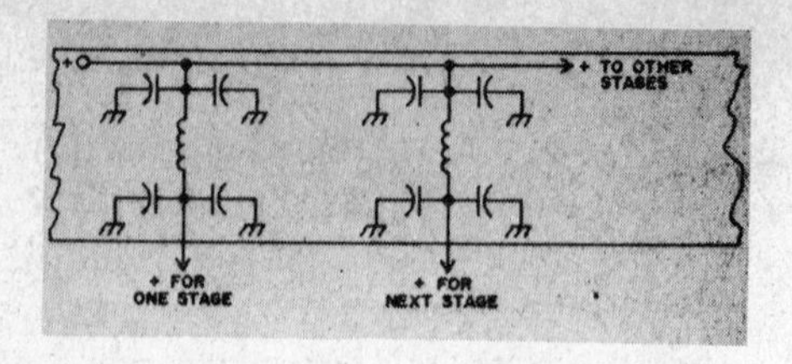

Fig. 5-54. UHF filters for interstage coupling.

Certain designs of the AF section of the phase modulator, its tuneup, and the connections to the phase modulator can be troublesome for the homebrewer, so considerable time was spent to make it as simplified and easy to adjust as possible. It also can be used in the receiver section as the AF amplifier because the frequency correction is done *outside*. The use of an 8Ω or 16Ω output connection into the phase modulator emitter circuit helps to stiffen the AF drive and keep it clean.

Phase modulation AF sections in commercial rigs are often qualified as "audio conditioning," or "processing" circuits, which they are of course, but don't let that bother you. Excellent FM quality can be otained by the use of an inductance of large value, placed outside of the AF amplifier, in the noncritical low-impedance output circuit. The inductance cuts down the extra high audio modulating frequencies caused by the phase modulator's tendency to make the FM deviation directly proportional to the modulating frequency, which emphasizes the highs too much unless corrected. Being outside of the AF amplifier, you can now use almost any good low-cost job and use it in the receiver also.

Figure 5-55 shows the simplicity of the method used. A four-transistor amplifier from Lafayette was installed, with a slight adjustment of the feedback resistor. This had nothing to do with the FM unit, it just happened that the Lafayette amplifier sounded and acted funny at first. And no wonder,

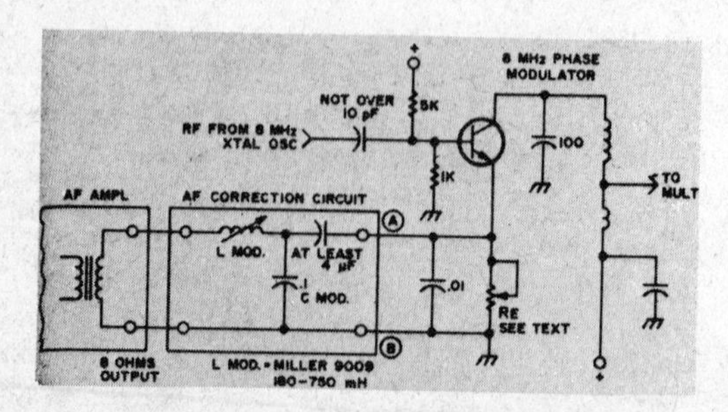

Fig. 5-55. Phase modulator interconnect circuitry.

it was oscillating up in the 100 kHz range! After trying to bypass and decouple almost everything. The printed lead going from the 8Ω output connection over to near the input was checked and sure enough, that was it: too much feedback! An additional 50 kΩ resistor in series with the one already there did the trick and from then on nothing but good AF came out.

The AF output needed to drive the phase modulator emitter is several hundred millivolts, and the low impedance allows the usual RF bypass capacitor of 10,000 pF to act simply as an additional AF filter, which it does.

As a result, the entire tuneup is done by adjusting the value of the emitter resistor and the phase modulator tank tuning coil. Neither are actually critical but should be adjusted while listening to the 146 MHz carrier on a good amateur narrowband FM receiver. The emitter resistor will be heard to kill the modulation when going much below 2 kΩ and to bring in distortion on large amounts of audio when going a lot more than 2 kΩ. This latter condition also causes a drop in the RF output. You may hit it right the first time with the 2kΩ value.

The actual phase modulation resulting from varying the emitter voltage with audio is adjusted by tuning, which is also smooth and noncritical. Use the tried-and-true method of listening to your own voice with plenty of audio on the receiver and a set of well-padded earphones which keep your voice from reaching your ears directly through the air. It also cuts down audio feedback.

Tuning with AF going into the phase modulator as per Fig. 5-55, you will notice good strong clean FM on either side of the peak tuning. These points occur *before* the 146 MHz carrier output starts to drop from detuning the phase modulator tank, so don't worry about that part. In any case, you are supposed to be following the phase modulator with enough saturated class C multipliers and amplifiers to prevent any variation in amplitude (otherwise known as AM!). You may have noticed an unduly large number of tubes showing in ads for surplus commercial FM sets. This large number is due to the designer's wish to get *all* the benefits of FM into his package. In one box if possible. You have to watch *very* carefully when using ICs for modulators, they tend to pick up RF and generate feedback with their wideband audio circuits and sometimes as many as 11 or even more transistors in one little can. Just a word on what to look out for. Also, don't put more than the specified voltage on IC amps. You can easily drop down with a resistor and a *large* bypass capacitor.

Almost any desired amount of highs and lows can be obtained or suppressed by the manipulation of the LC values in the modulator. If you use a Miller .9009 wide-range adjustable inductor, 180—750 mH, you can hear the difference as you adjust the core in and out.

You can start out with a large-scale layout for the parts, but you may wish to skip that and go right to a life-size layout as in Fig. 5-56. To make the life-size drilling template, lay out the components one after the other, "standing up" on a 1/2 in. strip of good-grade white cardboard and mark the component lead holes, which should result in something similar to Fig. 5-56. A nice feature of the cardboard method is the easy punching of the holes and the way it holds the drill as you go through the strip. Tape the template in place onto the insulating vertical strip. Do not use anything that melts under heat, though. Even if you ruin part of the strip, or want to make a large change of one stage you can just saw that out and make up another section and go ahead.

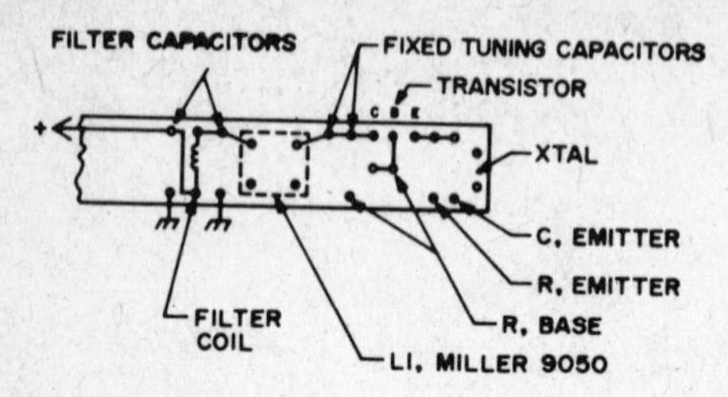

Fig. 5-56. Drilling layout (wiring side).

The 8 MHz Oscillator. Figure 5-57 shows the schematic of the crystal oscillator stage. Note the apparent use of negative feedback with the base return through the crystal to a tap on the inductance. It is only apparent though, as the crystal reverses the feedback phase, making it positive. It is a very powerful, sure-fire circuit.

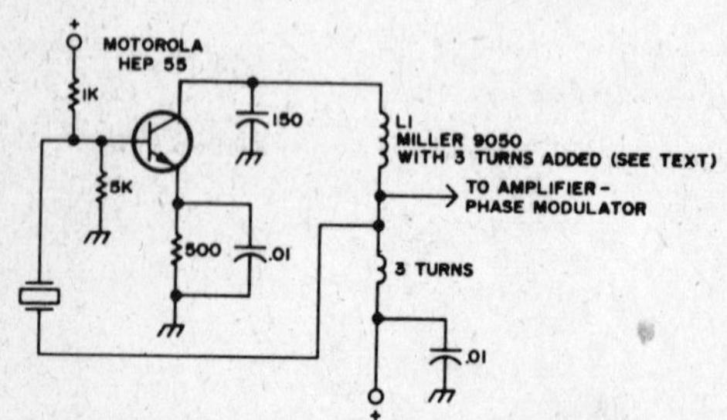

Fig. 5-57. 8 MHz oscillator schematic.

The tap on the coil also provides a good low-impedance match for the next base input. The coil itself is made from a Miller 9050 shielded coil which has magnetic as well as electrostatic shielding, and a good adjustable core that works well mechanically (which is more than you can say for some of those types of cores).

Remove the aluminum can by bending back the four holding tabs and wind on three turns of 30-or 32-gauge silk-covered wire onto the existing winding of the coil. Be sure and wind them in the same direction as the turns that are already there. The oscillator coil will then look like Fig. 5-58, and is ready to mount on the strip.

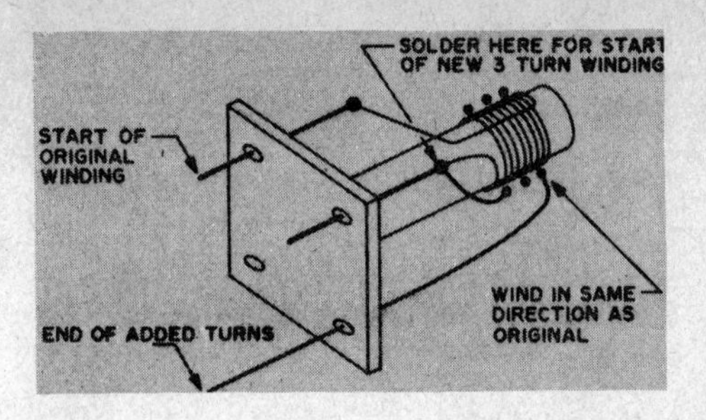

Fig. 5-58. Miller 9050 coil with added turns.

The wiring on the lead side of the strip is shown in Fig. 5-59, where most of the leads are seen to fall in place quite well.

Insert the component leads through the strip and bend them slightly in the direction they will go, such as the two base resistor leads which are bent towards the base lead, as shown clearly in Fig. 5-59. When all the leads to be soldered in one place are all touching each other, a final dressing can be done, followed by soldering. In the example mentioned, the base lead has three other wires soldered to it, a wire from the crystal, the 1 kΩ resistor, and the 5 kΩ resistor.

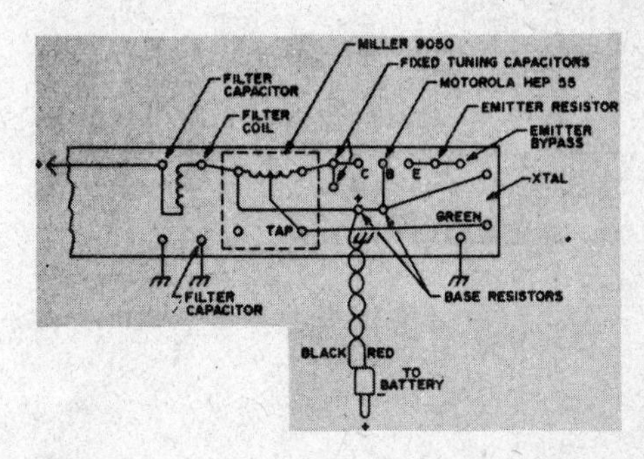

Fig. 5-59. Oscillator wiring diagram (lead side of strip).

The can of the 9050 coil has a tab which should be soldered to ground. The ground lead of some resistors (or all of them) is not routed through the strip but is soldered to the baseboard on the component side of the strip.

When the oscillator is assembled and wired, B+ can be brought in and the unit tested for RF. Some 5—10 mA of current should register and as soon as the oscillator coil is resonated to the crystal frequency, the oscillator should show RF output to the 8 MHz tuned diode test set connected to the output tap lead.

Check the oscillator carefully on a sharp receiver for its frequency-holding ability while tuning the slug in and out of resonance at 8 MHz. Actually this will be near 8.130 MHz. (With a multiplication of 18X, this should land on whatever 2 meter FM channel you're aiming for.) It should come into resonance on one side with a good "plot" and gradually build up on the other side as you tune.

Start with a large calibrated variable capacity at C3 (some 500 or 1000 pF) and then put in fixed values so that the iron core tuning slug in the 9050 coil tunes properly about 1/2 in. under the winding of the coil.

Power can be adjusted by the emitter resistor, and feedback by the number of turns between ground and the oscillator-coil tap (These are of course the number of turns added to the Miller 9050.)

A 48 or 49 bulb, rated at 2V and 60 mA, should light up with about 50 to 100 mW worth of RF with a 50—300 pF trimmer in series, as in Fig. 5-60. When the oscillator is properly tuned and under good power control via the test pot (in Fig. 5-60) and the plus voltage is checked for the voltages you expect to see, the next stage can be assembled. Of course if you wish, you can mark out the whole strip template, drill all the holes, and mount and solder all components except the coupling capacitor and B+ to the next stage. This allows you to test the oscillator by itself.

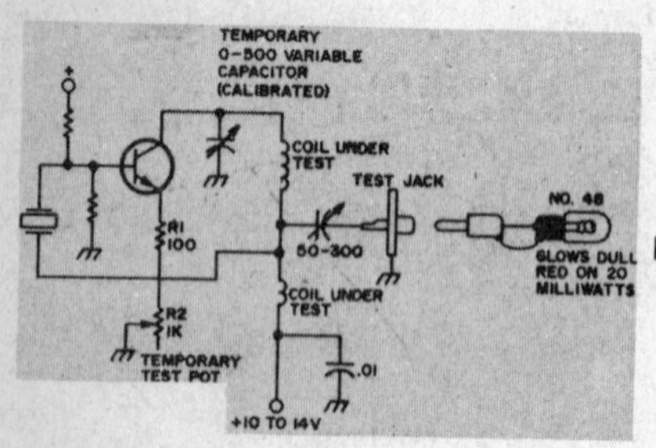

Fig. 5-60. Oscillator test setup.

The 8 MHz Amplifier—Phase Modulator. This stage (Fig. 5-61) is not critical, other than to keep the input base coupling capacitor at a low value to avoid self-oscillation. The only requirement is that the tuning should be correct for phase modulation.

Use the same methods of assembly, wiring, and tuneup for power output as with the oscillator stage. You do not need much gain, if any, in this stage.

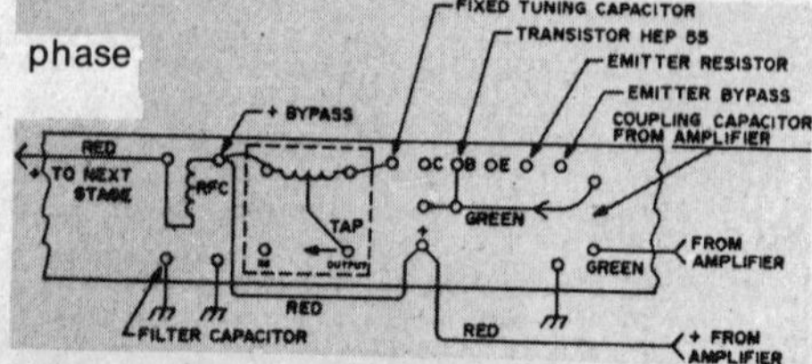

Fig. 5-61. Schematic of phase modulator/amplifier.

The 8—24 MHz Tripler Stage. A frequency multiplier has the advantage that generally (though not always) it is free from self-oscillation, due mainly to the output and input circuits being on different frequencies. The bias requirements are different in a tripler from those of a doubler or class C straight-through amplifier, but this can be adjusted simply by varying the emitter resistor during tuneup for maximum output on the desired frequency. Figure 5-62 shows the schematic of this stage, where the base input coupling capacitor is seen to be much larger than in the preceding stage. However, in spite of a small tap winding and low impedance in the preceding stage it is easy to cause superregeneration in the base circuit if the coupling capacitor is too large. 150 pF or slightly less is a good value.

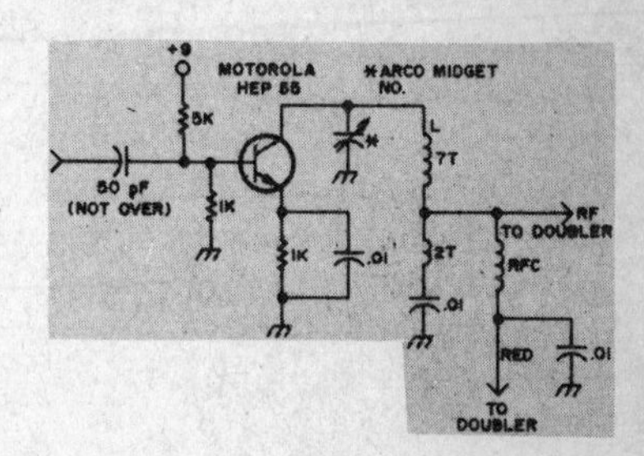

Fig. 5-62. 8 MHz tripler schematic.

The wiring side layout for this stage, which is typical of the multiplier circuits, is shown in Fig. 5-63. A logical wiring system is seen to prevail, especially as regards the emitter, base, and collector wiring and their components. Two extra wires are used, one red for the B+ and one green for the base input RF circuits, with a filter coil separating the plus of each stage.

A 24 MHz diode detector is clipped onto the RF output tap on the inductor Fig. 5-63, as was done in the preceding stages. Be very *sure* you're on 24 MHz, and not on 16 or 32. Here again you should be able to light a 48 bulb with RF with a 5—180 pF trimmer in series for matching. The collector tuning and power output curve with emitter resistor lowering should be clean and smooth.

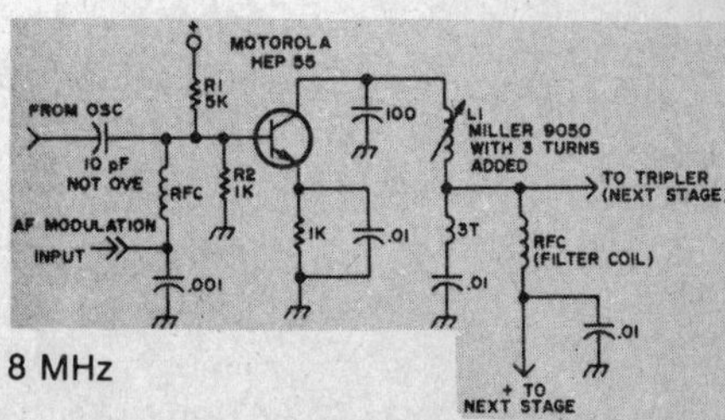

Fig. 5-63. Wiring diagram, 8 MHz tripler.

As mentioned in the test equipment section, it is a real *must* to *listen* to the carrier as you build it up in frequency. This can be done with a little AF amplifier continually connected to the diode detector output because the carrier has to be free of all spurious noise, squeals, frequency and power jumps, etc.

The Tripler to 73 MHz. This one proceeds in a similar fashion to the previous stage, except that now we begin to use capacitor tuning of the collector coils. The iron-core coils of the Miller 9050 series do not do a good job here, so you have to wind your own but that is very easy, as you will see.

Figure 5-64 shows the circuit and values obtained by tests here. Do not exceed the value of 50 pF for the base input capacitor. In case of any spurious noise, this is the first place to look.

When the stage is assembled and wired and under test as done with the previous stages, once again, look out for those *undesired* harmonics, especially the 64 MHz one in this case.

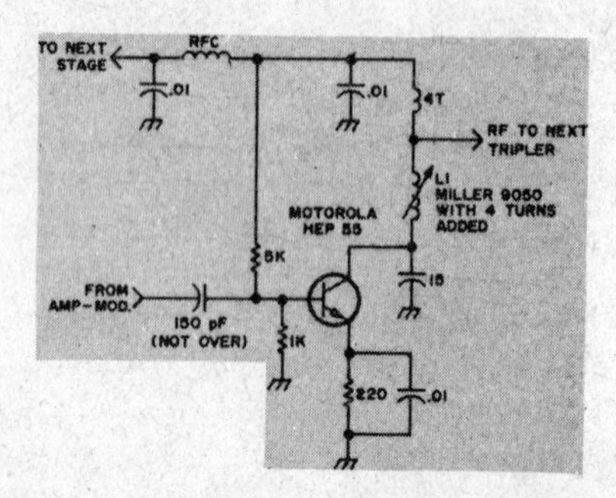

Fig. 5-64. 24 MHz tripler schematic.

The inductor may be fastened to the mounting strip with a nylon screw Fig. 5-65 for ruggedness. The variable capacitor does all right standing up on end with the fixed plates soldered to the baseboard, and the movable plates brought out through the strip with a piece of 16- or 18-gauge wire, where it is joined up with the collector, as can be plainly seen in Fig. 5-65.

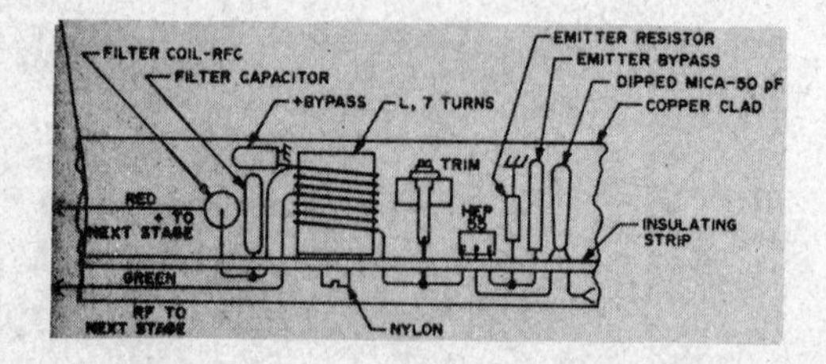

Fig. 5-65. 24 MHz tripler wiring diagram.

Clip on your diode for power checks and frequency. You can't check this latter too often. After testing for power control and noise, you are ready for the next stage, a doubler.

The Doubler to 146 Mhz. This stage uses a Motorola HEP 75 (2N3866), always a lively powerful one for VHF. The schematic, shown in Fig. 5-66, is quite similar to the others except for the different transistor and another coil tap. The tap base input capacitor worked out at 25 pF maximum, with a 39Ω emitter resistor to keep the power up for maximum input into the final stage. The collector lead is cut off and the collector connection is made by soldering a 1/8 in.-wide soft and thin copper strap, which increases the heat-sinking as well as RF conduction, directly to the HEP 75 case.

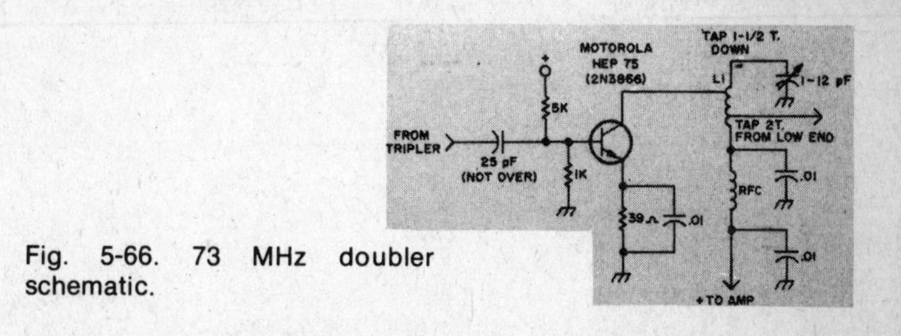

Fig. 5-66. 73 MHz doubler schematic.

Clean the case well by scraping at the place to be soldered. Use small solder, a small iron, rub the iron gently on the case two or three times for about one second only to effect a good joint for the collector strap over to the coil. The inductance is not actually critical but should be correctly tuned up and tapped for the collector as well as the output tap. After all you *do* need all the power you can get into that final.

You could use a little larger emitter resistor, for a little less current, but here again power is a point to watch, about 50 to 120 mW of RF at 146 MHz at the output tap, depending on the plus voltage. Of course, you can play around with up to 18V if you want to push out a bigger signal. Before buttoning up this stage, check it once more for frequency, please.

Power Amplifier. Refer to Fig. 5-67. Everything went along nicely with this one also, as in the previous stage. You will note a nice feature, true of most electronic circuits that are good and foolproof, that when everything is tuned up correctly and matched properly the whole stage becomes less critical all around. That is, the tuning is not touchy, the power goes up and down nicely, and even the output tap is not too critical. Note that with small emitter resistors of under 50Ω the collector current can get pretty high, so always keep at least 10Ω in series, as you test for the best emitter resistor value. Don't forget that 1W is 100 mA at 10V. And for a watt out you will need more than 100 mW even at 15V. "Big" transmitters (50 watters) use as low as 0.1Ω at times in

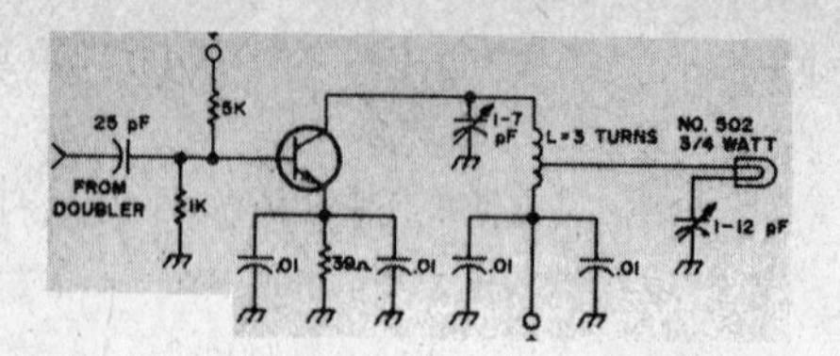

Fig. 5-67. 146 MHz amplifier schematic.

this place, and in some of the new heat-sunk jobs the emitter goes directly to ground.

Allow no currents of over 100 mA for this stage. The arrangement shown uses about 50 to 60 mA, depending on drive from the tripler, B+ voltage, and output loading.

Two bypass capacitors are used in the collector circuit. A test bulb (5V at 150 mA), in series with a 1—12 pF trimmer to ground, indicates RF output and loads the collector circuit. Without the test bulb or any antenna loading you can expect self-oscillation as the HEP 75 gives plenty of action on 146 MHz. The output tap can be led into the small but very useful 50Ω cable (RG-174/U). This cable will then go to your changeover switch or relay for the final assembly.

As a final note on the transmitter, each tuned circuit of the multipliers and finals should also be adjusted while listening to the carrier modulation.

Chapter 6
Radioteletype
Slow Scan Television and Specialized Communications

A Simple, Effective RTTY Terminal Unit

This basic unit consists of four functional blocks: the isolation amplifier, ratio detector, the voltage comparator, and the high voltage switch. The isolation amplifier and the voltage comparator are both low cost integrated circuits, whose use keeps the cost of this unit to a minimum.

The secret of this terminal unit is in the UA-710 integrated circuit. The "710" is a high gain, high speed voltage comparator. Its operation can be summed up by stating that a reference voltage is applied to one input and the signal in question is applied to the other. Any time the incoming signal is greater than the reference, the output is in one state, and when the signal drops below the reference point, the output will switch to the other state. There is a very narrow threshold which gives this unit its extreme sensitivity.

The detector portion of this TU is a ratio detector, such as is used in an FM receiver. This type of detector is very insensitive to amplitude modulation which allows the FM (FSK) to be detected when it is down in the noise.

The incoming detected signal is integrated across C5 which is divided by R7 and R8 to form the reference voltage for the comparator. Since this voltage is proportional to the detected signal, it automatically tracks and eliminates the need for a variable reference control.

T1 and T2 consist of the standard 88 mH toroid telephone chokes which are available as surplus. They have been modified into transformers by the addition of 50 to 100 turns for a secondary. The number of turns used is not at all critical, as long as the same number of turns are on each transformer.

A-1 is used as an isolation amplifier to drive T1 and T2 differentially, and is no more than a Darlington differential amplifier. Two different units were tried with good results, the Westinghouse "WC115" since it cost less than the MC1429G, but either unit can be used if attention is paid to the different pin connections.

The output keying transistor is a high voltage type, so the standard 150 volt loop supply could be used. Again, most any type of high voltage transistor can be used, but the MJE-340 which Motorola sells is recommended and can handle the power easily. The neon bulbs are a must to protect the MJE-340 from kick-back spikes.

Layout is not at all critical and modifications are easily added for tuning indicators, reversing switches and the like. The best way to put in a reversing switch is to switch the "hot" end of C1 and C2 between respective places. Wired as shown in Fig. 6-1, the mark frequency is 2125Hz, with space being 2975 Hz.

With this unit, very narrow shifts are possible. Copy of a 200 cycle audio shift when it is below the noise level and voice communication is impossible. Narrower shifts should present no problem.

Exact shift frequencies are not mandatory for copy since the ratio detector is an FM type. If the received station's tones are a bit off frequency, you probably will not notice any increase in error rate unless the received signal is very weak or the tones are considerably off. The operation of this unit can be vastly improved when utilizing narrow shift, by adding a good audio bandpass filter, designed for the shift in use, between the receiver audio output and the terminal unit input. If the signal being copied fades below copy level, or there is no signal at all, the Teletype machine will not run open; it will sit quietly until a proper signal is tuned in.

A SIMPLE SCOPE FOR RTTY MONITORING

Tuning in an RTTY signal is virtually impossible unless some form of tuning indicator is used. The most common types are the tuning eye tube, meter, and cathode ray tube, meter, and cathode ray tube. Of these, the most flexible by far is the cathode ray tube. In addition to providing tuning information it can be used to identify interference and check on proper transmitter operation.

Many amateurs shy away from building oscilloscope indicators, but they are no more difficult than any other electronic project. This little indicator is about the ultimate in simplicity. This is due to the availability on the surplus market of the 902-A cathode ray tube. It has a deflection sensitivity of 90 V/inch making it possible to obtain adequate deflection without amplifier stages. In addition, its high voltage requirements are modest, alowing the use of a small cheap power transformer (300-0-300 v) in a half wave rectifier configuration.

The scope is constructed on a 6 × 9 × 2 inch aluminum chassis. The circuit is divided into two sections (Fig. 6-2). The power and oscilloscope control components are mounted on the panel and chassis. The input transformer—tuned filter networks are constructed on a 2 × 5 inch piece of vector board which is mounted underneath at the rear of the chassis on

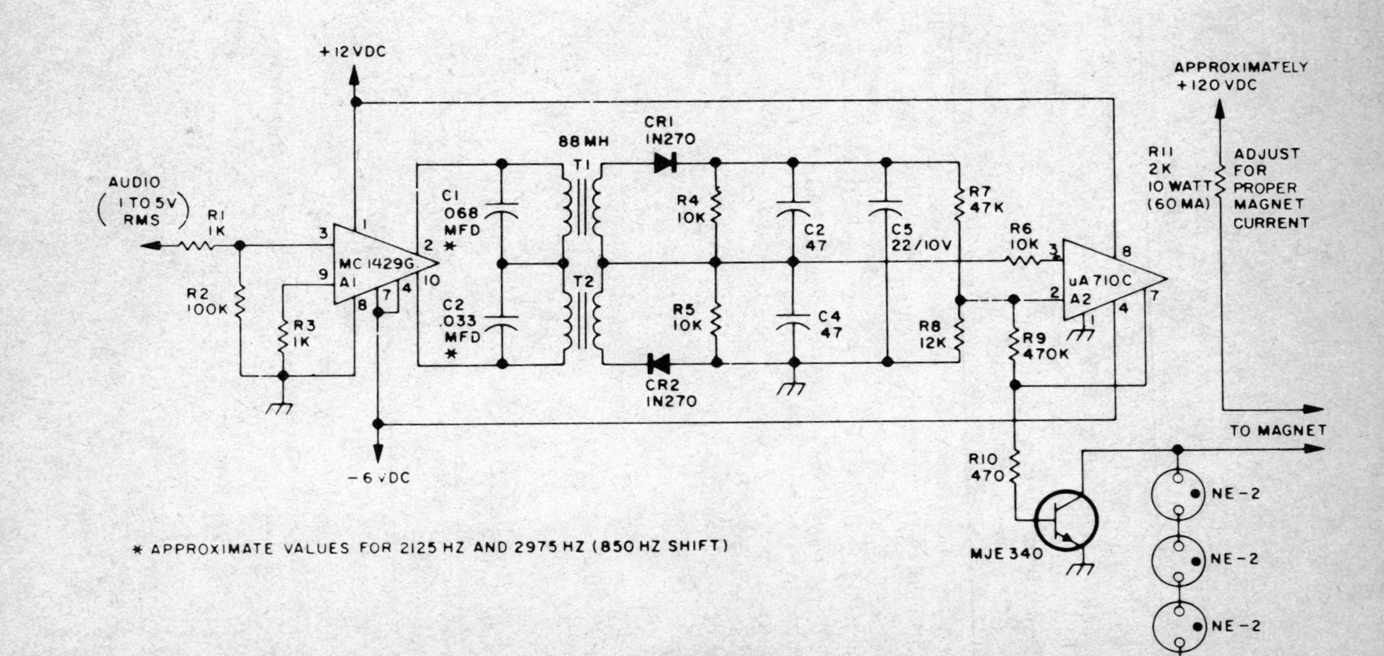

Fig. 6-1. Schematic diagram of the RTTY audio terminal unit.

stand-offs. Leads to this board should be long enough to allow it to be slid out of the chassis for tuning.

Due to the small size of the cabinet it is necessary to mount the transformer with the core parallel to the axis of the cathode ray tube. This produces an unwanted deflection. The cure is a tube shield made from a 4-inch length of 2-inch diameter galvanized water pipe. After cutting, the shield can be de-burred and painted black.

Many receiver output lines have an 8 ohm impedance whether they are from the HF communications receivers, or the VHF FM systms. A standard 88 mH toroid is used for a combination tone filter and step up transformer. To match the 8 ohm line a primary winding of 35 turns of No. 22 enameled wire is wound over the existing turns of the toroid.

After checking the wiring, turn the scope on, allowing it to warm up. Advance the brilliance control until a spot appears, then sharpen the spot with the focus control. Center the spot with the centering control. Connect an audio signal generator to the input. Remember, most audio oscillators have a 600 ohm output impedance, so a matching transformer should be used.

Now you are ready to tune the filters. There are several ways to do this. The results obtained by using a counter to check the frequency of your audio oscillator are well worth the trouble of obtaining the use of this instrument.

The easiest way to tune the filter is to use a value of 0.033 μF for the space capacitor. With your oscillator set at 2975 Hz tune the circuit to resonance by removing turns from the toroid.

About 4 turns per Hz is a good rule of thumb. Next tune the mark filter using a 0.066 mF capacitor to start with, and the same procedure. The audio oscillator should be set to 2125 Hz.

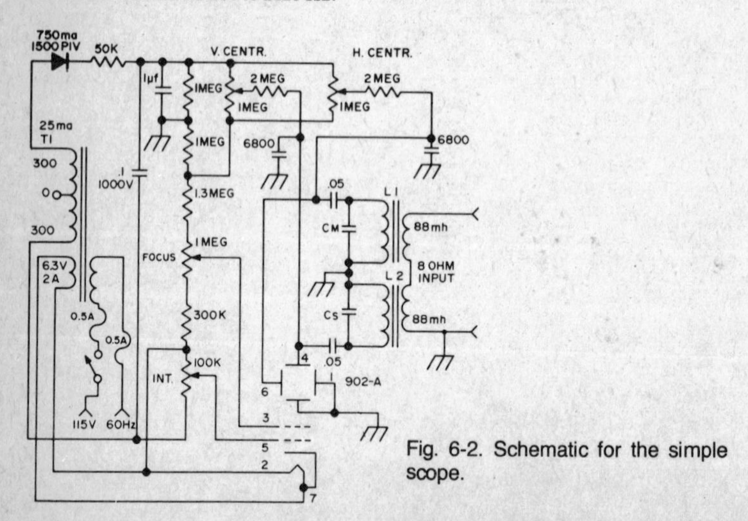

Fig. 6-2. Schematic for the simple scope.

An alternate method is to substitute capacitors until the desired frequency is reached. This is simpler than removing turns only if you have a large selection of capacitors.

Remember, in both cases, a good grade of mylar or paper capacitor should be used.

In operation the input line is bridged across the 8 ohm input to the TU. The desired RTTY signal is tuned in until a distinct cross is obtained on the CRT, then audio gain is adjusted to obtain the desired height.

UNIVERSAL AFSK GENERATOR

This section describes an AFSK tone generator which will work without modification in almost any local loop or with any terminal unit.

The two greatest shortcomings of most AFSK circuits have been cumbersome methods for adjustment of the tones to the correct frequencies and difficulty in adapting a particular circuit to an existing terminal unit. While some of the circuits based upon the Signetics 566 function generator have proved easy to adjust, the loop interface problems still exist. Many of these circuits do not produce a sinusoidal output waveform. This can mean severe adjacent channel interference if the transmitter audio bandpass allows transmission of tone harmonics. This circuit solves all of these problems and in addition is simple to construct. Circuit features include:

1. Plug-in operation in any RTTY loop independent of loop polarity or grounding.
2. Independent adjustments for each tone.
3. Constant amplitude sine wave output.
4. Excellent tone frequency stability.
5. 850 or 170 Hz shift operation with narrow shift ID.
6. Output adjustable from 20 millivolts to 2 volts peak to peak.

The RTTY Loop-AFSK Oscillator Interface

Most amateur RTTY loops are some form of Fig. 6-3. The machine keyboards and printer magnets are placed in series with the terminal unit output keying element and a high voltage supply. The current limiting resistor R1 sets the loop current to 20 or 60 mA as required by the machines. Normally the power supply voltage is 100V or so and most of the voltage drop occurs across the resistor. The drop across the printer magnets is only a few volts and the closed keyboard contacts have no voltage at all.

This arrangement is used for two reasons. First, the high voltage source decreases the time required for the loop current to build up in the printer magnets for each code pulse. This reduces the error rate of the printer. The second reason is operating convenience. Additional machines such as other printers, reperforators, or a tape distributor may be plugged into the loop

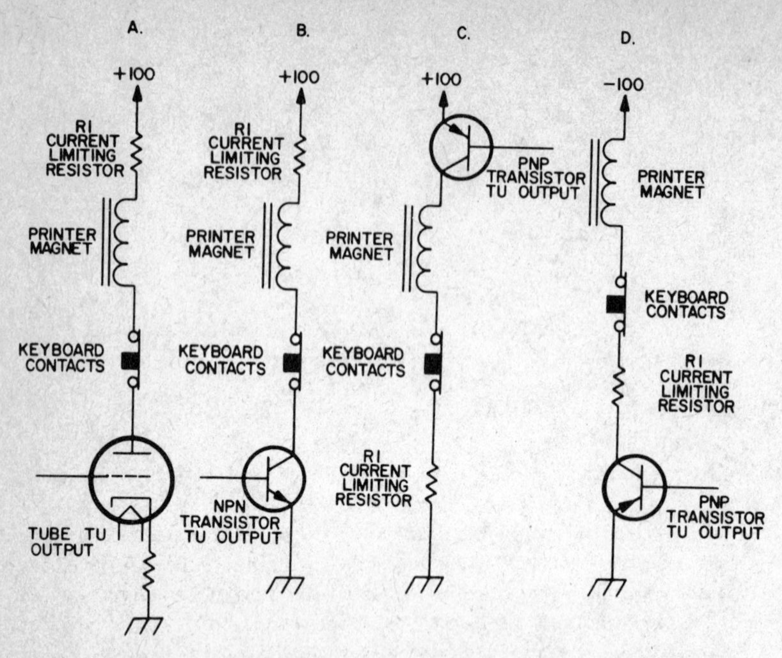

Fig. 6-3. Typical RTTY loop circuit: A—TU with tube output and positive loop supply. B—TU with NPN transistor output and positive loop supply. C—TU with PNP transistor output and positive loop supply. D—TU with PNP transistor output and negative loop supply.

without seriously affecting the loop current adjustment. The drop across any additional machine is small compared to the drop across R1 and therefore the loop current change is negligible.

The high voltage loop is all well and good for the machines themselves, but can cause real difficulty when an AFSK oscillator containing tender semiconductors must somehow be connected into the system. Further, each fellow's station is different. Depending on the particular operator, there may be one or more local loops and several terminal units. It is not uncommon to find that the serious RTTY operator has a Mark IV, a Mainline, an ST-6, plus some old military gadget such as a CV-57...all operational and powering one or more machine loops. Because of the differences in the various pieces of equipment, the AFSK oscillator design must contend with systems in which the loop voltage may be of either polarity. The machine end of the loop may or may not be ground referenced, and the loop supply may be anything from 24 to 200V. Thus, a special interface circuit is required to allow operation of the typical AFSK oscillator with each terminal unit.

This current, like any other piece of RTTY equipment, just plugs into the loop and works. In addition, it has good frequency stability and a constant amplitude sine wave output.

Two new integrated circuits, the IC optical coupler and a function generator with sine wave output, are used for this design.

The Optical Coupler

Optical couplers are a new type of IC in which a light-emitting diode and phototransistor are integrated in a DIP package. Figure 6-4 shows the internal circuit of a typical optical coupler IC. The LED and the phototransistor are electrically isolated, but placed such that light from the diode is focused on the phototransistor. Current through the diode causes it to emit light. This light causes the phototransistor to draw collector current. One specification for optical couplers is the current transfer ratio. A typical unit will have a current transfer ratio of 60%. For every 10 mA of current through the LED, 6 mA will flow from the emitter to collector of the phototransistor. Couplers are also specified for the maximum voltage allowed between the LED and phototransistor; 100V is not unusual. The optical coupler is the ideal device for coupling current pulses from the RTTY loop to the AFSK oscillator without any electrical connection between the two. All voltage, polarity, and grounding problems are eliminated with this simple method.

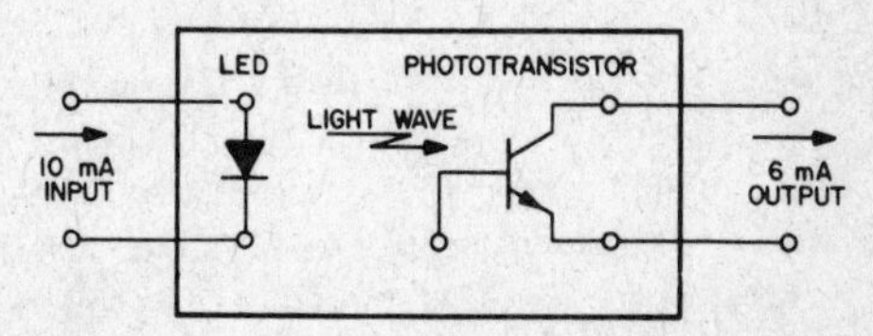

Fig. 6-4. Internal circuit of optical coupler IC. The input and output devices are completely isolated from each other. 10 mA through the light emitting diode will produce a current flow of about 6 mA from collector to emitter of the phototransistor.

The Intersil 8038 Function Generator

The tone generator portion of the AFSK oscillator uses the 8038 function generator IC from the Intersil Corporation.

Operation of this circuit is similar to the Signetics 566 function generator which has been used in previous amateur AFSK circuits. There is one important difference, in addition to the triangle and square wave outputs obtainable from other circuits, the 8038 also produces a *sine* wave output.

The basic RC oscillator portion of the chip generates a triangular wave form. This is transformed into a sine wave by means of a triangle-to-sine converter integrated onto the same chip. Special trimming adjustments are provided so that the distortion may be reduced by optimization of the converter circuit. Harmonic distortion is about 5% without trimming. Careful adjustment of all of the optional controls will reduce this to less than 1%.

If all of the trimming pots are not used, only three external components are required to make a basic voltage controlled oscillator. The VCO control range is such that a frequency shift of up to 100: 1 may be obtained by changing the control voltage input. The output amplitude remains constant as the frequency is shifted.

The chip is available in six versions which are graded on the basis of frequency stability and operating temperature range. The 8038BC used here has a maximum drift of 100 ppm/°C and is usable from 0 to 70° C. This version costs about $8.40 in single quantities.

AFSK Generator Circuit

The AFSK generator circuit is shown in Fig. 6-5. It is connected into the RTTY loop via a phone jack. In order to make the circuit truly independent of loop polarity, the current first flows through a bridge rectifier consisting of CR1-CR4 and then through 6V zener diode CR5. The bridge supplies the zener with current of the proper polarity such that a constant voltage drop occurs across the zener which is independent of both loop current and loop polarity. A 3/4W zener is used so that the circuit will operate with loop currents from 10 to 100 mA without damage. The bridge-zener diode combination then drives the optical coupler from an essentially constant voltage source.

The optical coupler is a Motorola MOC 1003 and is operated at an input current to the LED of 10 mA. This is set by R5. The phototransistor or output side of the coupler is grounded to the AFSK generator while the entire input circuit consisting of the bridge, zener and the LED are floating. The voltage difference between the loop and the AFSK circuit can be up to 500V before the coupler would be damaged.

The phototransistor drives the tone generator circuit via Q1 which operates as logic inverter to cause the frequency shift to be in the correct direction. Normally, the MARK (or closed loop) tone is 2125 Hz and the SPACE (or open loop) tone is either 2295 or 2975 Hz. The frequency shift is upward when the loop current is broken or keyed by the keyboard pulses. An upward frequency deviation is obtained from the VCO chip by decreasing the control voltage input at pin 8. This is done in the following manner: When the machine is at rest, loop current flows and the LED in the coupler is driven on. The light from the LED causes the phototransistor to be driven into saturation such that the collector-to-emitter voltage drop is about 0.5V. Q1 is turned off because its only source of base drive is via the 33K resistor, R6, and about 1.4V are required to overcome the forward drop of CR6 plus the turn-on threshold of Q1 itself. Because Q1 is off, the control voltage input to the tone generator IC is

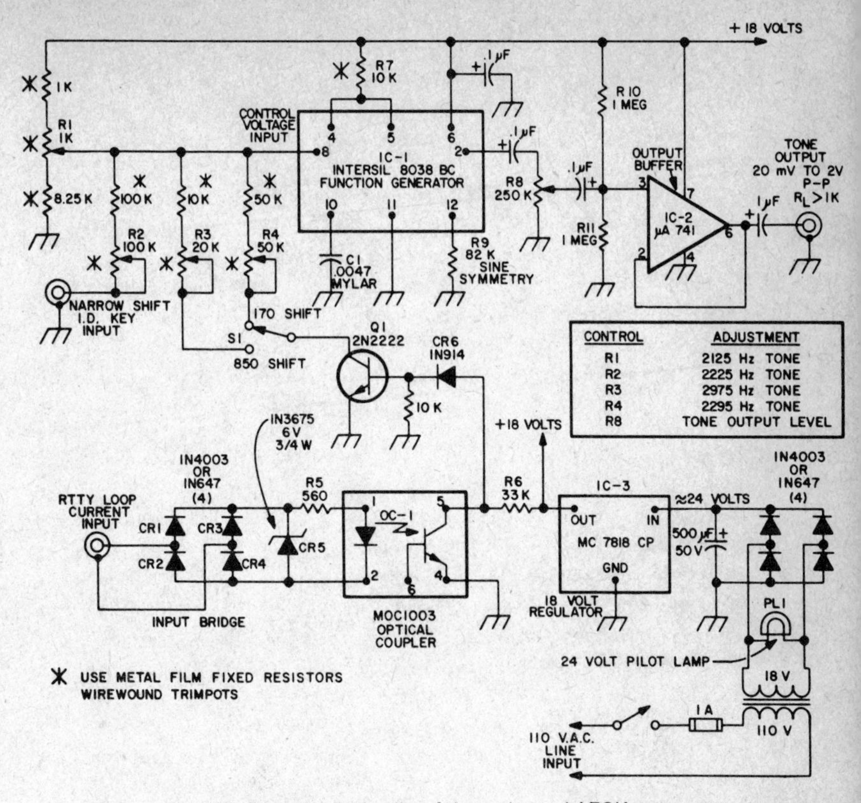

Fig. 6-5. Circuit schematic of the universal AFSK generator.

determined only by the voltage divider of which resistor R1 is a part. R1 sets the MARK frequency to 2125 Hz.

Each machine code pulse reduces the loop current to zero. The LED no longer has a source of current so the phototransistor in the coupler no longer conducts. The collector voltage rises until base current flows into Q1 causing it to saturate hard; the collector-to-emitter drop is less than 0.3V. The collector current for Q1 is obtained from the frequency setting pot R1 via resistors R3 or R4. This pulls down the voltage at pin 8 of the function generator IC and causes the frequency to shift upward. Switch S1 selects either R3 and its associated fixed resistor for 850 Hz frequency shift or R4 for 170 Hz shift.

Adjustment of the 100 cycle shift for CW ID is provided by R2, which is switched to ground by the key jack.

The basic frequency range of the function generator IC is set by R7 and C1. Other values may be used. The sine wave output terminal of the chip is pin 2.

The triangle is present across C1 at pin 10. The sine wave output is AC coupled to the output level adjustment pot, R8, and then to the output buffer amplifier which is a 741 operational amplifier. The 82K resistor, R9, is used to trim the sine converter in the chip for lowest distortion and is a nominal value.

The level set pot, R8, is 250K because the sine converter should not be operated into less than 100K if lowest distortion is to be obtained.

The output buffer, IC2, is operated from a single polarity supply so the inputs must be biased between the supply voltages. R10 and R11 accomplish this for the noninverting input at pin 3 while the inverting input, pin 2, is biased with feedback from the output at pin 6. This is a unity gain buffer connection which enables the tone generator to drive loads of less than 1K. The output voltage may be adjusted from 20 mV peak to peak to 2V peak to peak. Inclusion of the buffer is a bit of a luxury, but it does assure that the unit may be used with just about any transmitter and the low impedance output is nice if a long cable must be driven.

Power Supply

The power supply section of the AFSK generator is conventional with the exception of the voltage regulator. The transformer and bridge provide about 24V unregulated across the filter capacitor. The regulator is a new single voltage regulator IC from Motorola. This chip, type MC7818CP, is a fixed 18V regulator capable of up to one ampere of output current. It is one of a series of low cost regulators available in standard voltages such as 5, 12, 15, etc. Only three connections are required: input, ground and output. A bypass at the input is a good idea if the lead from the rectifier filter capacitor is long. While the one amp capability of the supply is much more than this circuit required, the ease of application and good performance of the regulator more than justified its use.

Construction

The circuit is packaged in a small sheet metal case and has its own AC power supply, but it may be built into an existing system from which power may be obtained. Anything from +15 to +20V regulated will work.

The majority of the circuitry is constructed on a small PC card sub-chassis. The most frequently used inputs and controls are located on the front panel and include the loop input jack, the shift selector switch, the AC power switch, and a pilot light. Rear panel controls include the narrow shift ID keying jack, the tone output jack, and the tone output level adjustment pot. Be sure to isolate the loop input jack from ground with insulated washers! The power supply is located on the main chassis.

Care should be taken in installation of the regulator IC in order to assure that it is not shorted to the chassis by the mounting screw or a sheet metal burr.

Starting from the bottom of the card, components are arranged in the following order: First is the input bridge rectifier and the zener diode. Just above is a DIP socket containing the MOC 1003 optical coupler which is in a 6 pin package. The pins are counted around the package starting with pin 1 at the index dot and are sequential as in the case of a standard 14 pin package. That is, they go 1 to 3 down one side, then jump across to pin 4 through 6 going back up the other side. Q1 is located just above the coupler socket while the frequency adjustment pots are just to the left of Q1.

All of the resistors associated with the frequency adjustments are stable metal film types such as the MIL RN60 series. Conventional carbons may be used but the stability just won't be as good.

Surplus wirewound trimmer pots are used for the frequency adjustments. The 20 turn resolution is a real help in setting up the circuit, but again, single turn pots will work if you have a steady hand!

The function generator IC is located above Q1. A 14 pin socket is required for this circuit also. The timing components are located around the function generator socket. I bring the pins on the IC socket through the board via wires to push-in type feedthrough standoff terminals. The components are mounted on top of the card and are easily accessible In addition, the larger terminals are easier to connect to than the tiny IC socket pins.

The 741 output buffer amplifier is located near the top edge of the card as shown in the picture. All input, output, and power connections are made to the feedthrough standoffs on the underside of the card. The card itself is mounted above the main chassis on half inch spacers.

Circuit Alignment

After assembly, check the wiring, particularly the connections to the two ICs. More chips are ruined from counting the pins from the wrong side than probably from any other cause.

Remove the 8038 function generator from its socket. Apply power and verify that the IC regulator output is 18V. Then check to see that this voltage is present at pin 6 of the 8038 function generator socket. Also check that R1 adjusts the voltage at pin 8. This voltage should never be less than three-fourths of the supply voltage. Remove power and install the 8038 chip in its socket. We are now ready to set up the tone frequencies. Provide a 20 to 60 mA constant current source such as a RTTY loop. Connect this to the loop input jack. Do not attempt to drive the loop from a constant voltage source such as a lab supply without providing a current-limiting resistor. A small increase in input voltage will cause a large current to flow through the bridge and input zener CR5, since the circuit contains no internal current-limiting resistor.

Set the output pot, R8, to maximum and connect a counter, TU, or other frequency measurement device to the tone generator output.

With loop current applied to the input jack, set the MARK frequency adjustment, R1, so that a 2125 Hz output is obtained.

Set S1 for 850 Hz shift and remove the loop current. Now set R3 for a SPACE frequency output of 2975 Hz. Leave the loop current disconnected.

Set S1 for narrow shift. Adjust R4 to obtain a 170 Hz shift or an output frequency of 2295 Hz. The narrow shift ID setting is all that remains. Reconnect the loop current and verify that the MARK frequency is still 2125 Hz. Short the ID keying jack and adjust R2 to obtain a 2275 Hz output. This completes alignment of the AFSK generator.

Optional Adjustments

The circuit performs quite well as described. The experimentally inclined can add optional adjustments to improve the output distortion so that the sinewave is nearly perfect. Reduction of the second and third harmonic to as much as 50 dB below the fundamental has been obtained through use of all of the trimming adjustments provided by the chip design.

Three pots must be added. They are connected as shown in Fig. 6-6. All other connections to the chip remain the same. The 1000Ω pot, R12, is used to adjust the time symmetry or duty cycle of the triangle wave generator by controlling the charge and discharge currents of C1. The pot should be set for a 50% duty cycle using a scope. If a wave analyzer is available, it may be set to null out the second harmonic. This null is quite sharp.

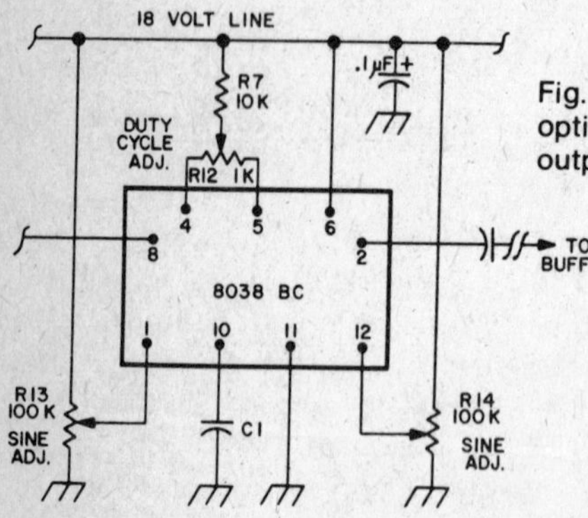

Fig. 6-6. Function generator with optional controls to reduce sine output distortion.

R13 and R14 are used to adjust the amplitude symmetry of the sine converter. One pot adjusts each half cycle. Start with each pot at center position and adjust carefully for best wave form symmetry or minimum third harmonic distortion. This circuit is recommended by the IC manufacturers, but should be used with care so that the pot wiper does not go completely to the end. It might be possible to damage the sine converter by sinking or sourcing too much current to the chip from the supply or ground. Used with care it works nicely.

Performance

The AFSK generator was subjected to several tests to verify performance. The frequency stability is quite good. The unit is about 5 Hz low when first turned on from cold, stabilizes within 1 Hz after a few minutes warm up in a normal room environment. The long-term stability is a function of the quality of the resistors and capacitors used with the function generator IC. I used MIL RN60D resistors with a mylar film capacitor for C1. Long-term drift after some six weeks is still within a couple of hertz after warm up as measured with a frequency counter.

The output wave form is a clean sine wave and no clicks or transients are present when the circuit shifts from MARK to SPACE. The scope camera photograph shows the input switching waveform on the top trace and the resulting frequency shift on the bottom trace. This particular picture shows the shift from 2975 Hz back to 2125 Hz. There are no transients present and the sine wave amplitude does not change at all.

Also measured was the harmonic distortion present in the output sine wave. Without adjustment of the 82K resistor, it is about 5%. Careful trimming will reduce it to below 3% which is good enough for most amateur RTTY applications.

The optically coupled input circuit works quite well. The shift from MARK to SPACE occurs as the loop current drops below 5 mA which leaves plenty of margin for these with 20 mA loop systems. The unit is truly universal. It does work with any local loop or TU, provided there is sufficient loop current

A DIGITAL TAPE DISTRIBUTOR FOR RTTY

This tape distributor has no motor or brushes and is about half the size of the conventional units. It is adaptable to other speeds. There are four integrated circuits and two transistors used so it does not cost much to build (Fig. 6-7).

A "clock" oscillator operates in bursts of 45.5 cps rate, toggling a binary counter connected to a decoder which gives accurately timed pulses 22 milliseconds long, each one the length of one RTTY "bit." The sensing contacts in the reader are either closed or open depending on the punched holes on the tape, so, if the outputs of the decoder are connected to the sensing contacts, a serial output is obtained. Diodes are added to form an "AND" gate to prevent outputs from the decoder from shorting together. The serial output then is connected to a keying transistor which may be used to operate a keying relay or key direct.

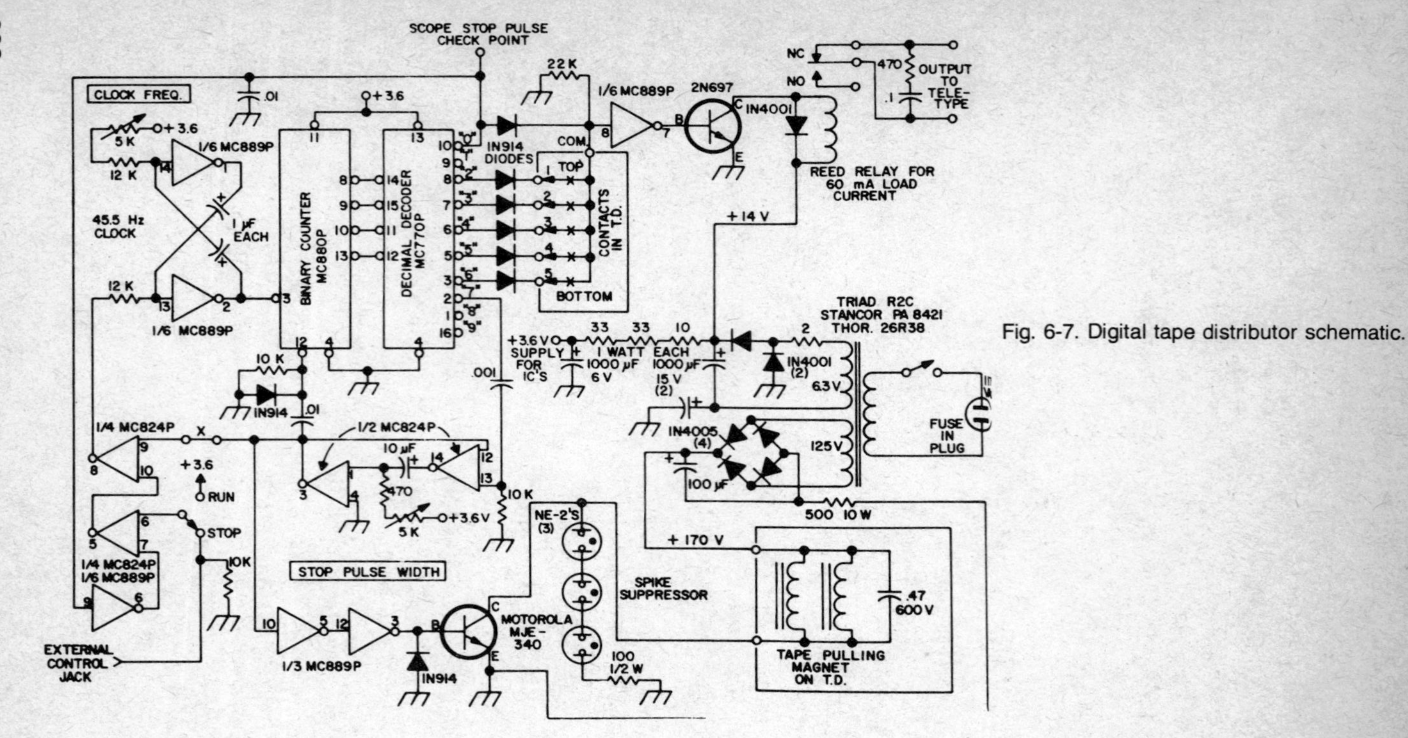

Fig. 6-7. Digital tape distributor schematic.

When all five bits are formed, the sequence reaches pin 7 of the decoder. The positive going edge of the pulse trips a monostable multivibrator which resets the counter to zero, shuts off the oscillator, and turns on the tape advance transistor, advancing the tape one notch. When the monostable pulse ends, the output of the NOR gate immediately goes positive, starting oscillation for seven more cycles. The adjustment of the pulse length of the monostable determines the length of time the stop pulse dwells at pin 10 of the decimal decoder. For 60 wpm, this would be 31 milliseconds. Later an easy method will be given to permit setting pulse widths without a scope. Note that the monostable pulse width and the actual stop pulse width differ slightly, so if you are checking pulses, the correct place to check is at the zero (pin 10) output of the decoder.

Gates are wired to the stop-run switch so that the device will complete the character in the gate of the TD before stopping. Normally the external control jack is unused, a positive voltage of from 1.5 to 3.0 volts on this jack will start the TD (Fig. 6-8).

The tape recorder has a large tape advance solenoid. Accordingly, it is pulsed with a high current for reliable operation. As mentioned above, the monostable pulse is actually about 20 milliseconds when the stop pulse is set at 31 milliseconds. The current through the magnet measures 200 mA peak but the duty cycle is so short that the average current is only about 25 mA. This permitted the use of a very small power transformer in the prototype, but other alternate members for substitutes are given. The solenoid advance transistor should not get hot in normal operation if you use the suggested types MJE340 or HEP244. An MJE340 could also be used instead of the 2N697 for direct loop keying.

NE-2 spike suppressors have been used across the tape advance transistor to hold the transients down below the rating of the transistor. Other methods have been tried including the use of zener diodes and ordinary OB2 regulator tubes. The NE-2s do the job, are small, and inexpensive. The tape advance solenoid would not operate at all when a silicon diode was connected across the coil as is usually done with relays to suppress transients.

A very simple power supply is used and was found adequate in the two models constructed (Fig. 6-9). It would be advisable to check the power supply voltage to the ICs and adjust the value of the series resistors until it is approximately 3.6V.

When the tape recorder is received, it is necessary to strip off all the parts on the side near the Blue Ribbon connector with the exception of the pulling magnet detent. The large Advance solenoid and the linkage to the gate is removed and the gate is secured with a collar-setscrew so it cannot slide back and forth. The parts may then be mounted in the clear space. While it is possible to get all the electronics in the TD, it takes a little planning. You might want to mount the TD on a small chassis and use this to mount the electronics,

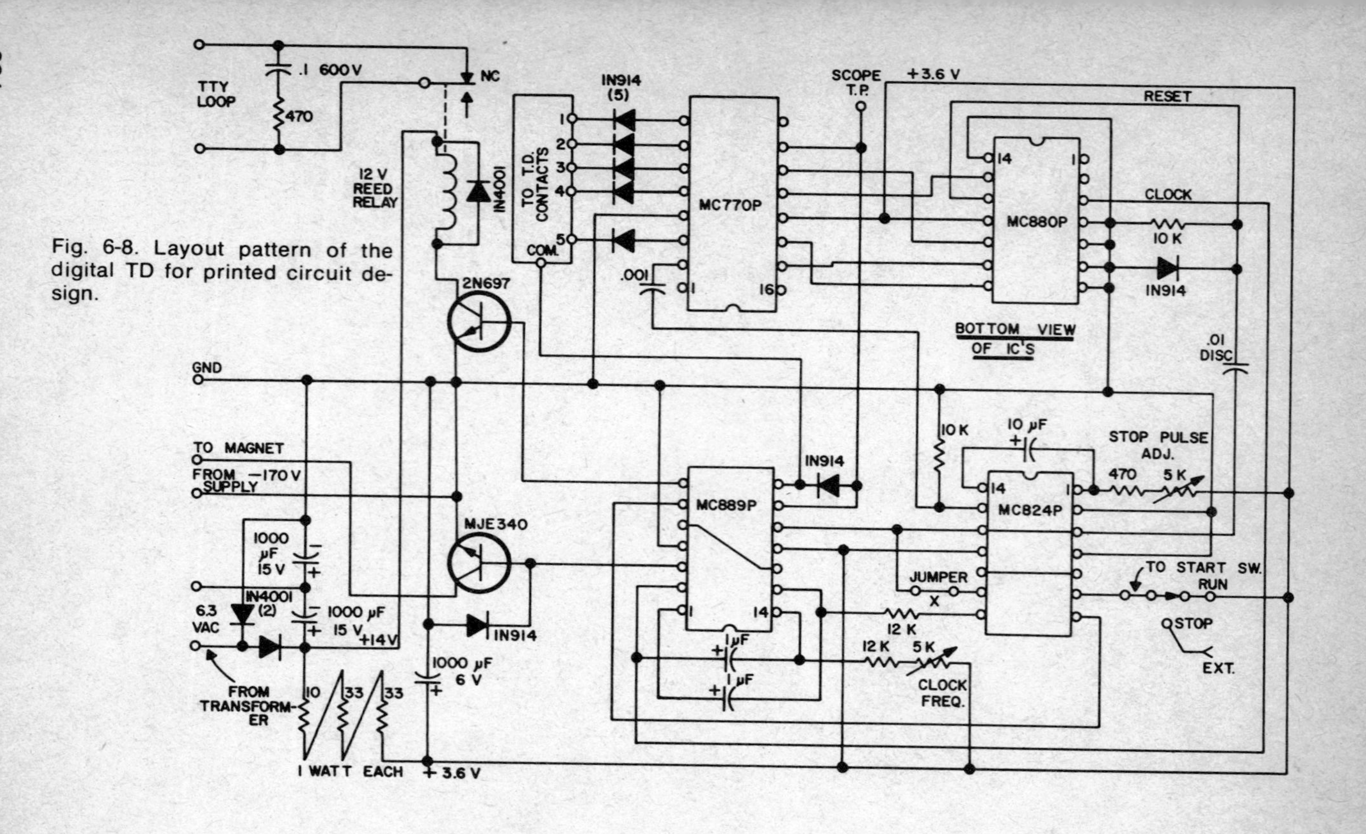

Fig. 6-8. Layout pattern of the digital TD for printed circuit design.

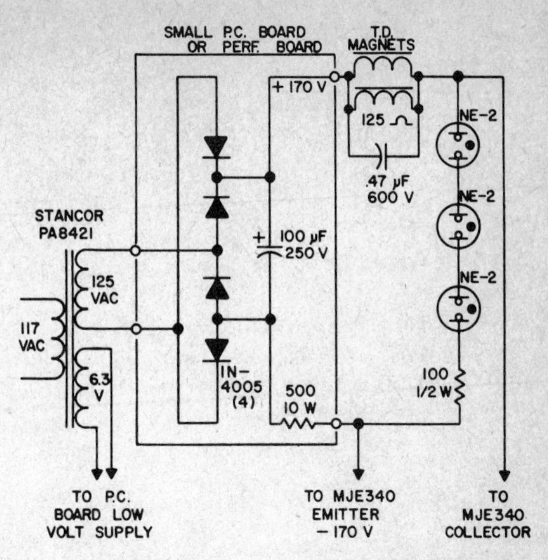

Fig. 6-9. Digital TD magnet supply board.

with the magnet supply in the head. An etched circuit board about 3″ × 6″ will hold all the integrated circuits and there is not too much wiring to do even if you want to hand wire the unit with small wire.

Two adjustments have to be made when the distributor is finished. The clock has to be set at 45.5 cps and the stop pulse width has to be adjusted. It should be pointed out that, due to the wide tolerances in electrolytic capacitors, it is possible that the clock will not adjust to 45.5 cps within the range of the pot. If this happens, merely change the value of the 12K resistor in series with the pot.

First, open the control line to the MC824 gate at the point marked "X" on the schematic. Then run a tape and adjust the stop pulse width trimmer so that the solenoid has a snappy action and pulls tape satisfactorily, but don't worry about how it is printing. Now, adjust the clock so a measured 39″ of tape goes through in one minute corresponding to 389 operations per minute. Restore the jumper at "X" and measure off 37″ of tape and adjust the stop pulse width until this length runs through in one minute. The TD is now set at machine speed, 60 wpm. If the teletype printer is misprinting, check for a 22 millisecond pulse at each of the tape sensing fingers. If OK, check for good contact at the fingers—sometimes they need cleaning. The TD can be checked with an ordinary oscilloscope if this is necessary.

It should be recognized by the reader that as a solid state device, this unit might pick up RF from the transmitter and malfunction. The cure is shielding and bypassing external lines. Shielding was all that was necessary in the prototypes.

RTTY TONE GENERATOR

The RTTY Tone Generator is a unique piece of RTTY test equipment. It is capable of simulating a variety of Teletype operating and QRM modes, since it can deliver all possible combinations of the 2125 Hz mark and 2975 Hz space audio tones. When not serving in the test equipment capacity, it functions as a conventional AFSK oscillator.

It may be keyed externally by a contact closure such as from a machine keyboard, or by a 3.5 volt peak-to-peak square wave signal. An internal keying generator provides a zero-bias 22 Hz keying signal. The internal keying signal allows the unit to be used for such things as adjusting TU polar relays and is also a great convenience when chasing signals thru the local system with an oscilloscope.

The unit is designed around transistors and low-cost integrated circuits. It requires 12 VDC (± 10%) at about 450 mA and delivers a maximum audio output in excess of minus 10 dBm into 600 ohms. Front-panel controls adjust the mark/space amplitude ratio and the output level, and select the desired operating mode.

System Operation

Figure 6-10 is a block diagram of the RTTY tone generator. The mark and space oscillator outputs are fed thru separate level controls to independent gated amplifiers. The gated amplifiers stop or amplify their respective signal inputs depending on the information they receive from the keying logic and keying selector circuits. The audio signals from the two gated amplifiers are combined at the output level adjustment potentiometer. The signal from the potentiometer wiper is connected to the mark/space output jack, J3. The particular combination of the mark and space audio tones present at the output jack depends on the settings of S3 and S4. The following possibilities exist:

a) Mark signal off, space signal off.
b) Mark signal on, space signal on.
c) Mark signal on, space signal off.
d) Mark signal off, space signal on.
e) Mark signal keyed on and off, space signal off.
f) Mark signal keyed on and off, space signal on.
g) Mark signal keyed on and off, space signal on.
h) Mark signal on, space signal keyed on and off.
i) Mark signal keyed on and off, space signal keyed on and off—mark signal present when space signal is absent and vice versa.
j) Mark signal keyed on and off, space signal keyed on and off—mark and space signal present and absent simultaneously.

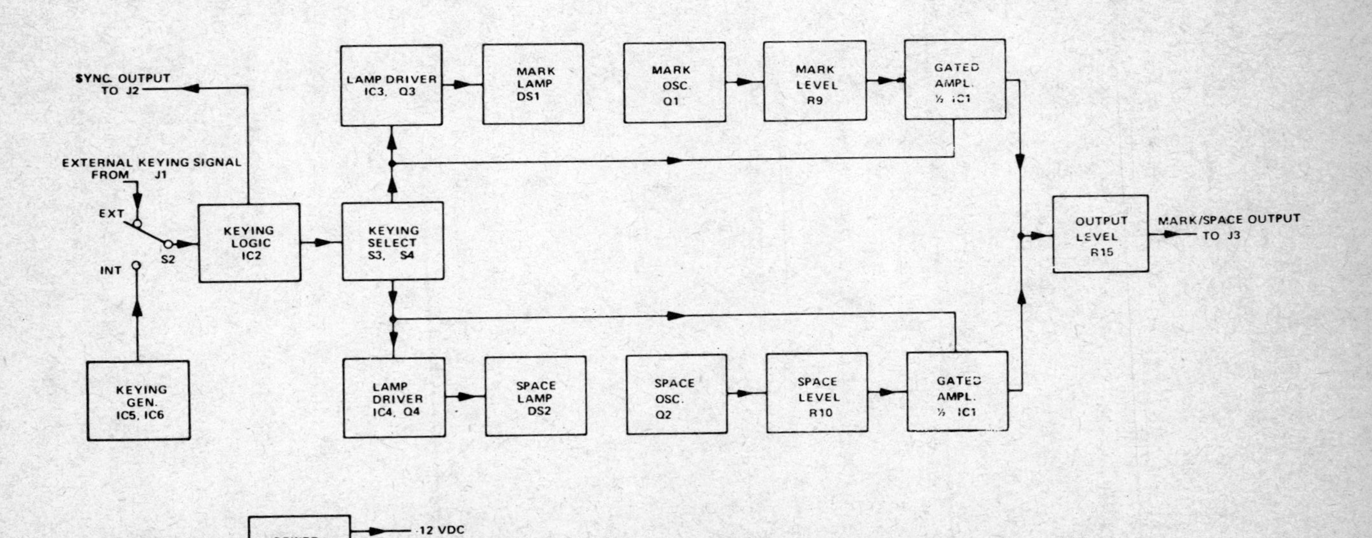

Fig. 6-10. RTTY tone generator block diagram.

The lamps (DS1 and DS2) indicate the state of the two gated amplifiers. They are extinguished when the amplifiers are in the "stop" mode, and illuminated when the amplifiers are in the "amplify" mode. The keying selector switch (S2) connects the keying logic input to either the internal keying generator or the external keying input jack. The power distribution circuits provide minus 12 VDC, minus 3.6 VDC regulated, and minus 8.2 VDC regulated to the various circuits.

Circuit Description

The majority of the circuitry is assembled on three 3″ × 4″ homemade component boards. The mark and space oscillators are contained on one of the boards (CB1), the gated amplifiers and lamp drivers on another (CB2), and the power distribution and keying circuits on the third (CB3) (Fig. 6-11).

The mark and space oscillators are conventional LC oscillators designed so that their operating frequencies are essentially independent of transistor case temperature. The two inductors (L1 and L2) are the usual 88 mH loading coils. Capacitors C1 thru C6 are mylar dielectric units having a ± 10% capacitance tolerance. Do not use ceramic capacitors in any of the frequency determining networks—ceramics are both temperature and voltage sensitive. Turns are removed from L1 and L2 to adjust the oscillator frequencies to 2125 Hz and 2975 Hz, respectively. This will be discussed later in the final adjustments. The capacitor values specified should resonate with any of the available "88 mH" inductors.

The mark and space gated amplifier and lamp driver channels are identical. The mark signal from R9 is fed to one input of a two-input NAND gate. IC1 is a *dual* two-input NAND gate. Potentiometer R11 biases this input (pin 5 of IC1) so that the gate can operate as a linear amplifier. The amplified mark signal appears at pin 6 of the gate as long as pin 3 is at or near minus 3.6 VDC. When pin 3 of the gate is at or near ground, pin 6 of the gate is essentially connected to pin 4 of the gate and the mark signal output at pin 6 disappears. The two-input logic gate thus provides a convenient means for switching (or gating) and amplifying the mark signal. The space channel functions in exactly the same manner, utilizing the two-input gate associated with pins 1, 2, and 7 of IC1.

When terminal 6 of CB2 is at or near ground potential, the mark signal output from IC1 is absent. Pin 3 of IC3 will also be at or near ground potential and the output of IC3 (pin 5) will be at or near minus 3.6 VDC. This voltage drives Q3 into conduction, extinguishing DS1. The whole procedure is reversed when terminal 6 of CB2 is at or near minus 3.6 VDC; the mark signal from IC1 may be present (depending on the setting of R9), the output of IC3 will be at or near ground potential, Q3 will not conduct, and DS1 will illuminate. The space lamp driver channel (IC4, Q4, and DS2) functions in exactly the same manner, responding to keying signals at terminal 5 of CB2. IC3 and IC4 are buffer amplifiers. Their sole function is to isolate the lamp driver circuits from the keying signals and assure complete switching of Q3 and Q4.

IC2 conditions the keying signal applied to terminal 14 of CB3. When the keying signal is at or near ground potential, terminal 17 of CB3 is at or near minus 3.6 VDC and terminal 16 is at or near ground potential. Conversely, when the keying input at terminal 14 is at or near minus 3.6 VDC (or open circuited), terminal 17 is at or near ground potential and terminal 16 is at or near minus 3.6 VDC. IC2 is a dual two-input gate operating as two inverters. Only one inverter is required to form the complementary outputs at terminals 16 and 17, but two are used to provide complete standardization of the keying signal.

One of the keying outputs is connected to terminal 15 and brought out to the front panel for use as an oscilloscope synchronizing signal. The keying mode switches (S3 and S4) select either of the two keying outputs, minus 3.6 VDC, or ground, and route the selected levels to the keying inputs of CB2. Figure 6-12 shows the presence or absence of the mark and space outputs with all possible keying/S3 and S4 combinations.

A third dual two-input gate (IC5) is connected as an astable multivibrator that forms the time-base for the internal keying signal. One of the primary requirements of the internal keying signal is that both halves of the cycle be of exactly the same time duration. *If* both sections of the gate were identical, *if* C11 and C12 were identical, *if* R18 and R19 were identical, and *if* the multivibrator were not connected to an external load, its output would be time-symmetrical and therefore suitable as the zero-bias internal keying signal. None of these "if" conditions are readily met in practice. There is however, a simple solution to the problem, as we shall see.

IC6 is a JK flip-flop. Connected as shown, its output (pin 7 in this case) changes state each time the input (pin 2) switches from 0 *to* minus 3.6 VDC. When the input switches back to 0 *from* minus 3.6 VDC, the flip-flop does *not* change state. Bear in mind that 0 and minus 3.6 vdc are only nominal values and that the flip-flop senses only HI (positive) to LO (less positive) transitions of the input signal. In each complete cycle of the multivibrator output, there is only one HI to LO transition. When the multivibrator output is connected to the flip-flop input, the flip-flop output changes state once for every complete cycle of the multivibrator output. The time duration of each half cycle of the flip-flop output is equal to the time duration of one complete cycle of the multivibrator output. If the flip-flop output is used as the internal keying signal, then the internal keying signal is time-symmetrical regardless of how unsymmetrical the multivibrator output is. The multivibrator must operate at 44 Hz to provide 22 Hz at the flip-flop output.

The internal 22 Hz zero-bias keying signal from the output of IC6 is connected to switch S2. S2 selects either the internal or external keying signal

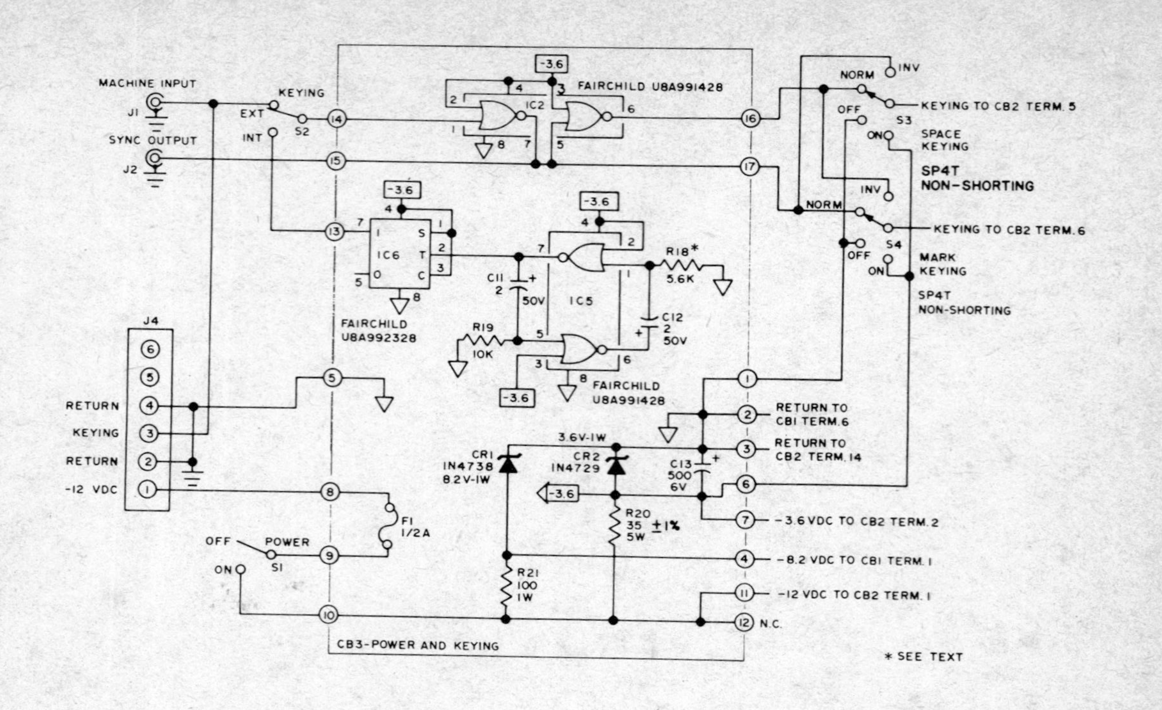
MACHINE INPUT
J1
SYNC OUTPUT
J2
KEYING
EXT
INT
S2
FAIRCHILD U8A991428
IC2
-3.6
IC6
S
T
C
FAIRCHILD
U8A992328
IC5
R18*
5.6K
C11
2
50V
C12
2
50V
R19
10K
FAIRCHILD
U8A991428
J4
RETURN
KEYING
RETURN
-12 VDC
CR1
1N4738
8.2V-1W
CR2
1N4729
3.6V-1W
C13
500
6V
R20
35 ±1%
5W
R21
100
1W
F1
1/2A
OFF
POWER
ON
S1
CB3-POWER AND KEYING
NORM
INV
OFF
ON
S3
S4
KEYING TO CB2 TERM. 5
SPACE KEYING
SP4T NON-SHORTING
KEYING TO CB2 TERM. 6
MARK KEYING
SP4T NON-SHORTING
RETURN TO CB1 TERM. 6
RETURN TO CB2 TERM. 14
-3.6 VDC TO CB2 TERM. 2
-8.2 VDC TO CB1 TERM. 1
-12 VDC TO CB2 TERM. 1
N.C.
* SEE TEXT

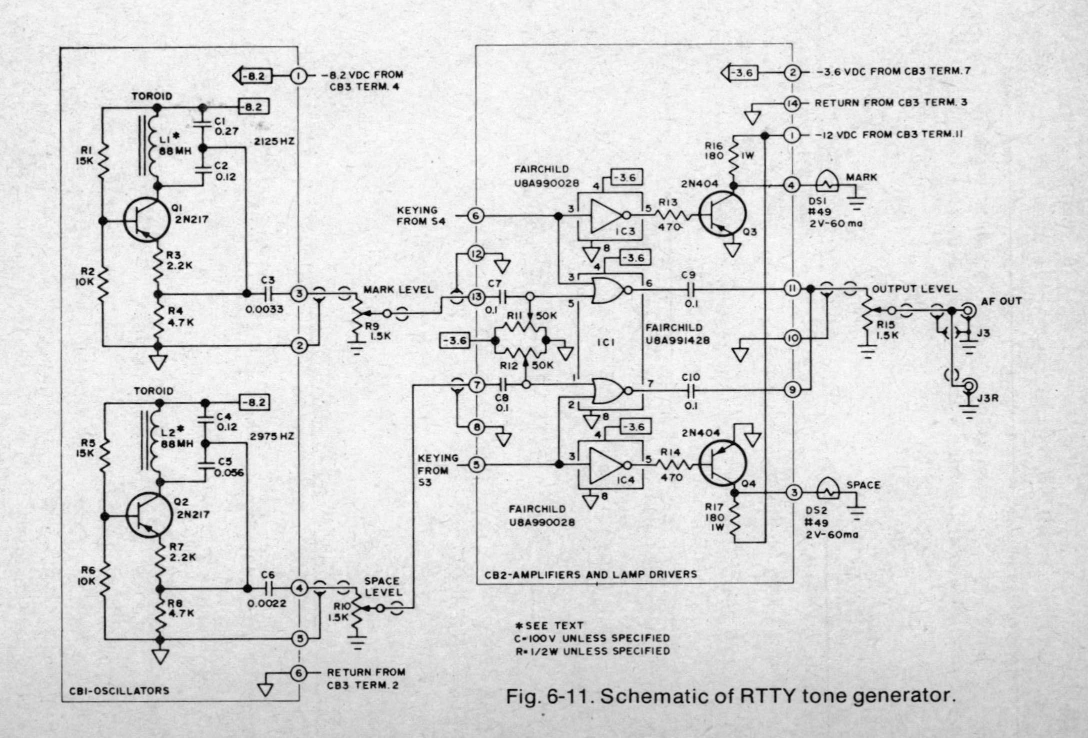

Fig. 6-11. Schematic of RTTY tone generator.

and applies it to the keying logic input. Referring to Fig. 6-12 the internal keying signal has the same effect as alternately opening and grounding the keying input. When the internal keying signal is at or near zero volts, the "keying input grounded" columns apply. When the internal keying signal is at or near minus 3.6 VDC, the "keying input open" columns apply.

The minus 8.2 VDC and minus 3.6 VDC sources are derived from the 12 VDC ± 10% input by conventional shunt zener diode regulators. Use the resistor and zener diode values and tolerances specified in the parts list.

selector switch positions		keying input grounded		keying input open	
mark	**space**	**mark**	**space**	**mark**	**space**
on	on	1	1	1	1
on	off	1	0	1	0
on	norm	1	0	1	1
on	inv	1	1	1	0
off	on	0	1	0	1
off	off	0	0	0	0
off	norm	0	0	0	1
off	inv	0	1	0	0
norm	on	1	1	0	1
norm	off	1	0	0	0
norm	norm	1	0	0	1
norm	inv	1	1	0	0
inv	on	0	1	1	1
inv	off	0	0	1	0
inv	norm	0	0	1	1
inv	inv	0	1	1	0

'0' indicates signal absent
'1' indicates signal present

Fig. 6-12. Mark and space outputs as functions of S3 and S4.

Component Boards

Each of the three component boards is made from a 3″ × 4″ × 3/32″ piece of micarta or phenolic. Brass eyelets 0.087″ O.D. × 1/8″ long are used for tie points.

The brass eyelets are inserted into all of the No. 43 holes in the component boards from the component side. Make certain the eyelets are pushed all the way into the board, so that the eyelet head is against the surface of the board. Turn the component board over, lay it on a piece of wood and funnel out each of the protruding eyelet "barrels" with a few gentle taps of a hammer on a 3/8″ center punch.

All wiring is done on the back of the component boards in point-to-point fashion with No. 22 AWG tinned bus-bar wire. Insert the wires through the eyelets and bend the ends of each wire over on the component side of the board to hold the wire in place. Clip each wire next to the eyelets on the component side. Insert the components and solder each eyelet from the wiring side. Clipping off the excess component leads completes the component board wiring. The leads of all the semiconductor components should be heat sinked during the soldering operation.

Checkout and Adjustments

After the construction phase is complete, two pairs of electrical adjustments are required to place the unit in service:

a) R11 and R12 must be set so that the gates operate as amplifiers.

b) L1 and L2 must be adjusted (turns removed) to set the exact mark and space operating frequencies.

Connect a 600 ohm (nominal) load and oscilloscope to J3 and set R11 and R12 (on CB2) to the approximate center of their range. Apply power to the unit and check the minus 3.6 and minus 8.2 voltage levels. Set the mark and space amplitude controls (R9 and R10) to about mid-range and the AF output level control (R15) fully clock-wise.

Place the mark keying selector switch (S4) to "on" and the space keying selector switch (S3) to "off." DS1 should be illuminated and DS2 extinguished. Set the mark amplifier bias by adjusting R11 for maximum amplitude of the mark signal as displayed on the oscilloscope. Maximum amplitude and minimum distortion occur simultaneously.

Place the mark keying selector switch to "off" and the space keying selector switch to "on." DS1 should be extinguished and DS2 illuminated. Adjust R12 for maximum space signal amplitude as observed on the oscilloscope.

Set the internal/external keying switch (S2) to "external." Key the unit at J1 or J4 and check each of the possible keying combinations listed in Fig. 6-12. The responses of DS1 and DS2 should follow the signal output. Place S2 to "internal" and observe that the internal keying signal keys the unit at about a 22 Hz rate. The synch output signal at J2 should be a 22 Hz square wave at this time. This frequency has no particular significance other than being at about the same rate as the keying frequency of a 60 speed machine. If it is too far off, bring it in by changing the value of R18. Bear in mind that each different set of components will have its own frequency vs. R18 characteristics.

Because of the capacitor and inductor tolerances, it is extremely unlikely that the mark and space frequencies will be correct. The frequencies will probably be too low, but can be set to within a few cycles by removing turns from L1 and L2. Go easy here—it's a lot easier to keep on removing turns than it is to start adding them back. The frequencies may either be compared with an accurate audio (or AFSK) oscillator or measured with a frequency counter. The mark signal should be at 2125 Hz and the space signal at 2975 Hz. Soldering iron heat conducted to C1, C2, C4, or C5 will affect the oscillator frequencies. Frequency measurements should be made only after the capacitor temperatures have stabilized.

THE SELCAL

The Selcal, an RTTY character recognizer, is sort of an electrical stunt box. It receives RTTY characters directly from the loop, with no machinery running. It recognizes four (or more) characters, in the proper sequence. An output relay closes to turn on your printer or other device. It then recognizes receipt of four letters "N," sent at the message end, to turn off your printer. While the characters must be received in the proper sequence, the Selcal does not distinguish between upper and lower case. Fig. 6-13 shows how the Selcal is hooked up.

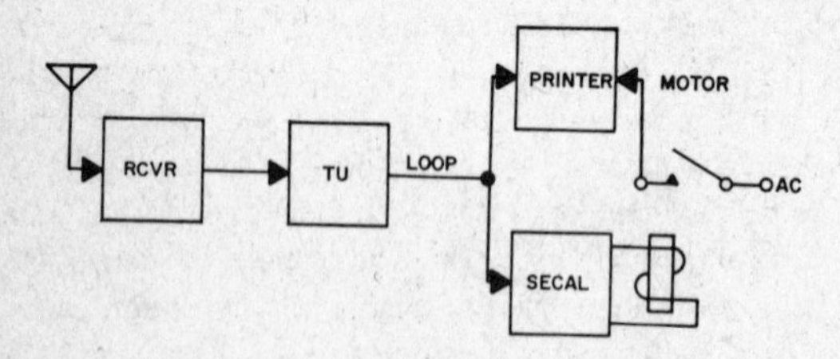

Fig. 6-13. Connecting the Selcal into your RTTY system to turn on your printer when your call letters are received.

The basic system is very versatile, and will be the basis of further RTTY logic systems such as regeneration, series-to-parallel conversion, and speed conversion.

The system is digital, using inexpensive Motorola integrated circuit (IC) logic blocks. This logic is designed to operate in practically any combination, with voltages, switching times, etc., figured out for you, eliminating much circuitry detail. Their cost is far below even junk box prices.

Logic

The Selcal is built entirely of three types of logic. Each will be described to allow the reader to follow the Selcal operation. See the reference list for more information on logic. This logic series operates on two voltage states: high (H) voltages—over 0.8—will turn on any gate; low (L) voltages—under .43—insure all gates are off. Levels between .43 and 0.8 would give erratic operation and are not used. The logic symbols do not show the B+ (3.6V) or ground connections (Fig. 6-14).

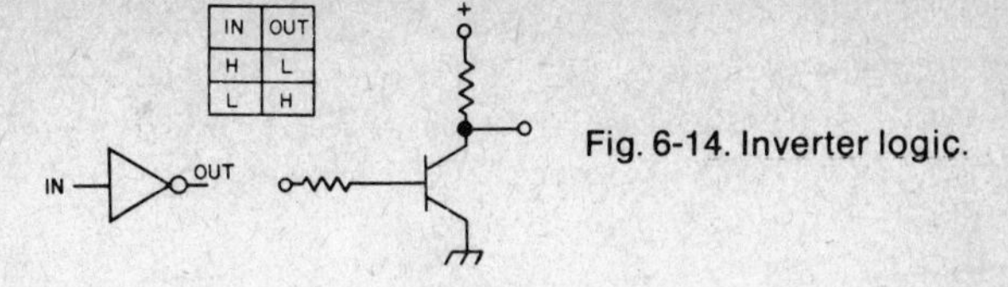

Fig. 6-14. Inverter logic.

Inverters

The simplest type of logic is the inverter, shown in Fig. 6-14. This is just a resistance-coupled amplifier designed so that in the "on" state the output is less than 0.43 volts. The inverter has a small "logic gain," or fanout, meaning one stage will drive several succeeding stages. A buffer is similar to an inverter but has a greater fanout capacity, and is available in both inverting and noninverting circuits. The MC789P Hex Inverter contains six independent inverter stages.

NOR Gate

The next logic type used is the NOR gate, shown in Fig. 6-15. It is obvious that if any input is high, a transistor will be saturated and the common output will be low. Only if *all* inputs are low can the output be high. The NOR gate is a most universal function, and nearly all digital computer circuits and systems can be built from combinations of this logic type. In the Selcal we will use the NOR gate as a coincidence recognizer. With varying high and low signals on all inputs, there will be an output only at the instant all are low.

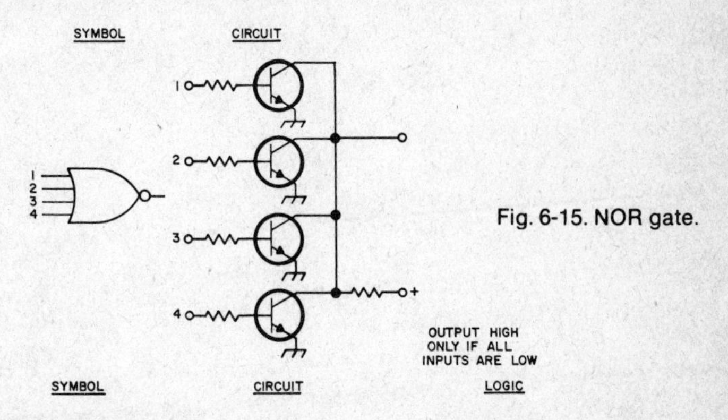

Fig. 6-15. NOR gate.

Flip-Flops

The J-K is an unusual but most versatile type of FF used in modern digital systems. It is also called a "master-slave," or "clocked" flip-flop. Its symbol and operation table are shown in Fig. 6-16. The inputs are : Set (S) and Clear (C) (sometimes called the J and K inputs), toggle or trigger (T) and preset (P). The outputs are (1) and (0), sometimes designated as ($\bar{Q}$) and (Q). These outputs are *always* in opposite logic states; that is, when one is high the other is low. The preset function is not shown in the truth table. When the (P) lead is high, the (1) output is forced low, regardless of the stages of the other inputs. While the integrated-circuit J-K contains the equivalent of 15 transistors, two independent circuits are contained in the Motorola MC790P.

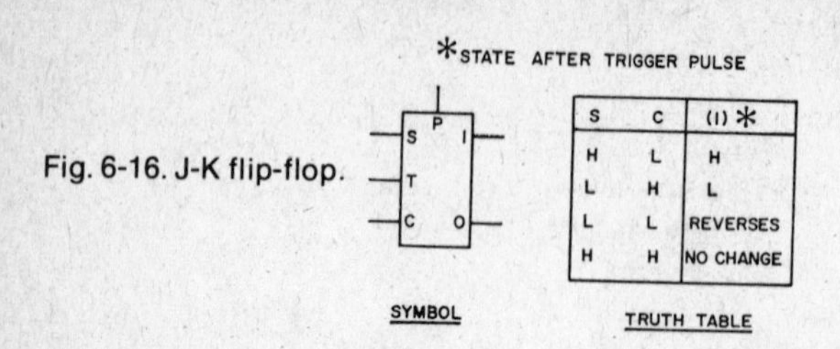

*STATE AFTER TRIGGER PULSE

S	C	(1) *
H	L	H
L	H	L
L	L	REVERSES
H	H	NO CHANGE

Fig. 6-16. J-K flip-flop.

The J-K can be connected for several different logic functions. Fig. 6-17A shows the J-K used as a common binary counter, or divide-by-two circuit. This divider will be used to count down the oscillator frequency in the Selcal.

Fig, 6-17C shows the clocked flip-flop operation. For this use the (S) and (C) inputs must be in opposite states, so an inverter is used as shown. The output logic states duplicate the input states *after the clock pulse*. This FF is seen to be timed, or "clocked." It will be used in this mode in the Selcal Shift Register. The truth table in Fig. 6-16 shows all modes of operation.

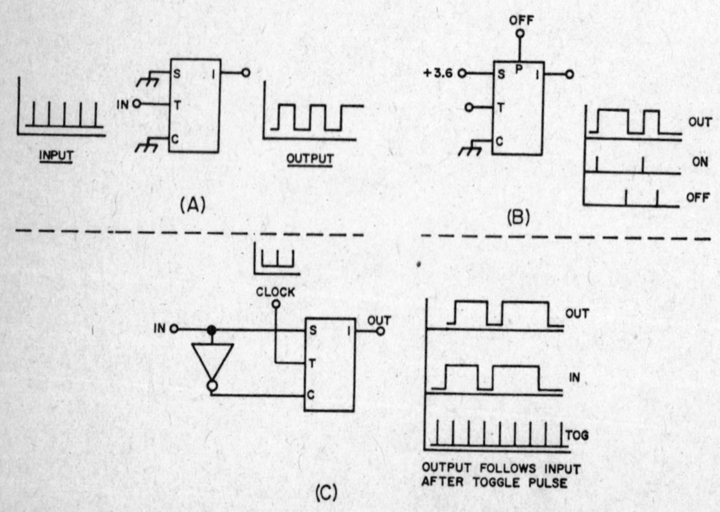

Fig. 6-17. Applications for the J-K flip-flop. A J-K divider is shown in A; a set-reset flip-flop in B; and a master-slave or clocked flip-flop in C.

Basic Operation

The Selcal is basically a series-to-parallel converter. The five character-information pulses, mark or space, are briefly stored in a five-stage shift register. The desired character is recognized by a coincidence circuit. The state of recognition is stored in a flip-flop. When all four characters have been received, the output relay is closed.

This is how the register stores the letter J (Fig. 6-18), which is Start-M-M-S-M-S-Stop. The first logic level seen by the register is the start pulse, a space. This (L) input is inverted to a (H) and applied to SR 5 lead (S). After the clock pulse, the (1) output also becomes (H), which we will define as the space condition of the flip-flop. The next signal pulse (one), is a mark, which makes SR 5 lead (S) low. The next clock pulse now does two things. At this point SR 4 sees the space condition of SR 5 and duplicates its output, making SR 4 (1) low. The start pulse has been passed from SR 5 to SR 4. Also the output (1) of SR 5 is changed to high, following the input signal. At the next clock pulse, the input is a mark (pulse 2). After this clock, SR 5 and SR 4 are in a mark condition, SR 3 in a space. The shift register now contains the start and first two information pulses of the letter J. These pulses continue to enter the register from the left. Finally the start pulse is pushed out the right end of the register, which then contains all of the five J information pulses.

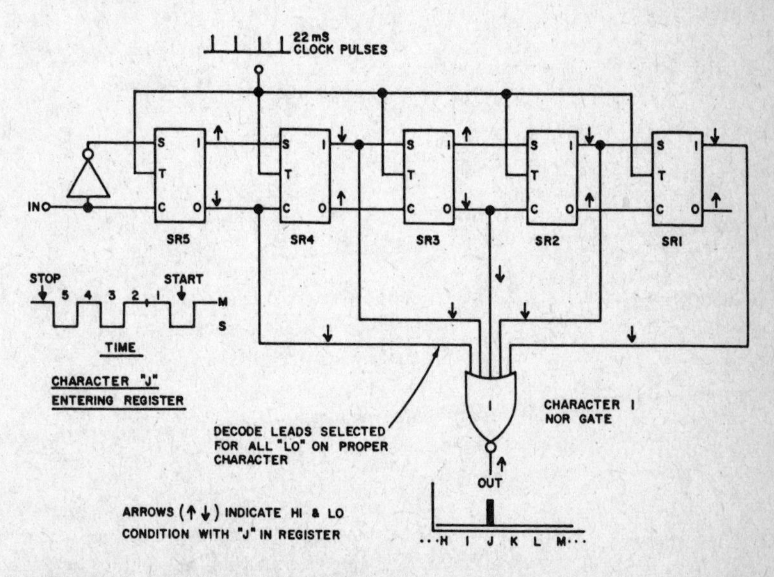

Fig. 6-18. A shift register connected to provide an output when an RTTY letter "J" is applied to the input.

Since both (H) and (L) outputs are available from each SR stage, we can select that lead of each SR that is low for a J. Only for this J (upper or lower case) will the all-low coincidence exist. These selected low outputs are now fed into a NOR gate. Recall that the output of a NOR gate goes high only when all inputs are low. It is the NOR gate that actually recognizes the J. The high pulse output is fed into a character-1 FF that flips and thus remembers that the J has been received. See Fig. 6-19.

To detect the next call letter, say K , another NOR gate is independently connected to the SR outputs that will give all lows with a K. The output of the character-1 FF feeds a low to the character-2 NOR gate so that the first character must be received before the second gate may look for its letter. This prevents the Selcal from responding to your call letters in an incorrect order (Fig. 6-19).

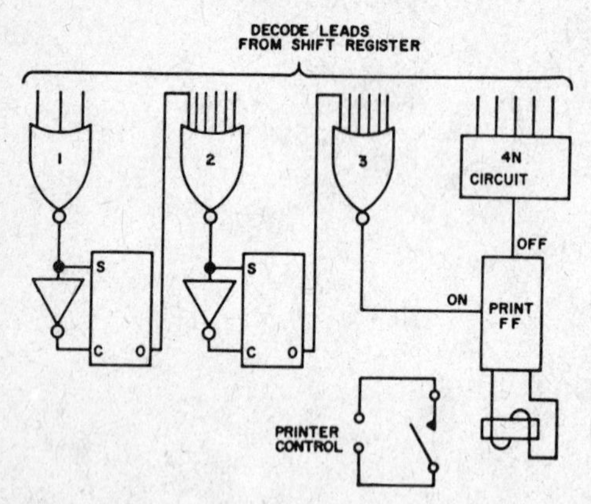

Fig. 6-19. The sequential selector. This circuitry is connected to the output of the shift register shown in Fig. 6-18 so that the letters of your call sign will turn the printer on only if they are in the correct sequence.

When both the J and K have been received, the third NOR gate is free to look for the third letter, say L. When received, the third gate gives a high output which turns on the print FF and the print relay. The printer is now on and receives your message.

To turn your machine off, the sender ends the message with "NNNN." The letter N is recognized just like the J, with a properly connected NOR gate. The gate feeds a two-stage binary counter which turns off the print FF when four Ns are received.

The Selcal circuit is complicated by the lack of the exact logic needed. Several NOR gates are paralleled to get enough inputs, and buffers and inverters are used to increase fanout or driving power. Note that the A-B signal lines carry the same pulses, the split being just to prevent device overload. The abbrevations listed in Table 6-1 will be used in the following detailed discussion of operation.

Table 6-1.

I	Inverter, or inverted.
SR	Shift register or stage.
N	Used in the 4N disconnect circuits.
C	Character; letter being recognized.
CH	Channel; memory for a character.
FF	Flip-flop.
Not	Circuit operating on all characters except ( ).
M-S	Mark-space.
Set	Pulses toggling SR5.
Shift	Pulses toggling SR4-SR1.
Hit	Non-RTTY pulses.
D	Divider stage (by two).
High	Voltage over 0.8.
Low	Voltage under 0.43.

Turn On

At the beginning of a start pulse, a high occurs on the m-s line, setting the start FF (Fig. 6-20). In this "set" condition, the start FF places a low on the divider preset leads, allowing them to operate. The oscillator inverter raises the voltage on the 6.8K resistor to high, starting the clock oscillator. The clock is a multivibrator that generates 5.5 ms (181-Hz) square waves. These pulses are divided by two, five times, by the dividers D1-D5.

Recall that a NOR gate has a high output only if all the inputs are low. By properly selecting the clock and divider outputs, a set of low leads can be found for each single clock pulse. As an example, let's see how the single decode pulse is obtained. At the decode time, (line 11), D2 and D3 leads (1) are low, but D4 and D5 leads (1) are high. By selecting the (0) leads of D4 and D5, we obtain all low inputs for the decode gate. D3 is not needed. Only at one particular time will the above conditions exist, so the decode NOR gate gives an output pulse only at the proper decode time.

In this way, NOR gates connected to the divider outputs produce properly timed set, shift, and end pulses. The end pulse resets the start FF, ending the Selcal sequence for one character. The "reset" start FF stops the clock oscillator and presets the dividers, making them ready for the next operation. The *hit* gate looks for a spacing signal partway into the start pulse. If a mark exists at this time (non-RTTY signal), the hit gate resets the start RF, terminating the operation. This resets the circuit after a false start from noise. The set-shift-not gate suppresses unwanted set and shift pulses.

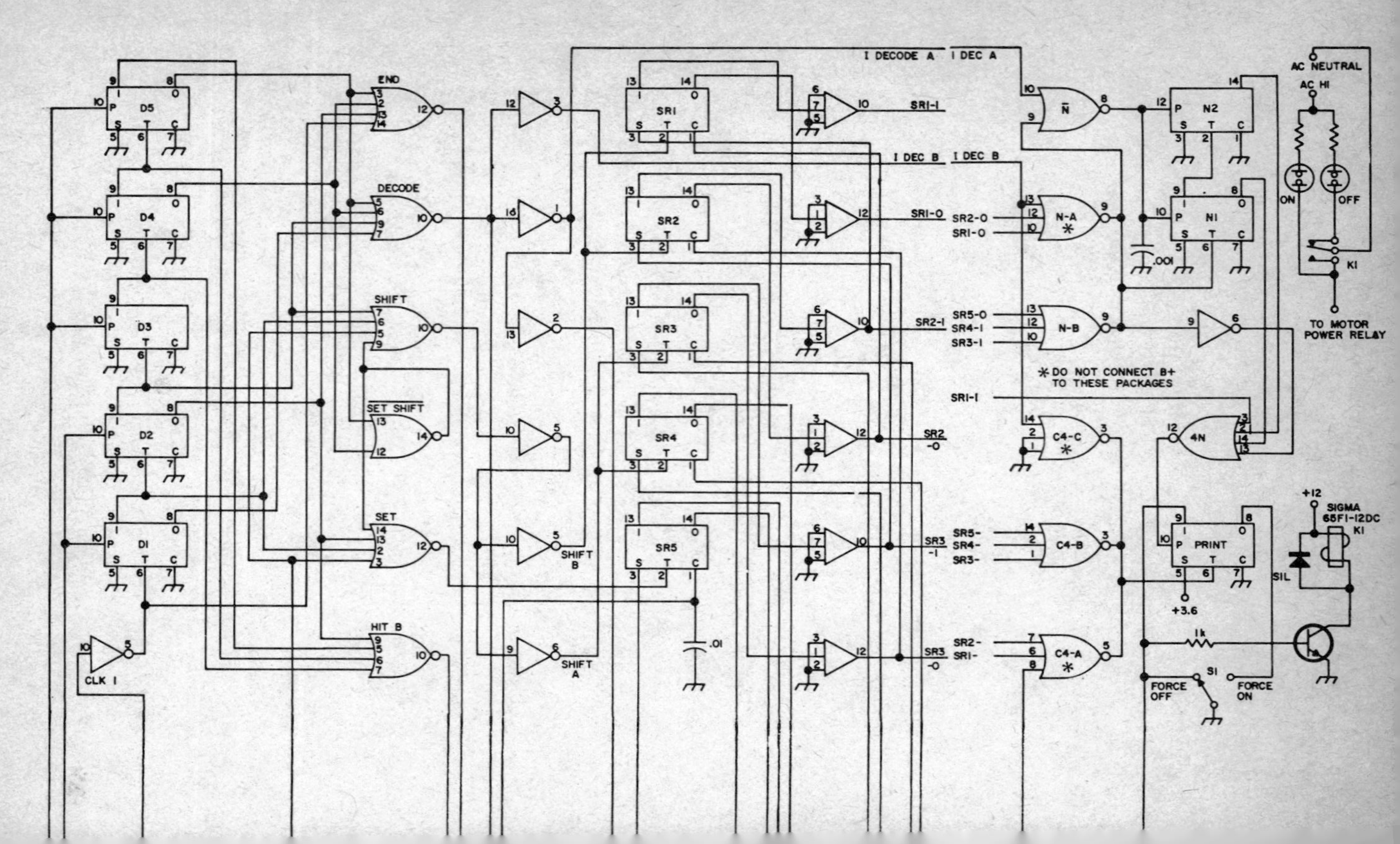
AC NEUTRAL
AC HI
ON
OFF
K1
TO MOTOR POWER RELAY
+12
SIGMA 65F1-12DC
SIL
FORCE OFF
FORCE ON
S1
PRINT
+3.6
1k
N2
N1
.001
4N
N-A
N-B
* DO NOT CONNECT B+ TO THESE PACKAGES
C4-C
C4-B
C4-A
I DECODE A
I DEC A
I DEC B
SR1
SR2
SR3
SR4
SR5
.01
SHIFT B
SHIFT A
END
DECODE
SHIFT
SET SHIFT
SET
HIT B
D5
D4
D3
D2
D1
CLK 1

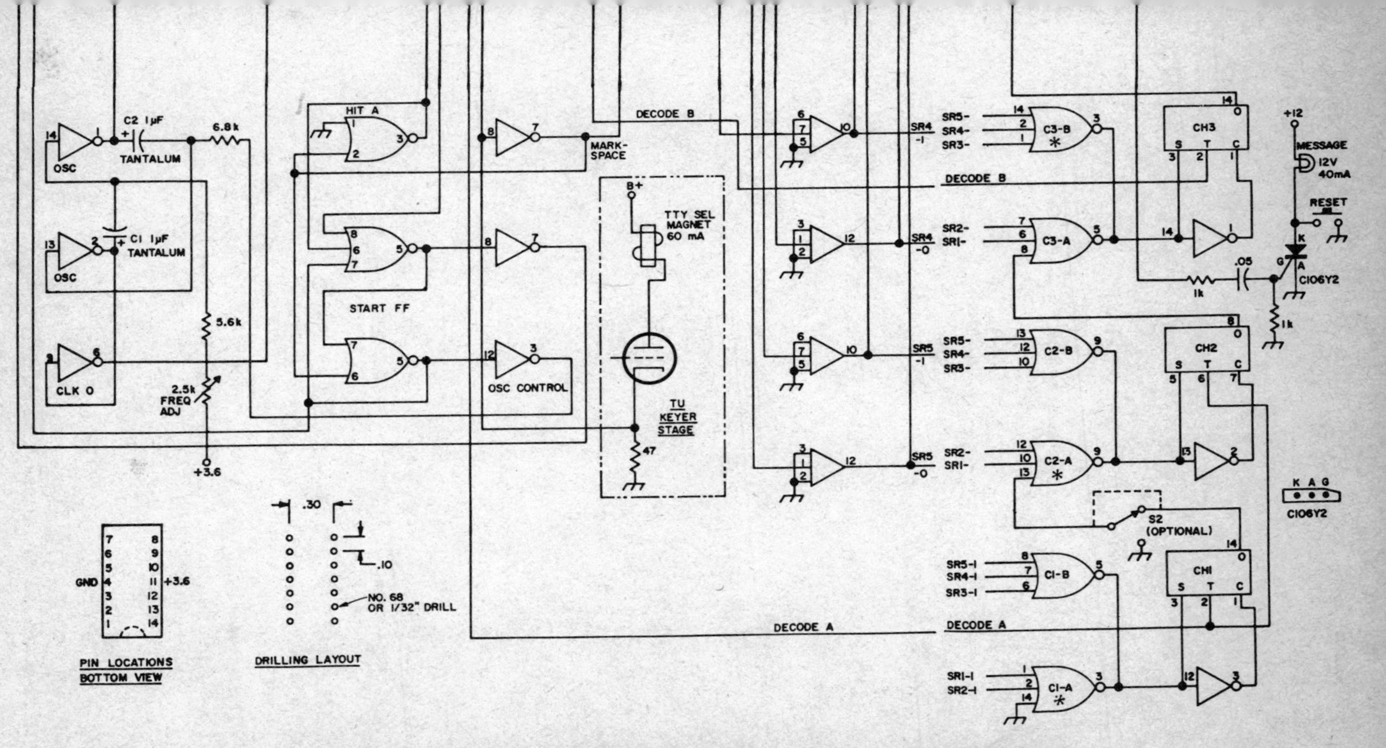

Fig. 6-20. Schematic diagram of the Selcal. Relay K1 is a 12V Sigma 65F1-12DC. Transistor Q1 may be any high-gain silicon transistor such as the Motorola MPS3393. The message light is a Sylvania 12ES. Switch S2 is an optional “omit first character switch.”

Now back to Fig. 6-20. Assume the call K8ERV is being received. To prevent casual copy reference to "ERV" from operating the printer, the code will be "ltrsERV," which already exists in the callsign. The first character "letters" (ltrs) enters the shift register as described earlier. At the time of the decode pulse, the ltrs marks and spaces are contained in the register. The C1 NOR gates are connected to the five SR output leads that will give all lows. The C1 gates will now recognize the ltrs character and give a high output to the CH1 FF. This high, with the decode pulse, causes the CH1 FF to set, remembering that character one was properly received.

The CH1 FF low output (lead (0)) is fed to the character-2 NOR gate, permitting it to look for, and recognize the next character, "E." As the decode pulse transfers the "E" recognition into the CH2 FF, it also resets the CH1 FF, which insures that characters will be recognized only in the proper sequence. The "E" makes the CH2 FF output (0) low, and the following "R" makes the CH3 FF output low. This low, plus the SR lows from the "V," and the inverted decode pulse (a low pulse) place all low inputs on the C4 gates. The high output from C4 sets the print FF, turning on the output relay and your printer. The Selcal has recognized the last four characters of the call K8ERV! Any wrong character will interrupt the sequence and reset the logic, preventing turn-on.

Turn Off

The print FF will now remain on until reset by a switch or by the reception of NNNN, a commercially used disconnect sequence. This section operates by recognizing and counting consecutive Ns. The fourth N received gives an output through the 4N gate which resets the print FF. Any character other than N operates the N-not gate which resets the FFs, destroying the count. The Selcal must see four consecutive N characters (or upper case equivalent) somewhere in a sequence to turn off.

All-Call

An important addition by K0OJV permits all Selcals to turn on with one particular calling code besides your selected call letters. Since recognition circuits exist for both "ltrs" and "N," an all-call code requiring a minimum of additional logic is "LtrsNLtrsNLtrsN." This code, besides being the easiest, will not occur in normal text. The use of six characters decreases the chance of false turn-on from noise.

Figure 6-21 shows the all-call addition. This is a counting arrangement similar to the 4N turn-off, except that the sequence "LtrsN" is counted until all three pairs are received, turning off the print FF. The counter is reset by any character other than "Ltrs" or "N."

Message Light

This circuit can be included to lock on a pilot light when a message is received. This alerts the operator to look at the copy. The print FF output

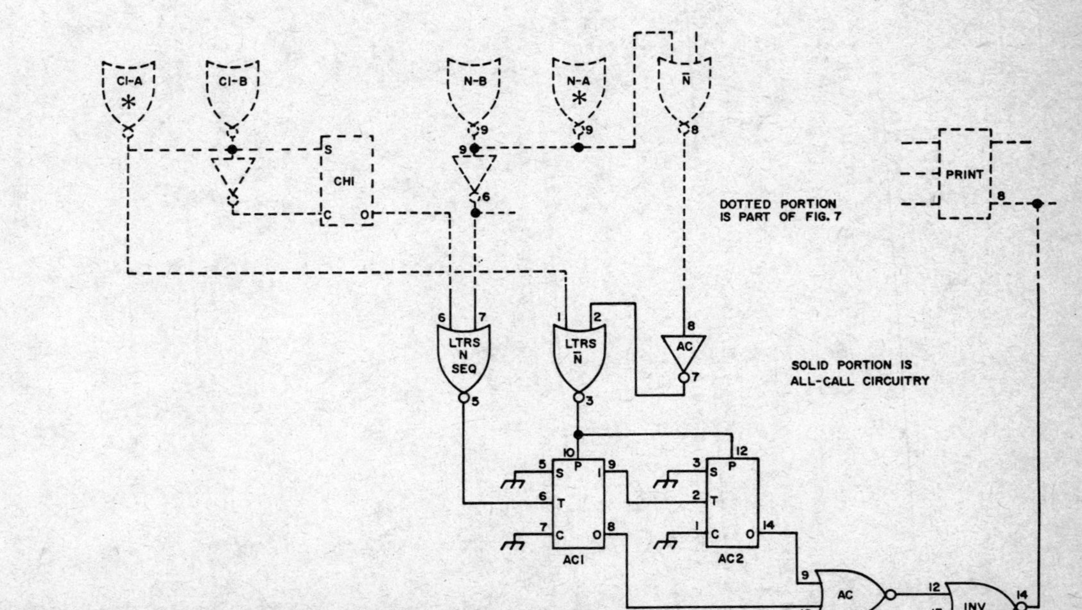

Fig. 6-21. All-call circuitry which may be added to the basic Selcal shown in Fig. 6-20. This circuitry permits the printer to be turned on by sending "LtrsNLtrs-NLtrsN." This is particularly useful for turning on all the machines of an RTTY net.

pulse is used to trigger a small SCR that locks on a low current lamp. The lamp is manually reset by a momentary, normally open push switch.

Construction

The integrated circuits used in the Selcal are the Motorola RTL (Resistor-Transistor-Logic) 700 or 800 series, in a plastic dual in-line package. These differ only in price and temperature range, the 700 types covering 15-55° C and the 800 types covering 0-75°C. Data sheets are available from Motorola.

These logic blocks may be laid out in any order. While IC sockets are available, they are expensive and unnecessary. One way to mount the ICs is to drill holes in a plastic sheet, insert the IC leads in the holes, and wire to the pins. Another way is to mount the blocks on their backs, using an adhesive, or double faced tape, and again wire to the pins. Leave plenty of room for the wires, there are several hundred of them! The easy way is to use the pair of circuit boards from K0OJV at $10.00 a set, undrilled. We strongly recommend small (#26) colored Teflon wire to prevent soldering iron damage in the rather cramped wiring space. The cheapest Teflon seems to be the Knight brand available from Allied radio at about 5¢ a foot.

The power supply must provide 3.6V ±10% at about 600 mA. The design shown in Fig. 6-22 has excellent regulation and negligible ripple to about 90 line volts. Its performance is better than needed but not expensive. Z_1 is a group of forward biased diodes of any silicon type, used as a low-voltage zener. Z_2 is optional, being a group of one-amp diodes used to limit the voltage in case of any type of supply failure. The Selcal can be operated from two flashlight batteries for testing, using a voltmeter in place of the output relay.

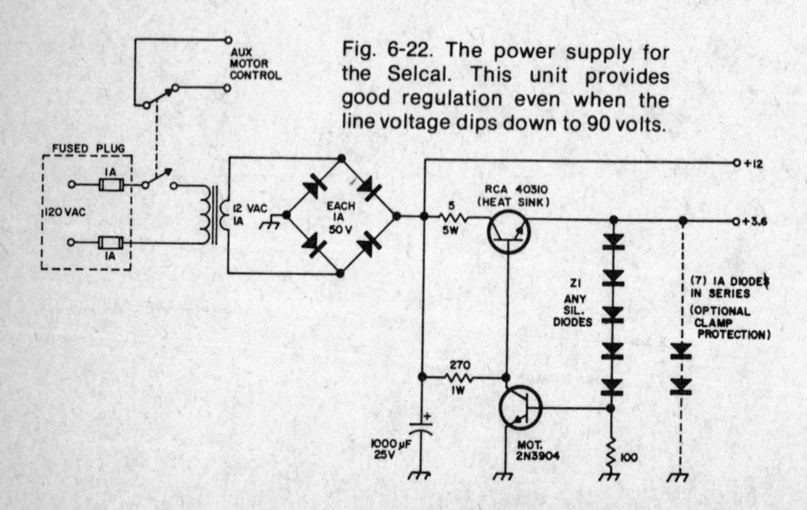

Fig. 6-22. The power supply for the Selcal. This unit provides good regulation even when the line voltage dips down to 90 volts.

The power supply and front panel layouts are not critical. The only controls really needed are the on and off switches, but all sorts of pilot lamps and other goodies can be added as described.

Decoding

Setting up the letters you wish to receive is done by hooking the particular character NOR gates to the proper SR outputs. Character 1 is shown set up for "Ltrs." The N gates are, of course, wired for Ns although any repeated character could be used. To construct the decode chart (Fig. 6-23) for any character, replace the character marks with (1), and spaces with (0), omitting the stop and start pulses. The first information pulse (after the start pulse) will eventually be in SR1, so the chart is actually reversed from the normal character construction. Since N is S-S-M-M-S, it becomes SR-1(0), SR2-(0), SR3-(1), SR4-(1), SR5-(0). Enter your letters in rows C2, C3, C4. Now transfer this decode to the C2, C3, C4 NOR gates in Fig. 6-20. C1 is done for you for "Ltrs." Connect each NOR gate lead to the indicated SR output lead. The SR outputs may feed more than one NOR input. This is the reason the non-inverting buffers are used.

DECODE CHART					
SR	5	4	3	2	1
LTRS	1	1	1	1	1
C2					
C3					
C4					
N	0	1	1	0	0

Fig. 6-23. The decode chart which is used in setting up the Selcal for receiving your call leters.

The simplest method of decode wiring is to permanently connect the decode leads. But two other methods are more versatile. Figure 6-24 shows how twenty inexpensive slide switches can be used to set in the four characters at will. This scheme permits fairly rapid changes in the decode set. A piece of cardboard with holes that accept the slide levers in a particular decode setup can be used to check the settings.

A still faster decode change can be obtained by using a multi-pin connector as a patch board. Each decode group is wired to a separate plug and inserted into the socket in the Selcal. Twenty-five pins are required for a three-letter decode, thirty pins for four letters.

The Selcal turns on the printer motor when its code set is received. If fed with continuous random noise, eventually the Selcal will receive its code and give an unwanted turn-on. A three-letter decode for commercial or experimental copy can be obtained by grounding the C2 output lead, as shown in Fig. 6-20. This is not recommended for unattended copy due to the increased

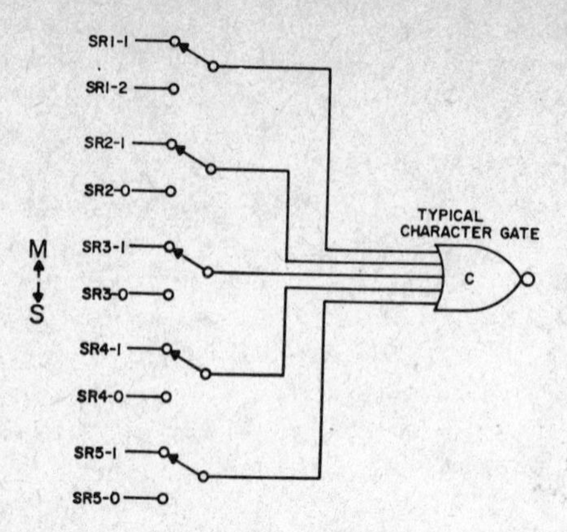

Fig. 6-24. By using slide switches in the input to the character gates, various turn-on codes may be used with the Selcal.

possibility of noise turn-on. We suggest the Selcal be teamed up with an auto-start system, such as in the TT/L to inhibit the noise fed to the input.

Adjustment

The only adjustment is the clock oscillator frequency. Temporarily turn on the clock by shorting the Selcal input. Connect a scope to either clock output. Using the line frequency for comparison, adjust the 2.5K pot for 180-Hz output. If a scope is not available, set the pot in the center of the range that gives proper Selcal operation.

The power supply output should be from 3.3 to 3.9 volts. It can be varied slightly by changing the 100-ohm resistor from 50 to 200 ohms. If greater shift is needed, change the number of diodes in Z_1. Caution, do not operate into the logic with Z_1 disconnected. A 6-ohm, 5-watt resistor can be used as a supply load to simulate the Selcal when "tuning up." If a Variac is available, run the line voltage down until the output starts to drop. This should be about 90 volts, but depends on the gain of the 40310. Lowering the value of the 270-ohm resistor will reduce the required input voltage, but too low a value will reduce regulation and may overload the 2N3904.

If wired correctly, the Selcal should take off when connected as in Fig. 6-1. Note that the Selcal relay will not handle a printer motor load, and must be used only to drive a suitable motor relay, such as the RBM 84-903 ($3.05).

Connect your printer into the local loop and send your call letters. The Selcal relay should turn on. If it by any chance does, you have made about 350 proper connections! Now send any letter except N to reset the all-call, and then send "NNNN," and hope it turns off. If not, don't despair, a troubleshooting guide follows.

Troubleshooting

First check to see if the start FF and clock oscillator are being keyed. Hook a scope or headphones through a 1000-ohm isolating resistor to the Clk-1 output. Sending any letter should produce a burst from the oscillator. Ground the Selcal input and check for proper outputs from each divider and from the set, shift and decode gates, as shown in the timing chart. Any logic block pin, except B+, may be connected to B+, or grounded, to force a circuit on or off, without harm to the logic.

Now check the shift register by sending a "letters" character. With a voltmeter see that all the SR-1 leads are low (less than 0.43 V), and that all the SR-0 leads are high (over 0.8V). Send an N and check for lows on SR3, 4-1, and on SR1, 2,5-0. Any letter should leave its proper pattern in the shift register. If the higher number SR stages work, but the lower ones don't, check the wiring and logic at the point of signal loss.

Now send any letter not in your code set. A meter should show highs on all of the character FF (0) leads. Your first code character (Ltrs) should make CH1-0 go low. The second code character will reset CH1-(0) to high and make CH2-(0) low, etc.

Check the 4N gate as follows. Send any letter other than "N." All four inputs to the 4N gate should be high. The first N will place a low only on 4N pin 14. The second N should make only 4N pin 2 low. The third N will make both pins 2-14 low. During the fourth N, only at the decode time do pins 3-13 go low, but a scope is needed to see this. The 4N gate output briefly goes high, resetting the print FF and turning the relay off. With the exception of the 4N system, most of the Selcal functions hold their states after decode, so that a voltmeter is all that is needed for testing.

Use

In operation, the receiver, tuning unit, and Selcal are left running continuously, or connected to a time clock. The sender should transmit your call several times to insure reception and turn-on. After his call, it is helpful to include the time in GMT, followed by an extra line feed to separate the messages. After sending the message, he should return the carriage to the left, and send 8-10 Ns. If conditions are poor, send extra Ns to insure turn-off, any not needed will not be copied. Automatic CR-LF systems are very convenient for any unattended autostart or Selcal operation.

While autostart is not useful for monitoring continuous commercial stations, the Selcal is, and can be used to select only those parts of interest to you. However, for 75 or 100 wpm monitoring, the Selcal clock must be changed.

A SOLID-STATE SLOW SCAN TELEVISION MONITOR

The first requirement for a newcomer to SSTV is to acquire a monitor. The unit described here is a monitor with good performance which keeps the component count and circuit complexity to an absolute minimum. The resulting circuit uses a magnetically deflected cathode ray tube (CRT) for bright picture display and is relatively simple in concept and construction. The circuit is designed around readily available over-the-counter parts to minimize procurement problems which often plague the home brew fanatic these days. If minimum cost is a factor, extensive substitution of surplus and bargain parts is possible and guide lines are provided following the circuit description. A well-stocked distributor should have virtually all of the components called for with the exception of the CRT which will probably have to be special-ordered. If the monitor is built in the modular form recommended, it is possible to experiment with new circuit ideas as they become available, thus gradually updating the circuit. As described, the circuit will recover excellent pictures under any but the worst conditions of noise or QRM.

Circuit Description

The SSTV picture is transmitted as an audio tone of varying frequency. Sync pulses to start each line and frame are sent as bursts of 1200 Hz while light intensity values between black and white are transmitted by varying the tone between 1500 Hz (black) to 2300 Hz (white). A monitor circuit must separate out the 1200 Hz sync pulses and use them to control the scanning of the monitor CRT and must also utilize the video frequency shift to bias off the CRT to provide brightness variations on the monitor screen. A block diagram of the monitor circuit elements is reproduced in Fig. 6-25 and will be referred to in the following description.

Since the picture information is transmitted as a tone of varying frequency, we want to remove all amplitude variations in the incoming signal caused by factors such as fading, variations in input level at the transmitter, changes in receiver gain controls, etc. IC1 functions as an audio frequency limiter to perform this function and is followed by a fixed gain audio amplifier (Q1 and Q2) to provide sufficient signal level to drive the other circuits. Video information is recovered by passing this signal through a 2300 Hz tuned circuit (L_1) which acts as a video discriminator. The black and white frequency limits will be used for the purpose of illustration but a linear response between black and white may be assumed in visualizing the reproduction of intermediate grey values. The output of the discriminator is a signal whose amplitude varies from a relatively high value for black (1500 Hz) to a relatively low value for white (2300 Hz). This amplitude "modulated" signal is fed to Q3 and is further amplified. Variable gain is incorporated in this stage for contrast control. The output of Q3 is detected, providing a relatively higher DC voltage during black portions of the picture and a low DC output when white is begin transmitted. This varying DC level is used to drive a high voltage transistor, Q4, which is connected to vary the grid bias on the CRT. At

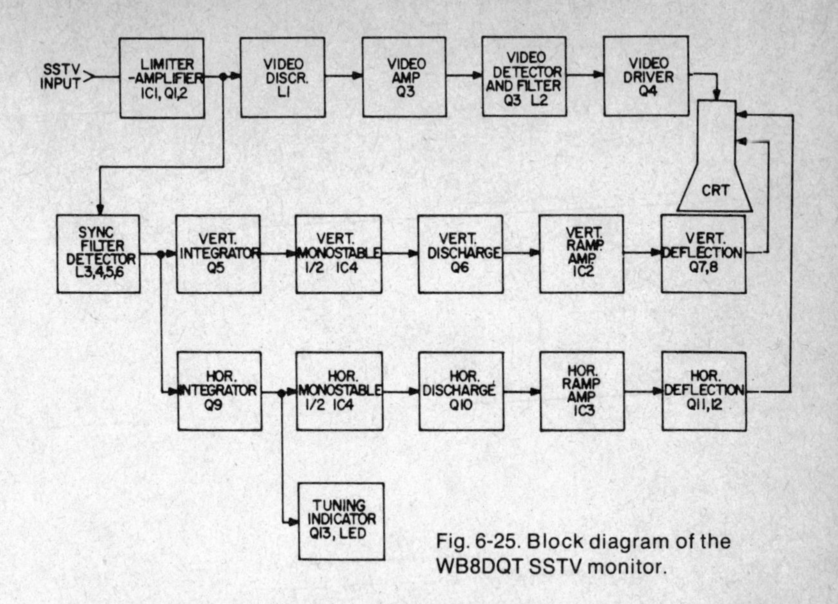

Fig. 6-25. Block diagram of the WB8DQT SSTV monitor.

the optimum setting of the contrast control, the relatively high DC output of the video detector will bias off the CRT during black portions of the picture, causing the CRT beam to be extinguished and producing a dark display on the screen. The low output of the detector during white portions of the picture has a minimal effect on the CRT bias, the beam is almost at full intensity, and a white display (actually bright yellow due to the type of phosphor required) results. A long persistence (P7) phosphor is used so that the video information may be viewed in its entirety despite the extended frame time (Fig. 6-26).

Output from Q2 is also fed through a series filter (L3, L4) to a 1200 Hz tuned circuit (L5, L6) where the sync pulses are detected. Output of the sync detector drives Q5 which is set up to integrate the relatively long vertical sync pulse (30 ms) while rejecting the shorter horizontal pulses. The output of Q5 is a 30 ms DC pulse which is fed to one half of IC4 which is wired as a monostable multivibrator. The rectified vertical pulse, which may be noisy, distorted by multipath, or of incorrect length, depending upon the setup of the camera is used to trigger the monostable which generates a clean 30 ms pulse. The output pulse from the monostable is used to turn on Q6 for the duration of the pulse and effectively shorts the 50 μF vertical discharge capacitor to ground. At the end of a vertical sync pulse, Q6 goes "off" and the capacitor begins to charge through the 1 meg resistor and 1 meg size control in the collector circuit, producing a ramp voltage. This voltage drives IC2 which amplifies it and in turn drives a complementary pair output stage (Q7 and Q8) which provides deflection to the CRT through the vertical windings of a standard TV yoke coil.

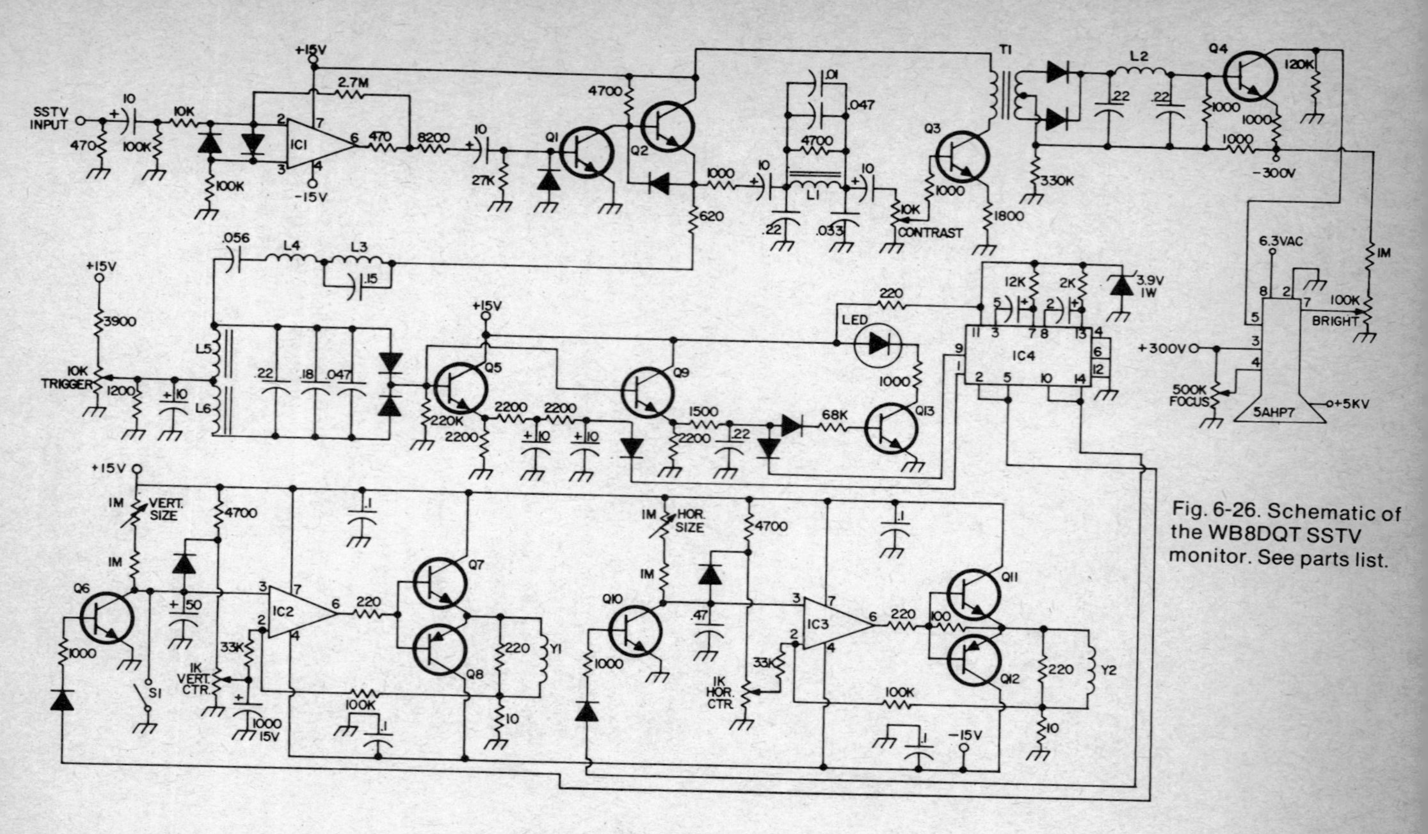

Fig. 6-26. Schematic of the WB8DQT SSTV monitor. See parts list.

PARTS LIST

Q1,2,3,5,6,9,10,13 – HEP 55; Q4 – HEP 240; Q7,11 – HEP 245; Q8.12 – HEP 246; IC1,2,3 – HEP C6052P or C60536; IC4 – HEP 570; All unmarked diodes – 1N457, 1N914 or general purpose rectifiers; all resistors 1/2W; All capacitors in mF – decimal values are 100V tubular Mylar, electrolytics are 25V unless otherwise designated; contrast. vertical and horizonal size and center pots are 1/4W PC pots; all other pots are 1/2–2W units; Y1,2 – both coils incorporated in standard 70–90° TV deflection yoke (Stancor DY-21 or equiv.); S1 – manual vertical reset – pushbutton switch; T1 – Triad TY-27XT (500:500Ω CT); L1,4 – Triad EA-100 toroid (100 mH); L2 – Triad C-27X filter choke (0.7H); L3 – Triad EA-070 toroid (70 mH); L5,6 – Triad EA-020 toroid (20 mH); LED – Industrial Devices Inc. 2190LI-12V – other LED's may be used if resistor values are altered to accommodate their voltage and current ratings; front panel controls – brightness, trigger, LED, manual vertical reset.

Fig. 6-26. Parts List.

The operation of the horizontal stages is similar except for the time constants involved. Q9 serves as the horizontal sync integrator driving 1/2 of IC4 which drives Q10 to generate the horizontal ramp voltage which is amplified by IC3 and used to drive the horizontal output stage consisting of Q11 and Q12 which in turn drives the horizontal windings of the CRT yoke assembly. DC pulses from Q9 are also used to drive Q13 which functions as a solid-state switch and turns "on" an LED tuning indicator whenever a sync pulse is present in the passband of the sync detector. This provides a convenient tuning indicator since the sync pulses will only appear in the narrow "window" of the sync detector when the station SSB receiver is tuned for proper carrier insertion.

The suggested power supply circuit (Fig. 6-27) is straightforward and provides regulated plus and minus 15V outputs for the solid state circuits and plus and minus 300V unregulated, for the CRT circuit. The CRT requires between 6 and 10 kV for acceleration voltage. This is provided by a HV module available from Robot Research which is powered by the low voltage supply.

Construction

The monitor can be assembled in almost any fashion that permits proper interconnection of the circuit elements, but certain points deserve mention. It is highly desirable to remove the power supply circuit from the monitor itself to prevent distortion of the CRT scanning beam by magnetic fields from the power transformers. The power supply can be interconnected to the monitor via a multiconductor cable. The HV assembly should be mounted directly on the monitor chassis.

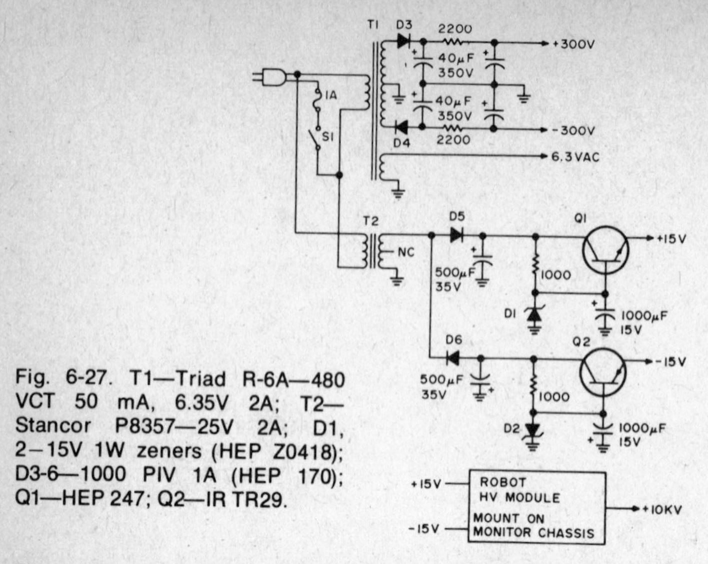

Fig. 6-27. T1—Triad R-6A—480 VCT 50 mA, 6.35V 2A; T2—Stancor P8357—25V 2A; D1, 2–15V 1W zeners (HEP Z0418); D3-6—1000 PIV 1A (HEP 170); Q1—HEP 247; Q2—IR TR29.

The brightness (plus on/off) and sync trigger controls are the ones which have to be mounted on the front panel. The manual vertical reset pushbutton switch and the LED indicator should also go on the front panel. The contrast control can be mounted on the chassis for easy access but need not be adjusted following initial setup. All other pots can be 1/4W trimmers and are best located on their respective circuit boards since they are usually not adjusted after initial setup of the monitor.

An LMB CO-1 cabinet makes an attractive package for the monitor, but you might wish to consider a cabinet styled to match existing gear in the shack. A sheet of yellow acetate plastic may be placed between the CRT face and the rear of the front panel to mark out the bright blue flash characteristic of the P7 phosphor, permitting easier viewing of the persistent yellow video display.

Parts Substitutions

The single most expensive item is the 5AHP7A tube specified. This is a modern, electrostatically focused CRT with an aluminized phosphor which provides excellent performance in this application. This tube is not usually available on the surplus market. The 5FP7, a common surplus item, can be used with only two modifications. First of all the focus pot specified should be deleted and, secondly, a permanent magnet focus coil assembly such as the Quam "Focalizer" must be mounted on the CRT neck in back of the deflection coil assembly. These units were used on many models of early tube-type TV sets and the junkers in local TV service shops are your best (and cheapest) source.

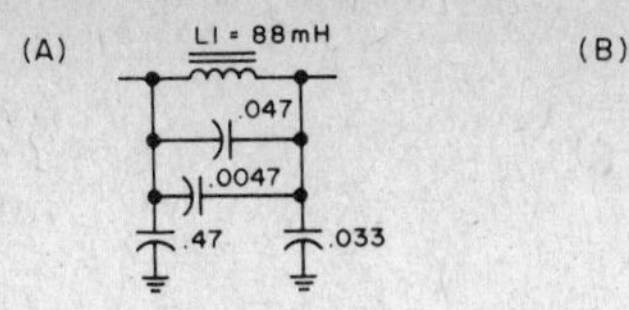

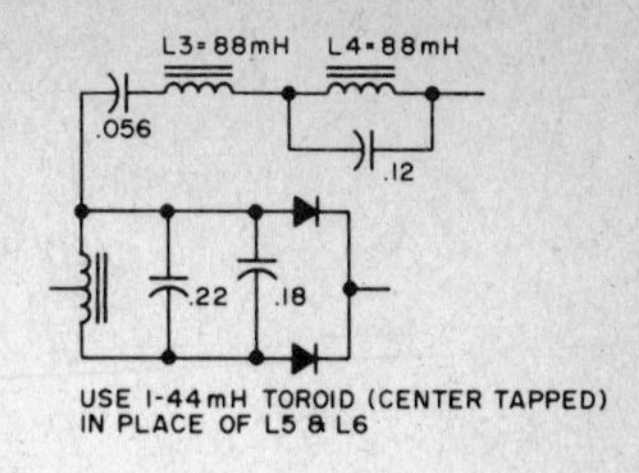

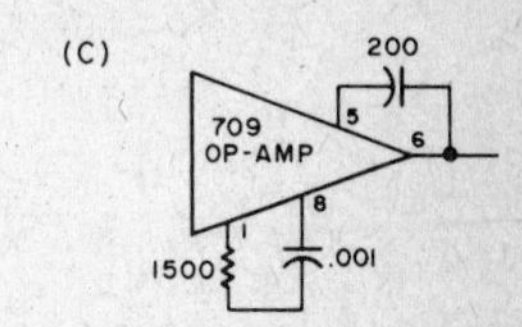

Fig. 6-28. Modification of monitor when using surplus components. (A) Modifications of discriminator when 88 mH toroid is used for L1. (B) Sync circuit modifications for use of 88 and 44 mH toroids. (C) Additional components required for frequency compensation when 709 op amps are substituted for IC1, 2, and 3.

Deflection coil values are not at all critical and virtually any deflection coil assembly can be used as long as it will physically fit the neck of the CRT.

Most of the transistors specified (Q1-6,9,10,13) are general purpose NPN types (2N718, 2N3391, etc.),so surplus bargains are fine. Be sure to test them first, however, if your source is suspect. Q4 is an NPN unit with a 300V rating—otherwise it is not critical. The complementary pairs (NPN-PNP) in the output of the deflection stages can be any complementary pair types rated at at least 30W. The NPN and PNP transistors in the power supply regulators are general purpose power types in TO-3 cases. They drop approximately 10V in the circuit shown and will get very hot unless properly heat-sinked.

The HEP ICs specified for IC1 3 are op amps in 8 pin TO-5 or mini-DIP packages. 741 op-amps in either the TO-5 or mini-DIP packages can be substituted with no change in pin numbering. These ICs are readily available and quite cheap. 709 op amps in the TO-5 package may also be used with no basing changes but these must be frequency compensated by the *addition* of a few more components. Figure 6-29 indicates what changes must be made. IC4 is a Quad 2-input gate which can be replaced with Motorola MC-824P.

The toroids specified are relatively inexpensive Triad types since the availability of surplus toroids becomes more variable as time goes by. Figure 6-28 also shows changes in the tuned circuits which can be made to incorporate surplus 88 and 44 mH toroids if you have a source of supply.

Initial Setup and Adjustment

This monitor cannot adequately be set up using a tape you record off the air. It achieves its resistance to QRM by using high Q coils in the sync section and tuning accuracy is so important that the chances of your getting the correct tuning operating blind are minimal. A recording obtained from another slow scanner or the test tape which Robot Research sells for a very

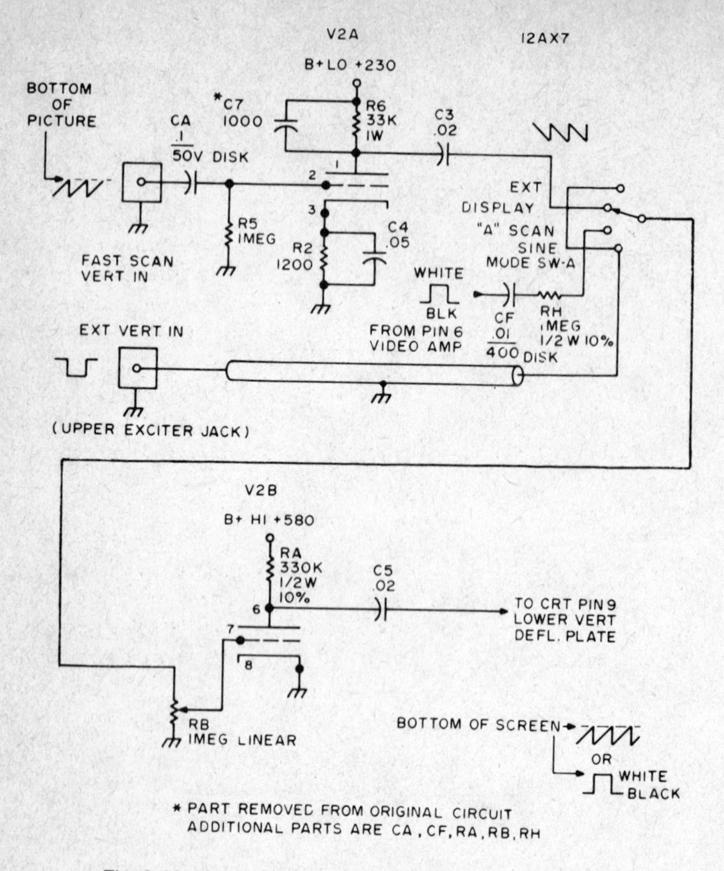

Fig. 6-29. Monitor vertical amplifier modification.

modest cost is required. The latter tape is ideal in that it contains a gray scale pattern which is ideal for setting up the contrast control.

Start by setting contrast, vertical and horizontal size, and vertical and horizontal centering controls to mid-scale. Turn on the monitor but keep the brightness control turned down. As the CRT warms up adjust the triggering control through its range as a slow scan signal is fed into the input. You should note the LED tuning indicator flickering in time with the horizontal sync pulses over at least part of the range of the control. The indicator will brighten momentarily during the vertical sync pulse due to the longer duty cycle of this pulse. Turn up the brightness until the trace is just visible and use the horizontal centering control to move it into the viewing area. The horizontal size control is used to expand the trace to fill the viewing area. You should be able to get horizontal triggering over much of the range of the triggering control. Adjust the trigger control to the point where the picture will begin to scan downward but not past the point where it will fail to reset when a vertical sync pulse arrives. If the picture scans upward or the trace jumps to the bottom of the screen when the reset button is pushed you must reverse the vertical deflection leads. All of the size and centering controls should be optimized for a square picture that just fills the viewing area. The brightness can now be set at a comfortable viewing level, usually half scale or just a little beyond, and the contrast control adjusted for optimum gray scale rendition. With proper adjustment of this control the beam will be completely extinguished in black areas of the picture, white areas will have maximum brightness, and there will be a smooth gray scale response with all bars of the gray scale pattern clearly resolved. You will find that there is comparatively little interaction between brightness and contrast, so the brightness control can be varied without shifting overall video response. Since the monitor employs triggered sweep circuits the trace will move off the screen if no signal is present.

Actual operation of the monitor in conjunction with the station receiver consists of tuning the receiver for maximum brightness of the flickering sync display of the LED. If the receiver is tuned this way the monitor will trigger properly and the gray scale reproduction will be identical to that obtained with closed circuit tapes. The LED will brighten with any audio signal that falls in the sync detector passband as the receiver is tuned, but video information, noise, splatter and other interference sources will produce a random brightening quite unlike the normal sync display. Even a very weak fading signal can be tuned by peaking the indicator at the top of the fade cycle, thus assuring that the monitor will be locked on the signal whenever it rises out of the noise. This tuning indicator has made it possible to track SSTV signals from OSCAR 6 despite rather pronounced Doppler shifts on overhead passes, thus assuring optimum video display at all times.

FAST SCAN MONITOR FOR SSTV

When operating SSTV the use of a fast scan monitor provides the means to adjust the scene at a 15 frames/second rate rather than the standard 8 seconds/frame rate—a ratio in round numbers of 120:1. Commercial fast scan display units sell for about $250. Oscilloscopes now sell in the $120 range and are usually large and bulky. Who wants to tie up a scope for display only? This article describes modifications that can be made to a standard Heathkit monitor scope, found in many stations, to provide the fast scan display feature for the popular Robot 80A Camera.

The modification is based upon the assumption that the trapezoidal display found in the HO-10 is seldom used.

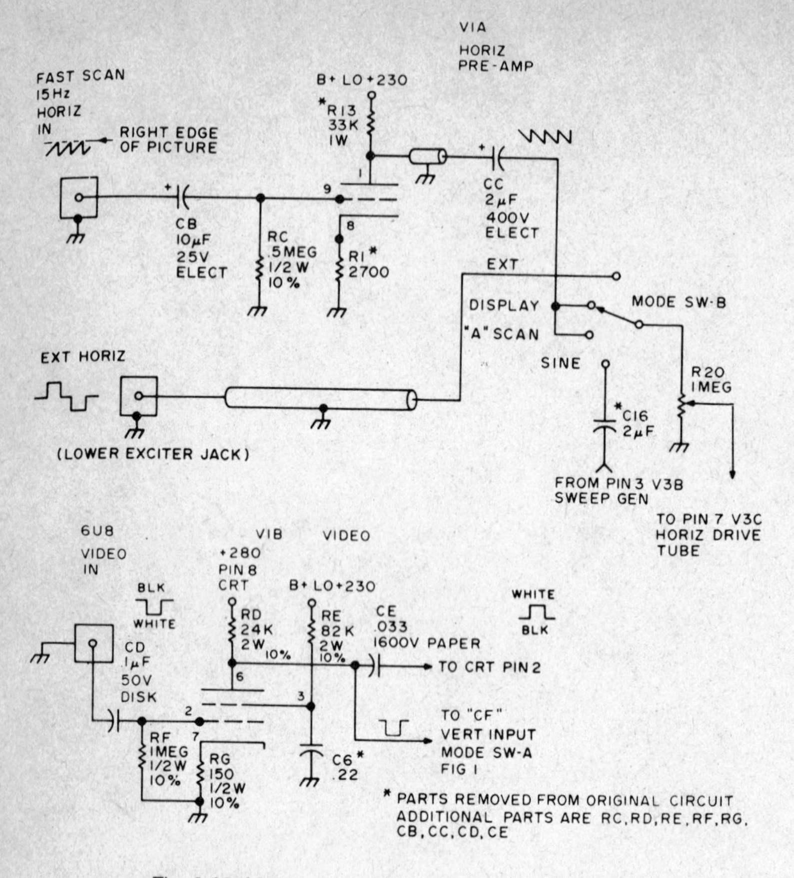

Fig. 6-30. Monitor horizontal amplifier modification.

Circuit Descriptions

Vertical Amplifier. The vertical amplifier, Fig. 6-29, was modified to satisfy two criteria:

1. Maximum sensitivity to a positive sawtooth at V1A.
2. No voltage transients on external vertical input line when mode switch is operated from display to EXT.

The vertical fast scan signal, from the 80A camera, is taken directly from the vertical yoke. This is a positive going sawtooth with a negative magnetic overshoot. C7 is used to minimize that overshoot. Notice that the coupling capacitor C3 is located on the input side of mode switch-A to prevent voltage transients from being applied to the external vertical input line when the mode switch is operated. Such voltage transients could damage external semiconductor circuits.

Horizontal Amplifier. The horizontal fast scan signal from the 80A camera is a positive going sawtooth. The amplitude of this signal is about the same as the fast scan vertical signal and requires two stages of amplification, Fig. 6-30. V1A the horizontal preamplifier must pass a 15Hz signal which accounts for the large capacitances (CB,CC). Notice also that both coupling capacitances have been moved to the input side of mode switch-B. This was done, as in the case of the vertical amplifier, to prevent voltage transients from being applied to the external horizontal input line when the mode switch is operated.

Video Amplifier. The HO-12 has no provision for video, or Z input. A video amplifier, Fig. 6-30, described by ROBOT in their Field Service note #4 was installed in V2B. This video grid resistor, RF, or the bias resistor, RG, can be made variable for gain control. I found 150Ω for RG satisfactory using the intensity control to vary the display. If the display is too contrasty, RG can be increased to 240-300Ω as required.

CRT. Three basic modifications were made to the CRT, Fig. 6-31, circuitry to:

1. Increase the vertical deflection.
2. Provide Z or video input.
3. Provide means to shut off the RF signal vertical input.

The 1MΩ resistor connected to the vertical deflection plate pin 9 was increased to 3.3μM. C8, through which the lowest level RF signal is applied to the vertical deflection plate, was removed and the switch point grounded to provide a means to shut off the RF input. This allows the display to be viewed while the picture is being transmitted—good if you're prone to move. In most cases it is not necessary to view the RF unless setting up or in case of troubles.

Mode Switch/"A" Scan Feature. The mode switch was changed from a 2-pole 3-position to a 2-pole 4-position switch, see Figs. 6-29 and 6-30. The currently installed mode switch has 2-poles 5-positions with only 3 being used, and can be modified with some difficulty. Use a new 2-pole 4-position switch.

The extra position allows amplified fast scan video to be applied to the vertical input and fast scan deflection to the horizontal input. These two signals result in an "A" scan, allowing synchronized video to be displayed. An "A" scan display is used for precise adjustment of the camera brightness control (see operation).

Physical Modification. In order to keep the fast scan signals together, the video input was entered via the two tone jack. The two tone signal was relocated to the lower antenna connector. Since there was no room to mount a terminal strip near the two tone tube, V4, the junction of R30, R38 and R52 was soldered to a short length of stiff wire, the other end of which connects to the lower antenna connector. The junction is, therefore, self-supporting. In order to prevent the junction from shorting to pot R29 a strip of plastic tape was

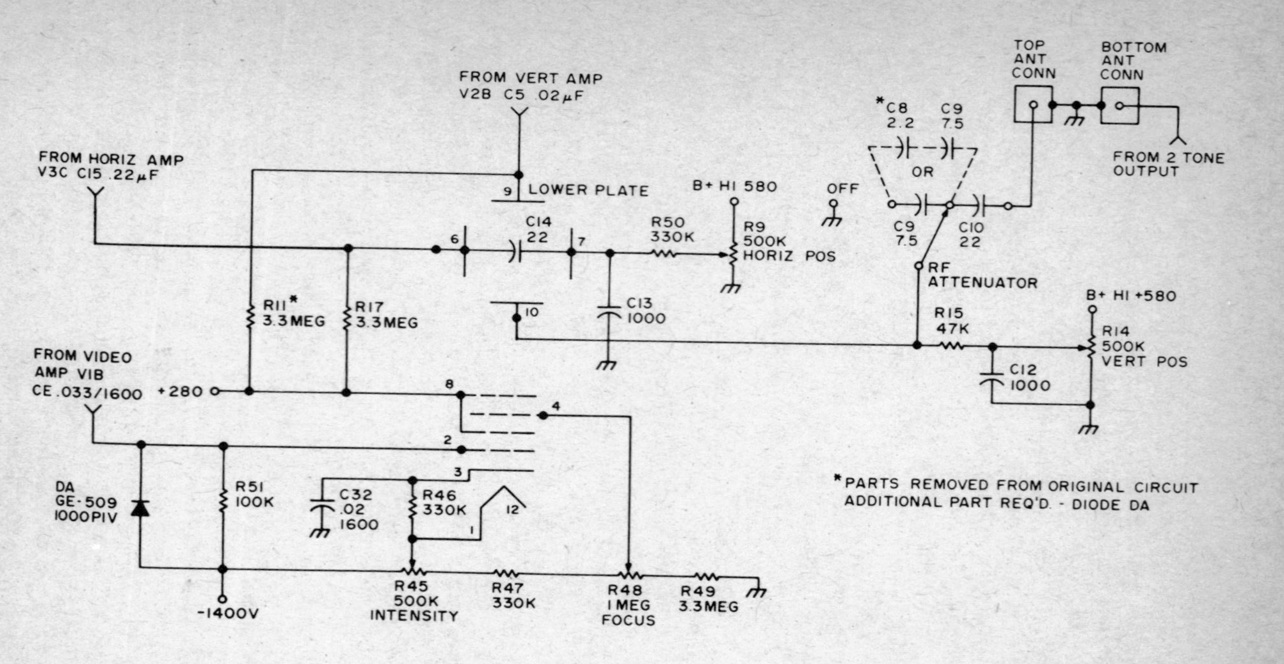

Fig. 6-31. CRT circuitry modification.

placed on the pot. The antenna is connected to the single connector via a coax "T" fitting.

The two exciter connectors, now no longer needed, were used for the external inputs; the top for vertical and the bottom for horizontal. The location of these connectors is convenient for connecting shielded wires to the mode switch.

The .003μF 1600V video coupling capacitor CE and the 24K 2W RD are connected via a 3 lug terminal strip. This strip is mounted on the side of CRT neck that is closest to the bottom side of the chassis. It is mounted using the same screw that mounts a terminal strip already in this location, Fig. 6-32. The strip currently mounted must be positioned such that its body is towards the CRT neck.

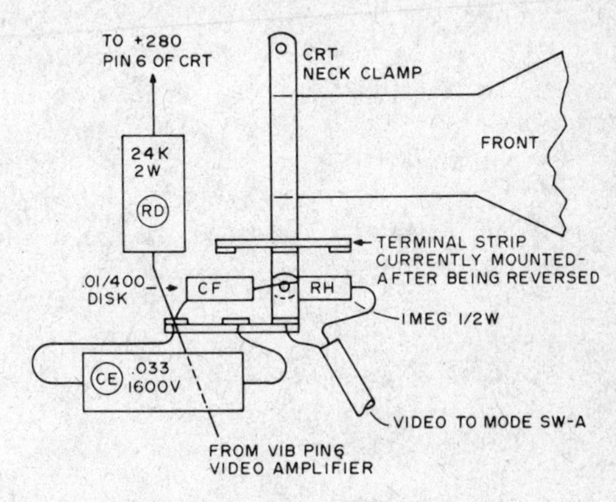

Fig. 6-32. Monitor physical modification.

Coupling capacitors 2μF 600V CD and .02μF C3 are mounted in the area of the mode switch.

The .22μF C6 is relocated from pin 8 to pin 3 of V1. Do not remove the ground end.

Don't physically remove the wire connecting pin 7 of V1 to pin 7 of the CRT. This wire is later used from pin 6 of V1B to the .033/1600 CE located on the CRT neck.

The shielded lead form V1 pin 2 is relocated to V1 pin 1.

External Inputs. Two external inputs, vertical and horizontal, were wired to the mode switch to permit attachment of an audio frequency spectrum analyzer or other device. The sensitivity of the vertical input is too negative pulsing. The sensitivity of the horizontal input is either polarity. Note that both external inputs are direct, not AC coupled. This was done in order to keep these inputs flexible. This allows the coupling capacitors to be matched to the external signal source frequencies and voltage polarities.

Power Supply Considerations Under No-Signal Input Conditions. The 6BN8 required 600mA of filament current. The 12AT7/12AX7 requires only 300mA, a net reduction of 1.9W.

The original V2B required 7.2mA at 580V, the revised V2B requires about 1.5 mA at 580V—a net savings of 3.3W.

The new horizontal preamp requires 3.3mA at 230V or an additional .76W.

The new video amplifier requires 4.9 mA at 230V or an additional 1.2W.

The net high voltage current drain is plus 1.5mA.

These values will vary, of course, under signal conditions, but the net change is towards less power consumption.

Chart. In order to adjust camera brightness (Fig. 6-33), set mode switch to "A" scan. Black will be towards the top of the screen and white will be towards the bottom of the screen. Adjust camera brightness for maximum difference between black and white signal levels. Notice that any further increase in brightness causes an increase in the overall down position of the vertical spikes (background shading), but does not increase the white to black ratio of the picture elements. This can be checked by switching to the display mode and "rocking" the brightness control either side of the optimum setting. The contrast can now be adjusted by looking at both display and "A" scan positions of the mode switch and touching up the brightness as required.

Conclusion

The display obtained with these modifications completely fills the screen and then some. The linearity of the display is not of TV quality but is more than sufficient for this application.

Operation

		Mode Switch		
HO-10	Sine	A-Scan	Display	Ext.
Intensity	Adjust to level to eliminate "Z" modulation	Normal	Black looks black	As desired
Focus		As required for intensity.		
Vertical gain and Position		As required but below ovarload level.		
Horizontal and Position		As required.		
Sweep Frequency	As required	N/A	N/A	N/A
Rf Attenuator (in rear)	As required	Off	Off/On When on, SSB rf overlays display during transmission, but can still be seen.	Off

Fig. 6-33. Monitor operations chart.

The camera was operational for one week before the modifications to the HO-10 were completed. The difference in performance and picture quality and the ease which they are obtained has definitely established the need for Fast Scan Display. The estimated cost of $11 for new parts is, without a doubt, a good investment to increase the versatility of an already valuable piece of monitoring equipment.

A SLOW SCAN TELEVISION SIGNAL GENERATOR

Circuit Description

The SSTV Signal Generator provides selectable test signals of the Sync, Black and White Frequencies, a variable Gray Scale signal, a vertical Bar pattern and a variable Dot-Bar pattern. The sync signals are stabilized by synchronization with the 60 Hz power line frequency (Fig. 6-34).

The transistors Q1 and Q2 are used to shape the 60 Hz power line frequency sine wave into a square wave so that it will stabilize the horizontal frequency multivibrator (Q3 and Q4) at 15 Hz. The 8 second per cycle vertical oscillator (Q5 and Q6) is frequency stabilized by inter-connecting the horizontal oscillator through the IN-648 diode and the .047 coupling capacitor. Sync pulses are taken from the collectors of Q4 and Q6, combined and fed to the sync amplifier Q10 and modulator Q11. The vertical sync pulse is 30 ms long and the horizontal sync pulse is 5 ms long. The Dot-Bar oscillator (Q7 and Q8) output frequency is manually adjustable from approximately 7 Hz to 80 Hz. This signal is amplified by Q9 and is connected to the Modulation Switch Dot-Bar position.

The 1500 Hz SSTV sub-carrier oscillator (12 and Q13) output is amplified by Q14 and is connected to the low pass filter through the step down transformer T1. The low pass filter rounds off the square wave output from the sub-carrier oscillator and attenuates any spurious frequencies generated above 3000 Hz. Positive going voltages appearing at the base of Q11 will shift the sub-carrier oscillator up in frequency and negative going voltages will shift it down in frequency.

The sync pulses from the sync oscillators are of the proper polarity when they appear at the base of Q11 to cause the 1500 Hz sub-carrier oscillator to shift to 1200 Hz for the duration of the sync pulse. The positive 12 volts from the "White" frequency position on the Modulation function control drives the sub-carrier oscillator from 1500 Hz to 2300 Hz. In the "Gray Scale" position, the voltage is manually adjusted from a positive 12 volts to near 0 volts which in turn shifts the sub-carrier oscillator to any frequency between 2300 Hz and 1500 Hz (White to Black).

Several fixed vertical bars are generated when the Modulation control switch is in the "Bar" position. The 60 Hz sine wave from the 6.3V filament transformer is rectified to a positive half wave by the IN457 diode (D1). This

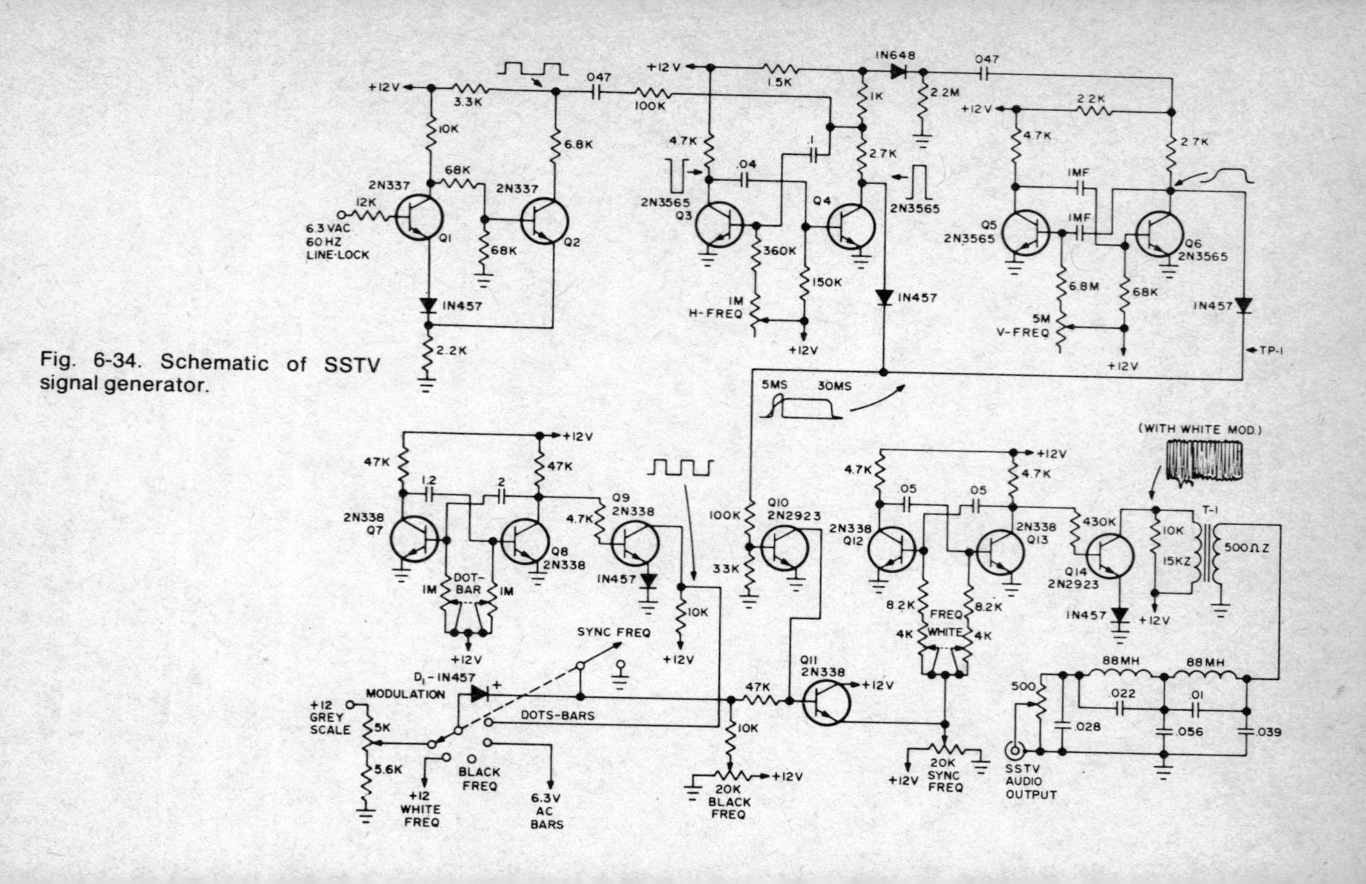

Fig. 6-34. Schematic of SSTV signal generator.

half wave DC voltage is used to modulate the sub-carrier oscillator thereby generating the desired stable vertical bar pattern (Fig. 6-36).

Construction

The major portion of the electronic circuitry is assembled on three plug-in perf-boards. One board contains the circuits for Q1 through Q6. The second contains the circuits for Q10 through Q14. The third board is used for the low pass filter and Q7 through Q9. The chassis is of hand-formed aluminum and is 5″ D × 6 1/2″ W × 1 1/2″ H. The cabinet is 6″ D × 8″ W × 6″ H. Point to point wiring was used throughout the unit. All resistors are of the 1/2 watt size unless noted otherwise. All transistors are of the NPN silicon type used for audio or switching applications. Many different types of NPN transistors were tried and all seemed to work satisfactorily in these non-critical circuits.

Adjustment

Connect a low capacity probe from an oscilloscope (Fig. 6-35) to the test point TP-1. With the scope sweep set to 15 Hz and the horizontal sweep sync at 60 Hz the signal generator horizontal frequency oscillator control should be adjusted for 15 Hz. There will be one 5 ms pulse appearing on the screen of the scope. The adjustment of the vertical sync pulse is more time-consuming because of the long time lapse between pulses (8 seconds). Slowly adjust the vertical frequency control until the start of the vertical sync pulse coincides with the start of the horizontal sync pulse at each 8 second interval.

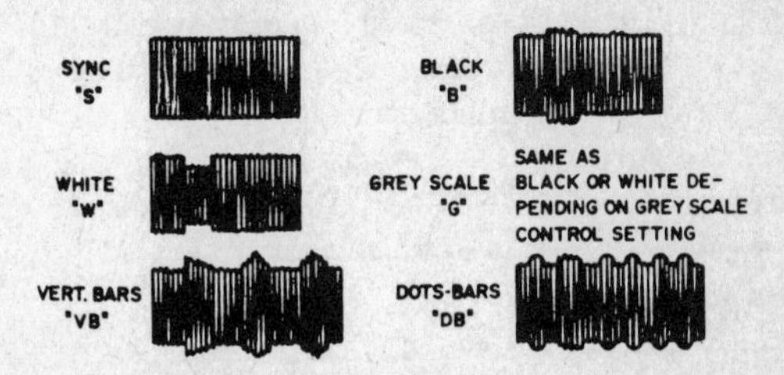

Fig. 6-35. Oscilloscope patterns at audio output jack.

An audio frequency meter or audio digital frequency counter is connected to the SSTV Audio Output Jack. Preset the Black, White and Sync frequency controls to mid-rotation. Select "White" on the Modulation control and adjust the white frequency control to 2300 Hz. Then turn the Modulation control to "Black" and adjust the black frequency control to 1500 Hz. With the modulation control at "Sync," adjust the sync control to 1200 Hz. Repeat these adjustments several times as there is some interaction between the three controls. Typical oscilloscope patterns as observed at the SSTV Audio Output Jack are shown in accompanying diagrams.

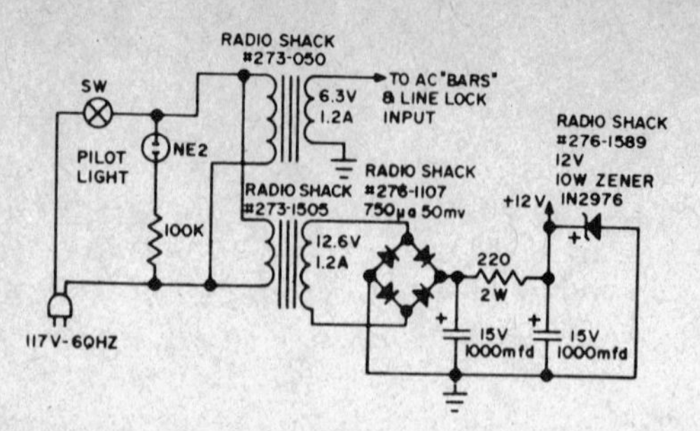

Fig. 6-36. Power supply schematic.

A FAST SCAN FACSIMILE SYSTEM WITH SSTV COMPATIBILITY

Mechanical systems of picture presentation popularly refrered to as facsimile are not new to amateur radio. Such systems were used extensively back in the 1920s when scanning discs produced the poineer equivalent of SSTV on the 150-200 meter band, where it was then legal. The relegation of all facsimile to the VHF placed most interested experimenters in a communications vacuum and has probably been the major factor in discouraging amateur facsimile.

The presently legal system known as SSTV, though facsimile, seems limited to TV methods; however, the parameters of the system are readily adapted to other systems of facsimile transmission and reception which are not limited by the 8-second useful persistence of the P-7 phosphor.

When so adapted, these other systems immediately become as legal on the DX bands as SSTV, thus broadening the potential horizons of those experimenting with these facsimile systems.

Most any system of mechanical readout may be adapted to the 15 line per second requirement of current SSTV. The drum system will produce beautiful pictures when carefully built, even when the drum is a rolling pin borrowed from the kitchen.

Little besides the drum and lead screw speeds and the video diode polarity need be changed to produce excellent slow scan pictures. However, such systems might not prove too attractive because of the necessity to retool between each picture, and the expense of the paper. Electrostatic paper under the same conditions is less expensive but has the same "retooling" disadvantage.

After several years experience with amateur APT weather satellite readout using several different systems, the writer adapted the continuous readout helix blade principle used in many commercial facsimile systems to home fabrication from readily available materials. The adaptation worked beautifully, using the inexpensive electrostatic papers available to experimenters in small lots.

The big advantage of such a system is to provide legal worldwide facsimile capability to those wishing to build their own facsimile equipment. As long as the transmitted signals follow the format of SSTV, these activities are obviously perfectly legal despite the equipment used for generating and receiving the signal. Furthermore, by the simple expedient of slowing the "vertical" scanning rate to send and receive the pictures in double or quadruple, the present 8-second limit of the useful brilliance of a P-7 phosphor, the number of lines per picture and, thus, the resolution can be doubled or quadrupled. (It is recognized that other parameters such as spot size must also be optimized in order to get the full improvement in resolution.) With the present wording of FCC regulations, such experimentation must await the development of amateur interest in such increased picture resolution.

The continuous readout helix blade principle allows the paper to roll continuously off the roll. This is scanned horizontally and then flows continuously out of the machine. Using electrolytic paper, the scan can be viewed immediately on completion of the individual line.

The basic principle of the system is at first more obscure than that of drum or scope readouts. It becomes readily comprehensible if one will get a cardboard tube, such as the core of a paper towel roll or a rolling pin from the kitchen, and wrap a single spiral turn of string or wire from end to end on it. Holding this tube parallel to the edge of a ruler, the tube is then rotated and the movement of the point of contact watched. One complete rotation of the tube will move the contact point the full length of the ruler, and represents one scan line when electrolytic paper is pulled slowly at right angles between this point of contact.

Construction Details

Construction may follow many forms. The box may be made of Masonite as in my APT readout apparatus, or in a 10.3 × 12.9 × 15.4 cm (4 × 5 × 6 in. Bud #CU-729-B) steel box as in the Fast Facsimile readout. The paper compartment or humidor could be in a plastic box with a slit for paper exit at an appropriate level and the mechanical printing complex arranged outside on suitable supports. The sole requirement in such case would be that the paper must not dry out while passing from the slit to the printing blade/helix combination—not probable at a rate of 6.5 cm in 8 seconds, but a situation to be reckoned with, nevertheless, where inter-picture waits take place.

If the cabinet is to be made of Masonite, one may well use 0.7 cm (1/4 in.) tempered Masonite held together by 4-40 machine screws 2 cm (3/4 in.) long screwed into holes drilled to the exact original diameter of the screw. To further firm up the cabinet, the screws should later be individually removed, "Will-hold," "Glue-bird" or "Elmer's" glue placed in the hole, and the machine screws again screwed into place.

In the metal box of the fast fax readout, the partition was placed at the approximate midpoint with the slit for the paper 2 cm (3/4 in.) from the top. Both the top and the bottom of the cabinet were removed and cut so that one portion of each covered the paper compartment. The bottom cover was then attached permanently with self-tapping screws and epoxy cement. The support for the paper was fashioned from two galvanized angles, two arms of which were superimposed to form the base of a U-shaped bracket. A drill and hacksaw aided by a rat-tailed file enabled the formation of the necessarily slanted holes for the reception of the alfax paper spindle (Fig. 6-37).

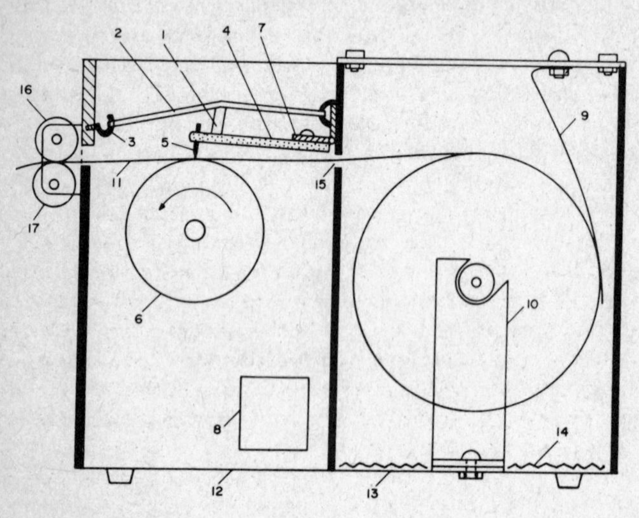

1. tension regulator
2. rubber bands
3. eyes for rubber bands
4. blade support
5. blade
6. drum supporting helix
7. hinges
8. relay
9. paper tension regulator
10. roll support
11. paper
12. front bottom plate
13. rear bottom plate
14. wet paper towels
15. paper exit slit
16. driver paper puller roller
17. idler roller

Fig. 6-37. Diagram of the mechanical layout of the Fast Facsimile recorder.

An area approximately 9.6 cm (3 × 3/4 in.) deep was cut from the upper front area of the front of the metal cabinet with a nibbler, and a 1.3 cm (1/2 in.) width of 0.65 cm (1/4 in.) plexiglass was bolted in so as to leave a 0.65 × 7.7 cm (1/4 × 3 in.) slit for the printed paper to pass out to the paper puller.

One of the two compartments of the cabinet is a "humidor" for the maintenance of the wet paper. The other is the printing chamber. The latter was originally planned as a humidified chamber, but this has been found unnecessary and seems undesirable, as the humidity might affect adversely the metal components of the printing complex. The humidor, whether made of Masonite or in a steel box, was first water-proofed with two coats of marine fiberglass paint (epoxy resin—not monoepoxy) as was also the printing compartment. After these coats had set, the paper compartment was given an additional coating of hot paraffin because it was found that water in the bottom caused the questionable grade of epoxy to turn white. The top of the humidor was made of plexiglass (or the original metal cover) held firmly in place by removable screws. Some sort of device must be included to prevent paper runaway. The device indicated in Fig. 6-37, made of a 2.54 cm (1 in.) width of shim brass, was used for this purpose in the fast fax readout.

The upper part of the central partition above the paper slit is made of plexiglass, as is also the lower portion, and is held in place to the walls with brass angles. This upper portion of the partition in turn becomes the support for the blade electrode supporting shelf, which is held firmly in place just above the paper entrance in the printing chamber by two small brass hinges on the upper side of the 0.65 cm (1/4 in.) Masonite or plexiglass which supports the blade electrode exactly above the center of the drum.

The printing blade is a 1.3 cm (1/2 in.) strip of stainless steel three inches long fastened to the edge of the 3.9 × 5.2 × 0.65 cm (1 1/2 × 2 × 1/4 in.) plexiglass blade support. 0.3 cm (1/8 in.) holes are drilled into the edge, Duco cement inserted, after which 4-40 machine screws were forced in. After a few minutes these were removed and the cement was allowed to dry. This in effect threaded the hole. The blade was attached by these same machine screws threaded into the holes. The blade rests on the top center of the helix. Desired levels of residual tension are applied to the shelf by several rubber bands originating above the point of hingement and extending to the front of the compartment. These are held in place at both ends by small hooks, and tension is maintained and varied by a strip of 0.65 cm (1/4 in.) Masonite or plexiglass held passively in place between the hinged blade support and the rubber bands. Considerable tension is necessary in order to print smoothly on the paper; however, too much tension results in paper shredding and cutting.

The drum should be of insulating material about 3.9 cm (1 1/2 in.) in diameter. There is nothing sacred about this size, however, and available materials should dictate. End pieces with exact centering should be turned from most any available material. I used fiber glass obtained from war surplus items throughout, and cemented them with epoxy. The length of the drum is controlled by the width of the paper decided upon. The most logical size of paper is 7.1 cm (3 3/4 in.) wide and was chosen to take advantage of the $1.50 price break between this and the next larger size. The drum length was a little over 9 cm (3 1/2 in.) and the spiral helix occupied a total length of 6.8 cm (2 5/8 in.).

The spiral helix was made of #22 nichrome wire bought from a laboratory supply house. Any relatively nonstretchable wire of equivalent size and behavior may work equally well providing it does not rust and is not reactant with the chemical in the paper.

The drum must turn in the counterclockwise direction observed from the (right) end, while the spiral must be made clockwise as indicated in the diagram. This is an essential feature for correct picture orientation. The wire is passed through a short diagonal hole to the end of the drum, where it is anchored to the end with a self tapping screw. It is then pulled tight, arranged in position on the drum where it is caught in the strong grip of a husky pair of long-nosed pliers. By twisting, the wire is tightened and the end then secured around a second self tapping screw. One end must eventually be connected electrically to the 0.65 cm (1/4 in.) shaft which serves as the drum axis. The center of the helix must be exactly midway between the origin and insertion of the two ends and on the opposite side of the drum. It should be tight enough not to require cement to hold it in place.

The ball-bearing assemblies which support the drum are mounted in the two ends of the cabinet at a position dictated by the location of the other active components of the printing chamber. In the metal model the bearings were cemented in holes in small pieces of Masonite with epoxy cement, and these were then bolted in the appropriate position inside of the cabinet. The holes receiving the ball bearings must be immediately below the printing blade and so positioned that the top of the helix is on the same level as the paper intake and exit slots. The bearings should be snug and tight. They may be held firmly in place in Masonite models by drilling a small hole beside each and then inserting an overlapping machine screw and tightening it firmly with a nut. The spring blade from an old relay riding against the shaft provides the electrical connection to the spiral wire of the helix.

The drum motor is the Bodine KYC-23, 3500 rpm synchrons, 9.5 watt job used in several industrial applications. They are available as industrial surplus from Herbach and Rademan, Inc., 401 East Erie Avenue, Philadelphia, PA 19135. This motor requires a set of gears to give a 4:1 speed reduction from 3600 rpm to 900 rpm. Suitable items are Pic Design Corporation's G 57-12 along with their G 41-48, which will give the required 15 rps or 900 rpm. Their address is P.O. Box 335, Benrus Center, Ridgefield, CT 06877.

The same supplier as for the Bodine motor has numerous synchronous motors at 1800 rpm. These might be substituted with a 2:1 gear reduction system. Hurst also makes a 900 rpm motor which would obviate the gear problem, but the price is prohibitive in most cases and the motor is larger than needed.

The drum motor is mounted on the right hand end of the printing compartment. The Bodine motor is supported by three of the four mounting bolts, the originals of which have been replaced with 6.4 cm (2 1/2 in.) replacement bolts and spacers. The fourth position allows the large gear on the drum shaft to reach the small one on the motor shaft.

The paper puller assembly uses a 3 watt synchronous motor turning at 15 rpm, which with the described rubber rollers give the desirable square picture for an 8 second frame. Any other 3 watt synchronous motor of appropriate speed would probably work, though the motor is working very near its maximum capacity and a Hurst 10 watt model CA might be more desirable.

The paper puller consists of two lengths of 0.5 cm (3/16 in.) brazing rod covered with 7.7 cm (3 in.) lengths of rubber tubing. If these parameters are changed, then the motor speed must also be altered to maintain the correct speed of paper advance. The brass rods are supported either by brass angles or brackets. It is desirable but not mandatory to turn the rods down at the ends to fit 0.3 cm (1/8 in.) holes in the respective brackets. I used appropriate sizes of spaghetti as coupling to the motor.

The position of the motor was dictated by the position of the paper exit slit and the direction of motor rotation. The top roller is driven when a clockwise motor is used on the left end, while the lower rod must be driven under the same conditions if a counterclockwise motor is used.

This paper puller motor must run from the light mains, but must never be in operation when the drum is turned off and tension is applied to the paper by the printing blade.

Synchronization

It goes without saying that the 60 Hz AC controlling the drum motor must be synchronous to that controlling the camera, the flying spot scanner, or the drum at the transmitting station. This condition is usually true within the continental United States and Canada, but is not when a tape recording is being used at the sending end, nor is it true with most foreign countries. 60 Hz will not necessarily be in sufficiently perfect sync to print out pictures.

For good sync, the voltage on the drum motor should be around 150 volts. This may be obtained from the AC line by connecting a 24V filament transformer so that it operates an auto transformer (Fig. 6-38). The secondary must be reversed if the resulting output voltage is less than line voltage.

If the pictures are to be tape recorded before being printed out, then a dual track recorder is necessary. On the left track the 60 Hz line is taped simultaneously with the received slow scan signal on the right track. If the two AC lines are synchronous, then the picture will be synchronous for any later printout. When thus printing out from a tape, a 15W, 60 Hz AC amplifier or equivalent must be used to drive the motor from the left track of the tape recorder simultaneous with printout from the right.

A much better and more versatile system, however, is the tunable frequency standard. This consists of a very stable variable frequency

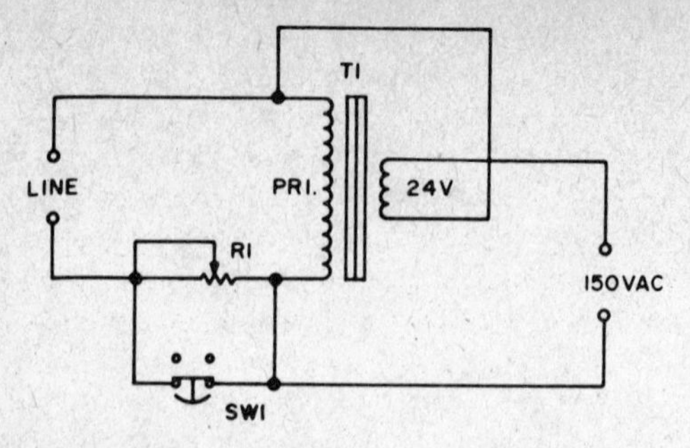

Fig. 6-38. Autotransformer to raise line voltage to 150 volts to obtain good synchronization from the Bodine motor. R1 should be adjusted so that when SW1 is pressed, the motor is only slightly slowed.

oscillator such as a unijunction or voltage sensitive integrated circuit as shown in Fig. 6-39, operating at, say 60 kHz, which is then "counted down" to 60 Hz with three decade counters, such as the SN7490s which are presently so inexpensive on the surplus market.

The 60 Hz obtained from the countdown is square wave and must be turned into sine wave, since the Bodine and Hurst motors definitely object to square

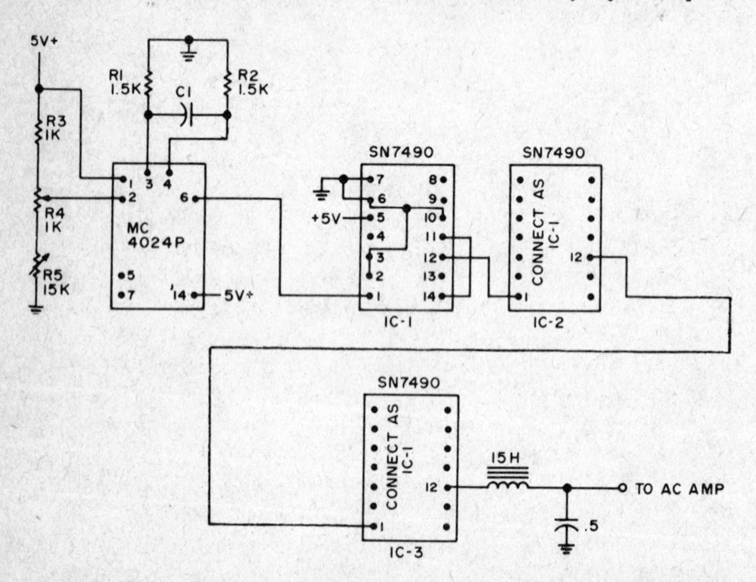

Fig. 6-39. The variable frequency synchronization system worked out by W6KT for his SSTV monitor, modified for use with the fast fax. R_1 and R_2 limit the range while C_1 and C_5 are used to set the frequency for exactly 60 Hz with R_4 at midscale.

wave. This is done by feeding the output into a 15H choke as inductance resonated at the output end by a condenser to ground. The junction of the choke and condenser shows perfect sine wave and is ready for amplification to the 150V at which the motors hold synchronization best. When the voltage on the motor drops below 130V, various synchronization anomalies begin to occur.

If carefully constructed, this makes a very stable standard which can be adjusted to exact synchronization with the received signal, generally even when AC operated tape recorders of the better types are used at the transmitting end. Replacement of a tape recorder motor with a synchronous one would probably be near optimum. Little can be done with the signal from the DC operated cassettes.

Of course, the ideal system would be for everyone to have absolute 60 Hz power to run their frequency sensitive apparatus. A standard frequency obtained from a crystal at, say, 6 MHz and divided down to 60 Hz by decade counters is extremely stable and probably represents the optimum. Briefly, the signal from a 6 MHz oscillator is fed at low impedance successively through three sections of a hex inverter to produce a square wave. This squared 6 MHz wave is then fed into five decade counters successively, coming out at 60 Hz. To improve the level it then goes into the last three sections of the original hex inverter. What comes out is square wave and the previous remarks hold in this arrangement as well.

In order to set the signal of the 6 MHz crystal to exact frequency, some signal from the crystal oscillator must be taken through a "divide by six" integrated circuit to give 1 MHz output. This should then be used to zero the oscillator with WWV at 10 MHz.

Video Circuitry

The video circuit feeding the blade and helix is shown in Fig. 6-40. It consists of a hard limiter feeding the classic SSTV discriminator introduced by McDonald. The output of the discriminator is fed to a transistor amplifier which is a transformer coupled to the demodulator diodes. These are connected to give a positive output, which then drives the Darlington paper driver amplifier. The whiteset potentiometer sets bias on the Darlington and thus should be adjusted to the threshold point with no signal. Black level is set by the contrast control. These parameters are the opposite to those of the usual SSTV monitor. The paper marking voltage should be of about 40V DC shunted with 200 μF of electrolytic capacitors to guarantee the high current capacity required during printing. Hard control of the voltage is necessary since the current varies from zero to 200 or even 300 mA. Be absolutely sure there is no AC hum in the power supply.

Notes on Operation

When operating the fast fax, unless you have a variable standard you should note if the transmitting station is sending live copy. If he is not, or

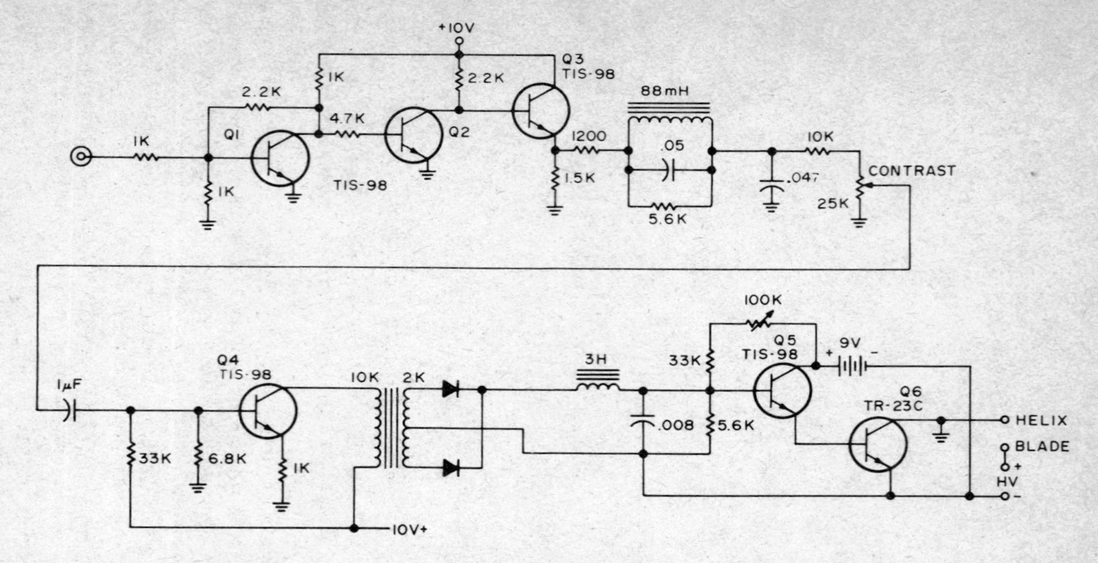

Fig. 6-40. The video circuit consists of the hard limiter $Q_1 - Q_3$ and a conventional McDonald L/C discriminator. This is followed by an amplifier, a rectifier, filter and the paper marking amplifier. Some of the former circuits may be eliminated by use of the video circuits of a conventional SSTV monitor, in which case one would probably connect in at the contrast control.

cannot do so, there is little use in attempting picture exchange. The signal is tuned in on normal voice, the motor is turned on, the contrast adjusted to produce copy, after which the margin is adjusted by the momentary loss of sync button (Fig. 6-40). The pictures should then roll out into your lap at an unnerving rate. Fortunately, experience will cure this initial adverse effect.

The printing surface of the blade requires frequent care. After every run when putting the fax away the blade should be lifted and the surface cleaned with water, and perhaps also with emery cloth. This will prevent a smudgy appearance when collected pigment accretions begin behaving as crayons.

As normal printing occurs, the blade deteriorates due to the necessary electrolysis which bleeds off positive iron ions into the paper. The iron ions unite with the chemical in the paper to produce the mark. After a time the blade may become uneven. Such a blade may be rejuvenated by grinding on a perfectly flat carborundum stone until any bumps are removed. The degree of printing deterioration determines the necessary frequency of this operation.

Threading the paper is a bit of a chore. Perhaps the simplest way is to expose the end until it dries, then glue on a thin strip of 12.8 × 17.8 cm (5 × 7) card to use as a "needle."

Pictures are considerably improved in appearance if they are ironed with "cotton heat" just before they are completely dry. They should feel dry to the fingers but still contain considerable water. When allowed to dry too much, the ironing will tend to wrinkle the paper.

In use, paper towels were placed in the bottom of the humidor compartment which were then saturated with water to keep the paper from drying out. Each time when beginning a run, a length of paper left in the front printing compartment will have dried out. This must be pulled out until the rollers make contact with the wet paper. The paper pulls easily when the dry is between the rollers, and it will slip until it is in contact with wet paper.

Electrolytic printing paper is available from Alden (Alfax Paper and Engineering Co., Inc., Westboro, MA 01581) in single roll quantities; there are several kinds to choose from. Alfax A2-41 gives much better depth without gamma correction circuits, operates more smoothly, and prints in black and white, but unfortunately fades and discolors in time. Alfax A paper is more sensitive, but does not make the clean, crisp picture that the A-2 paper produces. It requires 50% more drive from the video circuits.

Much improvement is yet to be made in the video circuits, the type of paper used, as well as in other parameters of the circuits and mechanics. Only the basic circuits and mechanics are presented here as an invitation to improve and perfect the system.

AUTOMATIC VERTICAL TRIGGER FOR SSTV

In order to improve the vertical sweep function of a slow scan TV receiver converter the circuit shown in Fig. 6-41 was constructed. The vertical retrigger

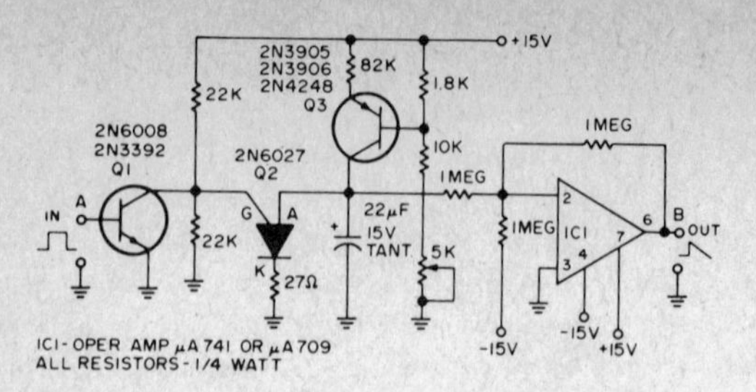

Fig. 6-41. Automatic vertical trigger circuit.

consists of Q1, Q2, and Q3. If output is taken from Q3 it would be necessary to reverse connections to the vertical plates of the cathode ray tube due to phase inversion. In order to eliminate the problem so that the oscilloscope used with the SSTV converter could be used without reconversion for normal response, the operational amplifier stage was added.

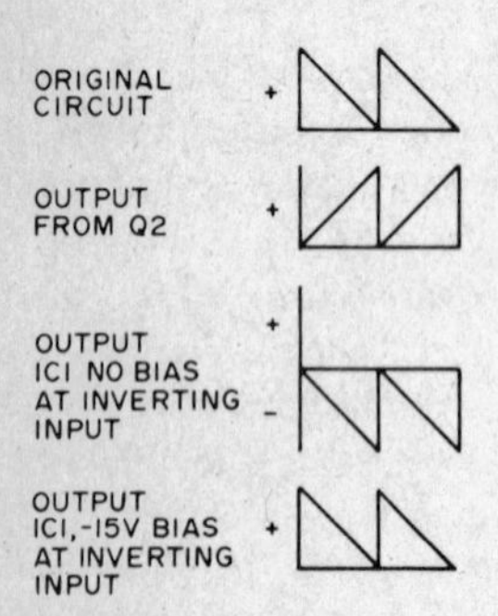

Fig. 6-42. Vertical output wave forms.

The wave forms shown in Fig. 6-42 indicate the output from the original vertical sweep circuit (non-recurring) output from Q2 and from the op-amp. The negative bias applied to the inverting input of the op-amp restores the polarity and makes the wave form congruent with the original. The 5000Ω potentiometer is set for a sweep period of 8.5 seconds.

Fig. 6-43. Circuit replacement points.

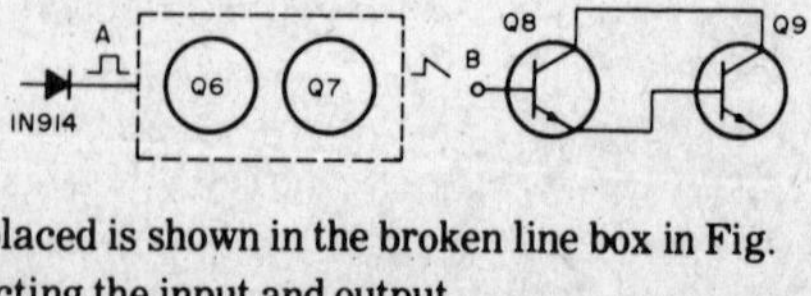

The original circuitry to be replaced is shown in the broken line box in Fig. 6-43 along with the points for connecting the input and output.

The entire unit is built on a 1 × 2.5 in. piece of perforated Vectorboard and mounted directly above the original components.

Chapter 7
Amateur Accessories

A DIGITAL READOUT FOR YOUR VFO

Are you on frequency? Are you within the band? Incentive licensing subbands have made this an increasingly difficult question to answer. Even if you have an Extra license, it is still nice to know where you are for net operation, OO work, etc. The declining price of integrated circuits and readout devices has made it unnecessary to continue to drool over the advertisements for that $1600 transceiver with the digital dial. A large percentage of currently manufactured gear can be fitted with a digital dial with greater accuracy, and very reasonable cost. Little modification of your gear is required,and what is required will not affect the resale value, since it is invisible and is easily removed.

As an extra added plus for the homebrewer, the digital dial makes unnecessary the greatest hate object of the electronic purist; the tuning dial, with its attendant impossibilities of getting linearity, accuracy, and above all, of inscribing, decaling, engraving, or calligraphing the dial in a neat, readable manner.

The device to be described is an adaptation of the basic electronic counter circuit. It takes the VFO signal of your receiver or transmitter and tells you what it is. Accurately. Not to the nearest kHz but rather down to the nearest 10 Hz, almost always, with no precautions or requirements other than a periodic check to zero-beat the oscillator with WWV. If it is installed in a transmitter, it can measure received signal frequency by zero-beating the transmitter with the receiver. It can also measure frequency shift of an RTTY signal by zero-beating the mark and space frequencies and subtracting the two readings.

Your rig should have crystal-controlled front end or conversion oscillator and a VFO covering a reasonable frequency range without odd number

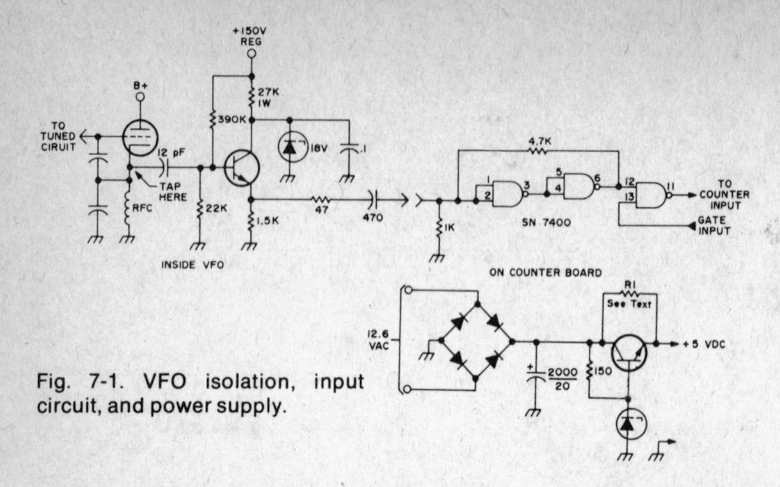

Fig. 7-1. VFO isolation, input circuit, and power supply.

kilohertz tacked on. Ideally, a range of from, say 5.000 to 5.500 MHz would be covered. The unit can be used with a unit whose VFO covers, for example, 5.300 to 5.800 MHz with only minor changes. Input frequency is unimportant—anything up to 15 or 20 MHz is okay with the ICs specified. What is important is that there be no odd numbers at the end. A VFO covering 5.455 to 5.955 would be unacceptable. Of course, the counter will measure such frequencies as well as any other, but they will be impossible to mentally relate to the operating frequency. Another requirement is that the frequency mixing scheme be either additive or subtractive, but not a combination of the two. A quick glance at your instruction manual will tell you the exact VFO frequency and mixing scheme. If your rig tunes in the same direction on all bands, you should have no problem with the additive/subtractive question.

The digital dial described herein is being used with an HX500 transmitter, whose VFO is 3.9–4.4 MHz, and which employes subtractive mixing. However, the principles and circuitry can be used with only minor modification in any transmitter or receiver fitting the above criteria.

The Circuit

The first step in going digital is to modify your rig. The digital dial needs +5V DC at about 500 mA, B+ at about 15 mA, and 1V of signal at the VFO frequency. To get the 5V, tap the filament supply (6.3 or, preferably, 12.6V) and connect it to the bridge rectifier shown in Fig. 7-1. If your filament supply is 6.3V, it may be necessary to connect another 6.3V, 500 mA transformer in series with the filament supply. Nixie tubes like 170V at about 3 mA across them, but since they tend to act as voltage regulators, current limiting is necessary. Measure your B+ supply, subtract 190V, and calculate a resistor that will give 3 mA per Nixie with this voltage across it. Be sure to make a power calculation also, as a 2W resistor may be required. Small series resistors are connected to each Nixie to equalize current; the resistor just calculated goes to the junction of the series resistors.

The purpose of the VFO buffer is to make sure that the VFO in the rig is not disturbed. An emitter follower is ideal for this application. The component values shown are for a supply of 150V, as this is the most commonly used. No trouble should be encountered if the pickoff capacitor (12 pF in the schematic of Fig.7-2) is at least an order of magnitude smaller than the capacitor from the cathode to ground. The actual capacity will be much smaller, since the pickoff capacitor is in series with the transistor base, but it doesn't hurt to be safe. This completes the necessary modification of the rig.

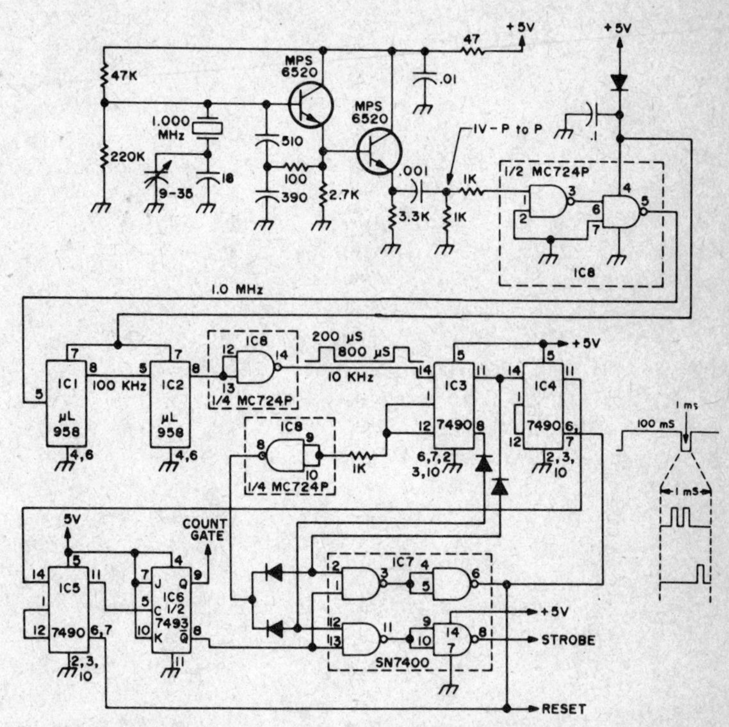

Fig. 7-2. Frequency-standard oscillator and counter control circuitry with timing diagrams.

Circuit Description

The theory behind the electronic counter is that the number of cycles of the signal in a given period can be counted. If the period, for example, is 1 second, the number counted comes out to be cycles per second. Deriving the period precisely is done by counting down an accurate reference oscillator and then using the counted down signal to gate the signal to be measured.

The circuit works as follows (see Figs. 7-3 and 7-4). The input signal goes into a divider chain, the first stage of which is the least significant digit. After 10 pulses, the first stage is reset and a carry pulse generated. Since the counter has five stages and the counted frequency is in the MHz range, you can see that the counter will overflow several times during each count interval. However, we know what the first digit will be, so there is no point wasting a counting stage on it. All the stages are identical except for the most significant digit, which uses discrete components since special decoding is required.

After the counters have counted, the SN7441AN decoder/drivers ground the appropriate Nixie cathodes and the corresponding numbers light up. Between the counters and the decoder/drivers, one additional stage is necessary—a buffer storage register. The reason for this is that the gate time is 0.1 second, which will cause blurring during counting. If you wish to read out once per second, this isn't too bad. However, the dial is more useful when it's responsive, and you can zero in a frequency much more quickly when you don't have to wait a full second to see where the dial turn took you. By using storage registers and a few logic gates, it is possible to make ten 100 ms measurements in 10.1 seconds with no digit blurring. How this is done brings us to the timing circuitry.

First, an accurate frequency must be generated. The circuit shown in Fig. 7-2 is a stable oscillator designed for 32 pF crystals. A trimmer is included for fine frequency adjustment. There is no reason why the 1 MHz output of another oscillator can't be used. Approximately 1V (peak-to-peak) is required. You can also use a 100 kHz oscillator and delete one divider stage. The MC724P quad 2-input gate shapes the signal into a square wave suitable for triggering the dividers.

ICs 3, 4, and 5 are dividers. At time zero, let us suppose that IC6, the gate flip-flop, is reset (gate at logic 0). The carry pulse from the last divider brings the gate high, starting the count. One carry pulse later (100 ms), IC6 is again reset, stopping the count. (The abbreviation "ms" means milliseconds; "μs" means microseconds.)

IC6's Q, being high, sets half of the *and* condition on the first two sections of IC7. IC3 is still counting from binary 0 to binary 9 at a rate of 1000 times per second. Its output states are decoded by the remaining inputs to IC7 in the following sequence: At binary 4 and 6, pins 12 and 13 are both high, causing pin 11 to go low, pin 8 to go high, and thus strobing data into the buffer register. The strobe is performed twice because doing so eliminates the need for an additional decode—the same data is strobed each time.

At binary 8, pins 2 and 1 of IC7 are both high, causing pin 3 to go low and pin 6 to go high, thus resetting the counter and resetting the last two 7490s in the dividing chain. However, these two dividers are reset to 9 instead of to 0. This means that just

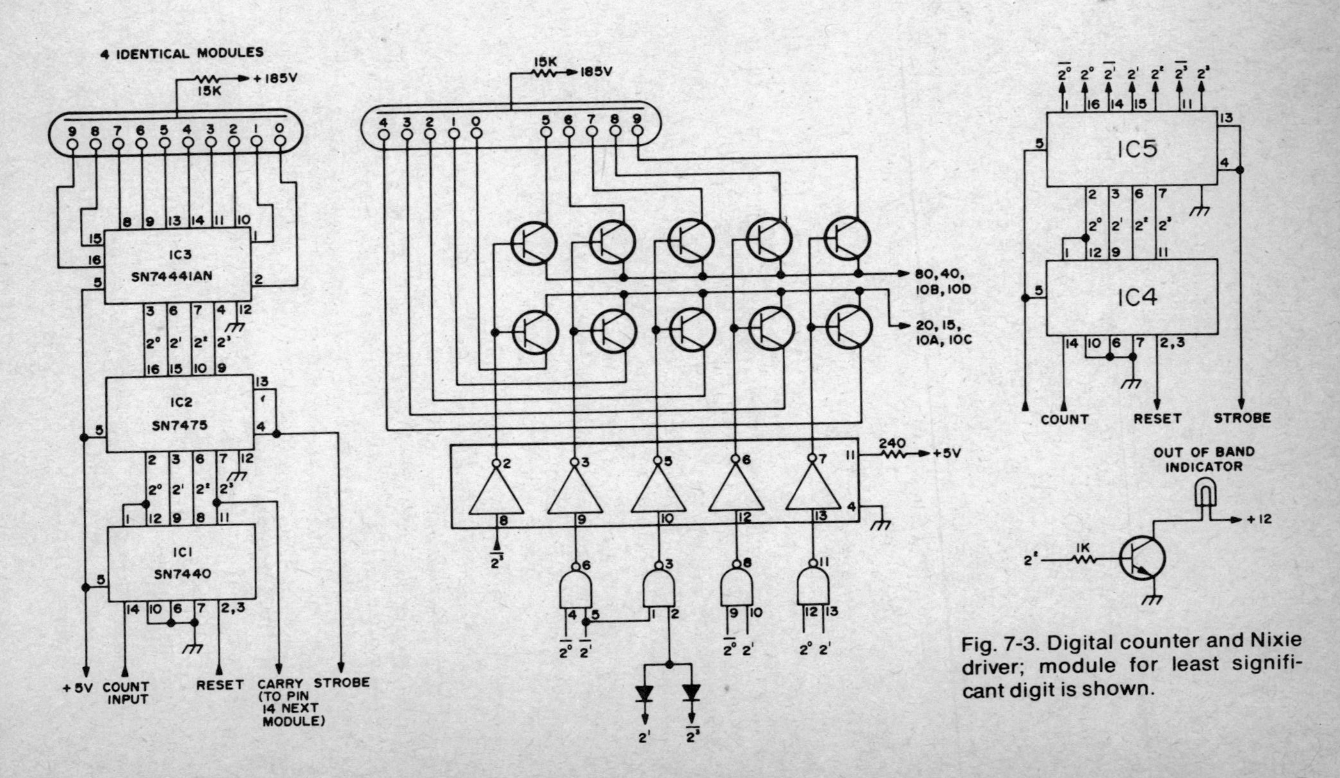

Fig. 7-3. Digital counter and Nixie driver; module for least significant digit is shown.

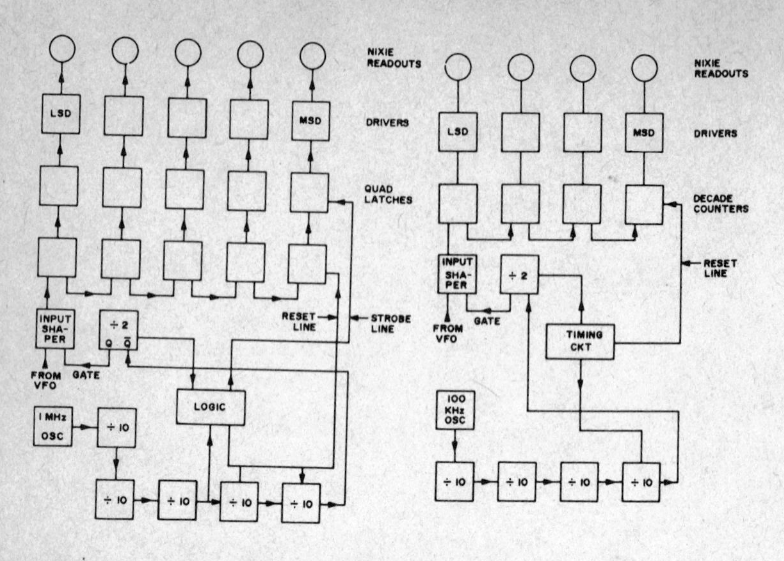

Fig. 7-4. Logic block diagram. The left block gives readout in 10 Hz increments at 101 ms intervals without blurring. The right block reads in 100 Hz increments; the update is adjustable and blurring occurs during 100 ms count.

0.1 millisecond later, when IC3 carries, IC4 and IC5 will also, starting the count again. Note that the total time between the end of the first count and the beginning of the next is just 1 ms, allowing almost 10 measurements per second.

Backwards or Forwards?

If your rig is of the subtractive mixing type, you are probably wondering how the readout frequency corresponds to the output frequency of the rig. It doesn't—everything is upside down. Of course, it will be accurate on one frequency in the middle of the band, but that's little comfort. Despair not. Instead, connect the Nixies to the counter backwards. (Switch 0 for 9, 1 for 8, 2 for 7, 3 for 6, and 4 for 5.) Through the magic of mathematics, for every 10 Hz increment of the VFO frequency, there is a 10 Hz decrement of the readout frequency. You can now see why the rig can be additive or subtractive, but not both.

Most Significant Digit

The one remaining readout problem is that of the most significant digit. If you want the digital dial to be accurate on all bands it must read: (3)745.92 kHz on 80 meters, and (7)245.92 kHz on 40 meters. One way this can be done is to have different drivers for the Nixie, and ground the appropriate one with an extra wafer on the band switch (see Fig. 7-5). By driving the bases in parallel and grounding the emitters of the appropriate group, a cheap and dirty "adder" is constructed.

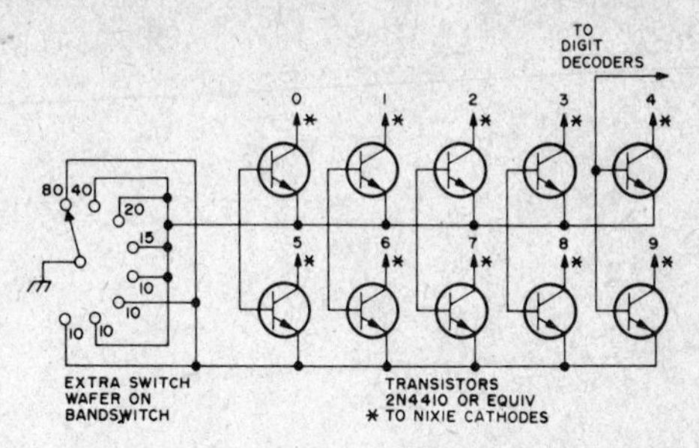

Fig. 7-5. Using a wafer switch to ground the appropriate Nixie drive for "adder" function.

If your VFO starts from some frequency other than an even megahertz, the decoder must be designed to match. If the VFO is, say 3.9 to 4.4 MHz (as is the HX500), where 3.9 corresponds to the top of the band and 4.4 to the bottom, then the decoding is as follows: 4.399.99 to 4.300.00 is 000.00 to 0.99.99 kHz on the dial. Thus, 3 must be decoded as 0; 2 as 1; 1 as 2; 0 as 3; and 9 as 4. To do this the gate that decodes a 3 binary state should drive the 0 (and 5) digit line, etc. The decoding section is fairly simple since only 5 states must be decoded by the SN7441AN. Figure 7-6 and Table 7-1 give the *general* decoding logic for all possible binary-coded-decimal states.

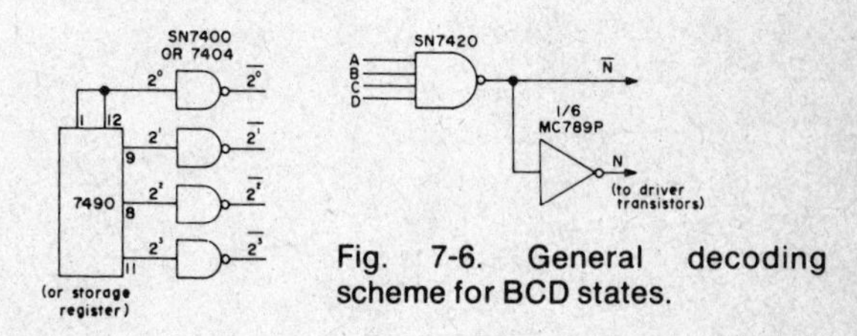

Fig. 7-6. General decoding scheme for BCD states.

If you wish, the unused states of the counter can be used to give an out-of-band indication (Fig. 7-7). None of the N outputs will be low if none of the proper states are decoded. This allows the transistor to be saturated. You can use it to light a light, ring a bell, disable the VOX circuit, or start a tape recording.

Thus far, all comments have been about a "deluxe" counter with storage and 10 Hz readout. If you wish to build a more austere model, there are ways to cut down on the cost. The best is to make a four-digit model with a 10 ms time base. Since

Table 7-1. Connection Chart for All Binary Combinations.

N=	BINARY	CONNECT A	B	C	D
0	0000	$\overline{2^0}$	$\overline{2^1}$	$\overline{2^2}$	$\overline{2^3}$
1	1000	2^0	$\overline{2^1}$	$\overline{2^2}$	$\overline{2^3}$
2	0100	$\overline{2^0}$	2^1	$\overline{2^2}$	$\overline{2^3}$
3	1100	2^0	2^1	$\overline{2^2}$	$\overline{2^3}$
4	0010	$\overline{2'}$	$\overline{2^1}$	2^2	$\overline{2^3}$
5	1010	2^0	$\overline{2^1}$	2^2	$\overline{2^3}$
6	0110	$\overline{2^0}$	2^1	2^2	$\overline{2^3}$
7	1110	2^0	2^1	2^2	$\overline{2^3}$
8	0001	$\overline{2^0}$	$\overline{2^1}$	$\overline{2^2}$	2^3
9	1001	2^0	$\overline{2^1}$	$\overline{2^2}$	2^3

the count interval is so short, you can sample once every tenth of a second, count for a hundredth of a second, perform housekeeping functions in 2.0 ms, and have a display visible for almost 99% of the time without using storage resistors. One Nixie state is eliminated, and IC6 is unnecessary.

To make the stripped-down verison, substitute the circuitry of Fig. 7-8 for the bottom half of Fig. 7-2. It works as follows: As the dividers count up from 0, the diodes decode a count of 88 and generate a 1 ms long reset pulse. At the transition to 90, the count gate decodes a binary 9 on the most significant digit

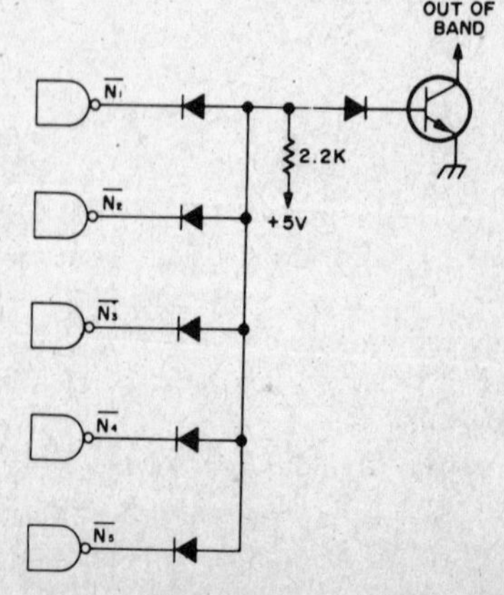

Fig. 7-7. An out-of-band indicator is constructed by ANDing the unused N states which feed a simple driver.

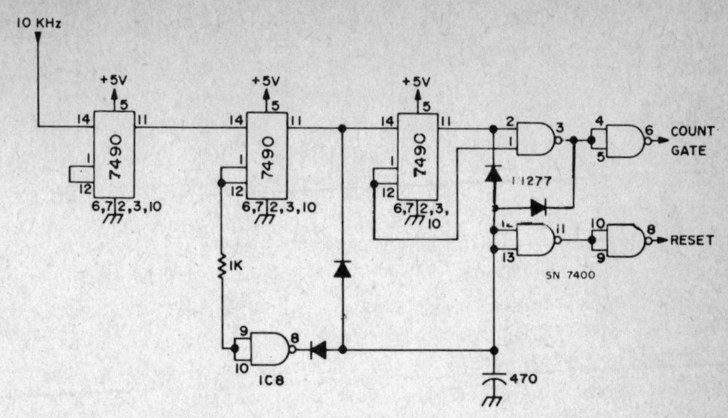

Fig. 7-8. By substituting this circuit for the bottom half of Fig. 2, a cheaper but less accurate digital counter results.

and goes high for 10 ms, forming the gate interval. Thus, the SN7490s essentially perform the timing functions.

An additional Nixie stage can be eliminated by deleting the most significant digit from the count chain. This also saves the trouble of figuring out the decoding circuitry. Since any VFO will be accurate to the nearest 100 kHz, a second of mental work will compute the correct number.

Construction

Unless you want to make a PC board, the easiest way to build the unit is to use a "universal" IC card. The ICs and the Nixies can be mounted on it, and no drilling is required. Since most of the wiring is repetitive, it can be accomplished quite rapidly. The counter board will take about 4 hours, the timing board about 2.

One suggestion to speed things up is to use the wire that can be found in multi-conductor telephone cable. A four foot scrap of this stuff has 200 feet of hookup wire in it, and the wire has a soft plastic insulation which is very easy to strip. It melts at a low temperature, but since you're using a low power soldering iron, there should be no problem.

The crystal oscillator is quite stable, but it can obviously be improved by installing the crystal in an oven. The TTL circuits used in this project are much better than RTL with regard to noise immunity. A significant amount of RF can be floating around before erratic operation occurs. However, it is recommended that the unit be carefully bypassed and shielded; first, because often there *is* a considerable amount of RF floating around, and, secondly, because the pulses in the digital circuitry are very fast and can radiate noise to the receiver.

Mechanically, the unit is constructed on two boards which are then sandwiched together. This saves about 10 square inches. There is no reason why the unit can't be built on one board. Final dimensions exclusive of case are 5 by 4 1/2 by 2 1/4 in. Something this size could probably be installed inside the cabinet of the rig. If you wish, the Nixie and their drivers can be placed on a separate board which should easily fit in the space taken by the typical dial drum or scale assembly.

The unit requires 5V and some convenient high voltage. The 5V supply is worth a few words. Regulation is not critical, but if the supply has lots of hum or noise, filter it out before it gets to the oscillator. The oscillator only takes 5 mA or so and an RC filter should be sufficient.

The resistor shown bridging the transistor bypasses some of the current without, theoretically, greatly affecting the regulation. Measure the voltage across the transistor—there should be a large AC component. Pick a value for the resistor so that when the voltage is at an instantaneous minimum, the current through the resistor is about 100 mA. If all this is too much trouble, don't bother. Almost any transistor you choose can handle the power even without the resistor.

Final Thoughts

This is a fairly simple and relatively cheap way of obtaining accurate frequency calibration. There are some inherent limitations. The main limitation is that the counter doesn't actually measure the transmitter output frequency. Since the assorted conversion oscillators are inevitably crystal-controlled, there is little cause to worry about significant errors. When you install the unit, it might be a good idea to set the VFO at 000.00 and zero-beat the transmitter output with your calibrator by adjusting the trimmer capacitors on the conversion crystals.

If you can't build the counter described because your rig doesn't fit the criteria, there is still hope. For instance, if it is additive and subtractive, you might consider using the bidirectional counters that are now available. Direction of count can be controlled by the band switch. These units are presently more expensive than the 7490s.

If your rig uses some odd VFO frequency, perhaps with a 455 kHz added or subtracted, you can try several things. One is to mix the VFO with a signal that will subtract the odd number. Another is to preset the counter with the reciprocal of the odd number and set to 9 all digits more significant than the odd number.

INTEGRATED CIRCUIT FREQUENCY COUNTER

Here's a useful transistor-IC counter that's sure to clear out a significant chunk of your junk box.

It's capable of 50 MHz operation with a 74196 in the first DCV stage, and the only test equipment required is a VOM to check voltages and a scope to troubleshoot possible band connections. To set the time base generator on frequency, make a comparison with a commercial counter or with WWV using a scope to get as close as possible to the exact frequency. The common practice of zero-beating by ear is just not accurate enough. This is especially true if you want the counter to have the best possible accuracy that this design will give. Calibrate a ten MHz crystal to WWV, then adjust the counter to read 10.000.000+. That is the best method available with limited resources.

Construction Ideas

The arrangement of the decode counting units (DCU) is straight in line with the SN74196 being first on the right as viewed from the front. The rest of the ICs are placed so as to keep the wiring as short as possible to minimize interference between sections. The main part of the counter is constructed on a 4″ × 6″ vector board with .1″ center spaced holes which match the spacing of the ICs. No sockets are used although all ICs are removable. This is accomplished by using integrated circuit terminals. These terminals come in strips of 56 each. This is a convenient size as the whole strip can be inserted in the holes, wired and then the top of the strip is broken off giving the equivalent of sockets with less cost and effort. In one counter, seven disply tubes which fit well on the board and come out just right for the integrated circuit terminals were used. AKZ type tubes and sockets fit the .1″ spacing on the vector board.

Input Amplifier

In the construction of the input amplifier (Fig. 7-9) several types of integrated circuits were tried before settling on the SN72733. Four 2N2369s were tried first and found to be unsatisfactory. The RCA CA3018 (Fig. 7-10) was then tried and found to be an excellent input when used with a 74H10 input gate (Fig. 7-11). It will go up to 32 MHz with no problem. The SN72733 from Texas Instruments will work up to 50 MHz with the SN74S10 input gate. You have to use the combinations together, the SN72733 with the SN74H10 for up to 32 MHz.

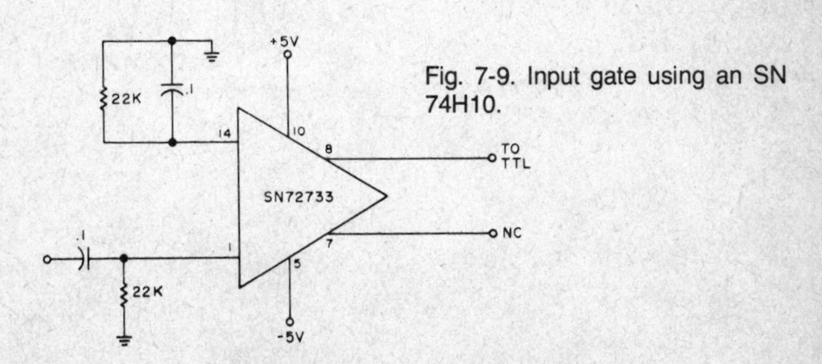

Fig. 7-9. Input gate using an SN 74H10.

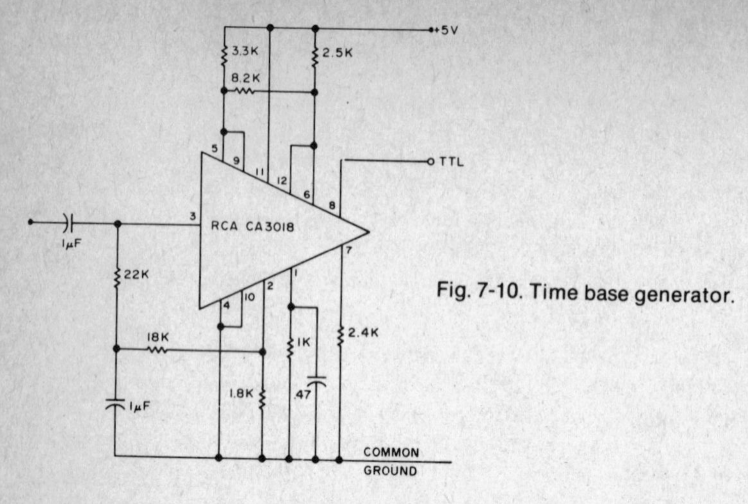

Fig. 7-10. Time base generator.

If you want lower frequency operation you can use the CA3018 with a SN7410 and by this route be able to use a SN7490 for the first DCU. This combination would be able to count to approximately 20 MHz. This is the only section of the counter where different designs were tried. The rest of the counter is designed and constructed with very few changes being made. From the input gate the signal is fed into the first DCU (Fig. 7-12). All of the DCUs are alike except the first one which uses an SN74196, SN7475 and SN7441 which gives the first digit. The rest of the DCUs are SN7490, SN7475 and SN7441. The SN74196 is used in the first DCU because it has a higher toggle rate (50 MHz) than the SN7490 which will only go to 20 MHz maximum. The SN74196 has to be reset through an inverter because it needs a negative going pulse to reset whereas the SN7490 needs a positive going pulse.

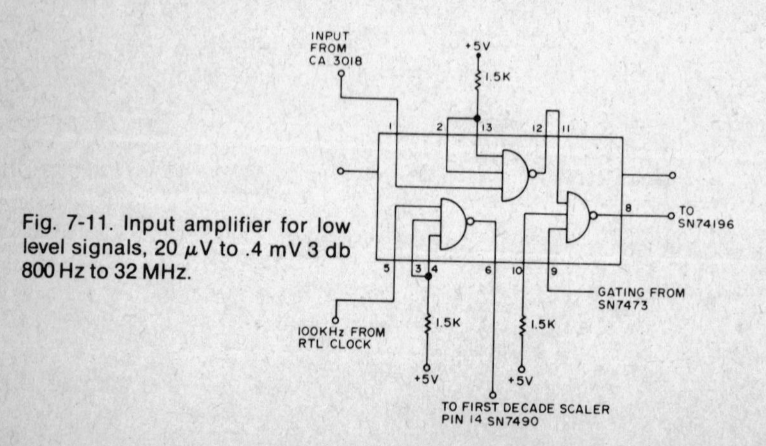

Fig. 7-11. Input amplifier for low level signals, 20 μV to .4 mV 3 db 800 Hz to 32 MHz.

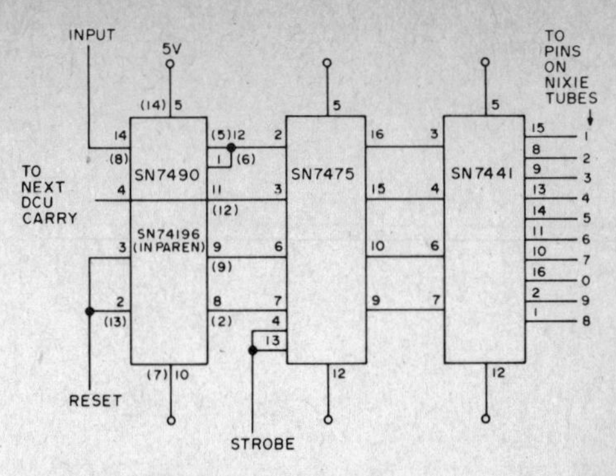

Fig. 7-12. The SN74196 requires an inverter on the reset. Pin numbers in bracket are for the SN74196. Pins 1, 3, 4, 10, 11 are open. Pin 7 Gnd, 14 + 5V.

The input gate is opened and closed by the Q output from the JK flip-flop SN7473 (Fig. 7-13). The $\bar{Q}$ output is connected to the four wide two-input *nand* gate SN7400 together with pulses from decade scaler number four to generate the strobe and reset. The diodes are used to decouple the set up pulses passing current in one direction only. The SN7400 is blocked from producing the strobe and reset except when the gate is closed by the $\bar{Q}$ output from the JK flip-flop. When the strobe is generated the count is transferred to the read out, then the reset is generated resetting the DCUs and scalers five and six to be ready for the next sequence in counting. This is a good scheme and gives a good ratio of read out time to count time.

Time Base Oscillator

The time base oscillator (Fig. 7-14) uses a 1.4 MHz crystal in a circuit using a 2N706. The padder in series with the crystal allows calibration to the exact frequency. The buffer is also a 3N706 that serves to drive the first μL 923 in the divide by seven circuit. Pin 5 of the third μL 923 is connected to another 2N706 in a half monostable circuit resetting the first μL 923 forcing a divide by seven. The last μL 923 divided by two giving 100 kHz output. This can be modified to fit any crystal as long as it can be divided by a whole number. The whole divider chain can be eliminated if you have a 100 kHz crystal. The 100 kHz from the RTLs is run through one section of the SN74H10 to improve the wave shape, that gate was left over from the input IC, so no parts were added. From the SN74H10 the 100 kHz is fed into the first of six decade scalers SN7490. Connected to the first three decade scalers is a four wide two input and-or-invert gate SN7454 together with four sections of hex inverter SN7407 making it possible to select one of four frequencies: 100 kHz, 10 kHz, 1 kHz and 100 Hz. These four frequencies, after going through the last three decade scalers, will give integration times of 0.01, 0.1, 1.0 and 10 seconds. The 0.01

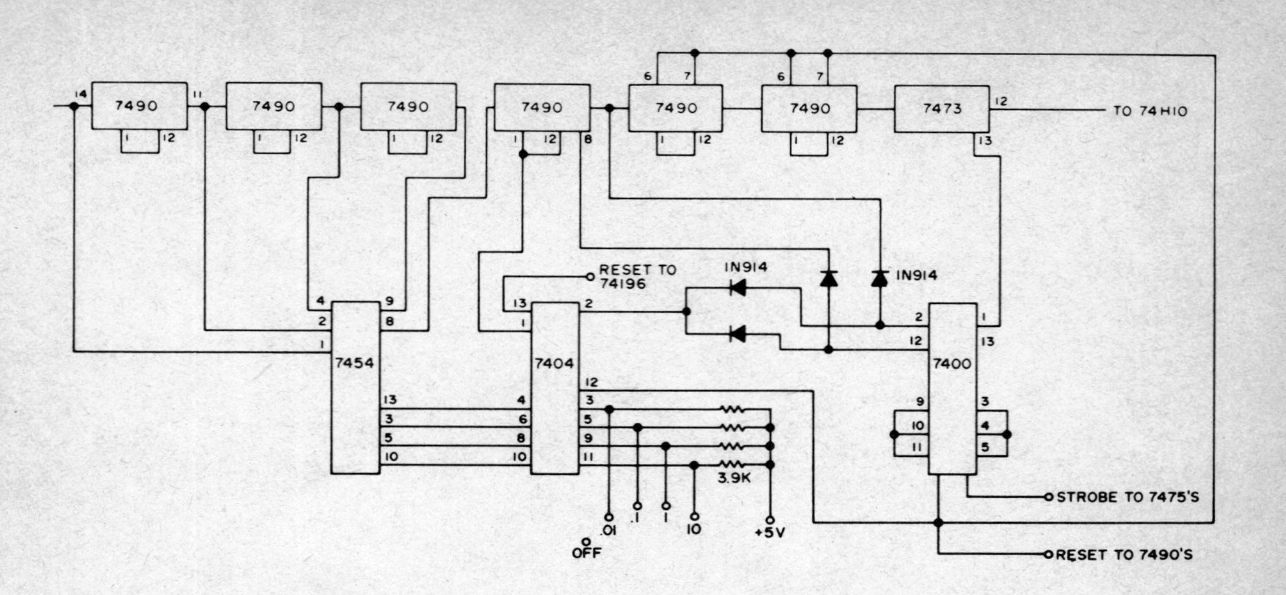

Fig. 7-13. Input amplifier circuit. Pins 3, 4, 11, 12 are open. Pins 2, 6, 9, 13 are NC.

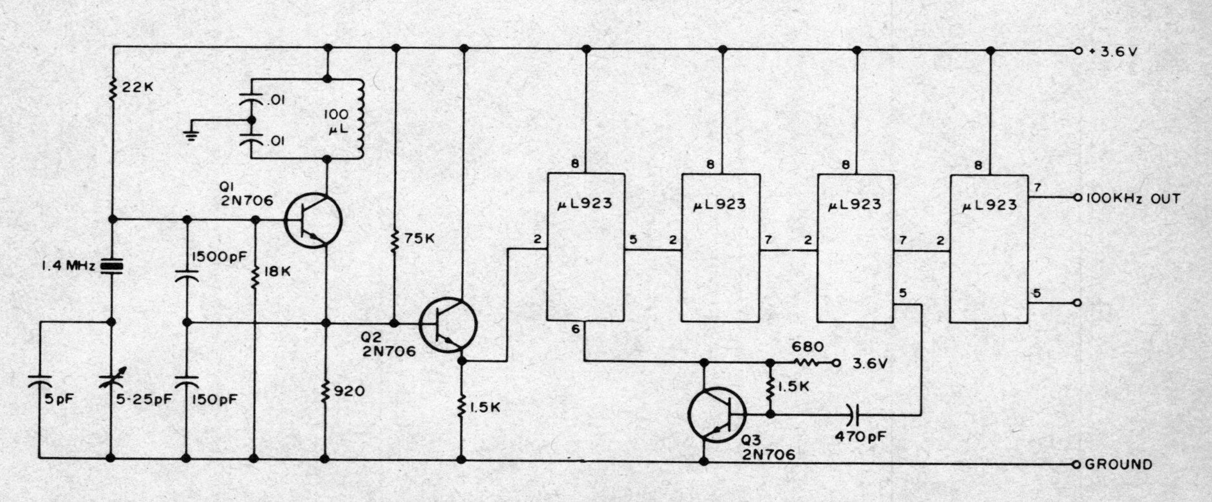

Fig. 7-14. Frequency counter timing.

second integration time has no real value in this design and need not be wired. However, if you use less than seven read outs, you will need it to read tens of MHz. It is also useful in checking and troubleshooting.

Power Supply

The power supply (Fig. 7-15) uses a transformer rated at 500 V centered tapped 50 mA and 6.3V 1.2A. Using a full wave rectifier on the 500 volts and 40 μF 450 volt capacitor gives approximately 300 volts DC under load. Three 27,000Ω two watt resistors in parallel drops the voltage down to the zener diode rated at 180 volts 10 watts. In series with each Nixie tube, there is a 33,000Ω half watt five percent resistor. The recommended voltage for the Nixie tube is 140 volts at approximately 2 1/2 mA. This arrangement can be changed to fit what you have on hand by changing the series resistor to a value that will assure that no more than 140 volts are on the Nixies. The low voltage supply is a full wave bridge rectifier with 40,000 μF filter capacitor which gives about 8 volts. This is regulated by an IC, LM-390-K, that has a rated output of 5 volts at 1 ampere. The counter draws a little over half of that and runs cold. The power supply is built right in the 12 × 8 × 3 inch chassis adhering only to an arrangement that saved as much space as possible. Not shown on the schematic is a minus five volt power supply for the SN72733 which is necessary if you decide to use this system. The SN72733 draws less than 30 mA, so not much is required. If you use the CA3018 this is not necessary. If you want to measure frequencies less than 10 MHz only, you could build a transistor amplifier using almost any fast switching type. In that case three or four readouts would be satisfactory. Then there would be no need for the different integration times so the SN7454 together with the switch could be left out. If you really want to cut it down you could use the power line frequency for the time base dividing by six to get .1 second integration; throw in a SN7490 to get one second.

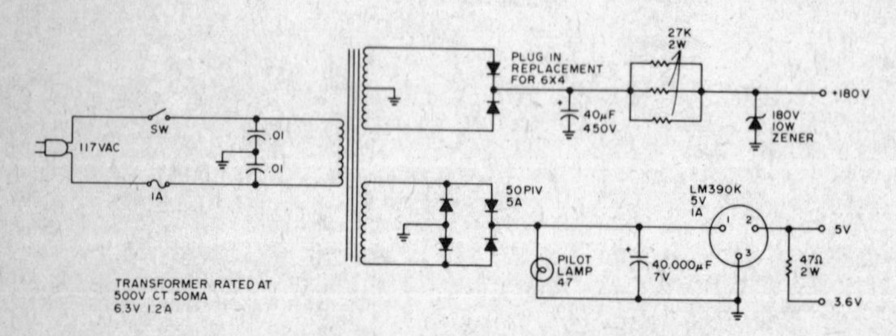

Fig. 7-15. Schematic of the regulated power supply.

Troubleshooting

Troubleshooting of integrated circuits requires a lot of pre-planning. First, with a piece of hookup wire, jump pin five of the last μL 923 to the input.

You should get 100 kHz on all ranges. Do not confuse this as an indication of the accuracy of the counter but as an indication that all systems are working. With a scope on pin 8 of the SN7554 you should get 100 kHz with the switch in the .01 second position, 10 kHz in the .1 second position, 1000 Hz in the one second position and 100 Hz in the ten second position. To check the rest of the decade scalers, put the switch in the .01 second position and follow the signal through the SN7490s by checking the pin 11 of each one, scaling down by a factor of ten for each IC. Finally, on pin 12 of the SN7473, check for the gate opening and closing, and on pin 13, for the timing pulses to the SN7400. Set the switch to the ten second position and measure the gate opening and closing with a VOM. The up's and down's occur pretty fast but it will show on the VOM. Pins 8, 1 and 12 of decade scaler 4 should be nice square waves with the frequency measured depending on the switch position. Pins 1 and 3 of the SN7404, pins 2 and 12 of SN7400 should all be square waves; however, the strobe and reset pulses are hard to lock on even with a triggered scope due to long down to up time.

Using your VOM, momentarily jump pins 3, 4, and 5 of SN7400 to ground and read a high on the reset line. Do the same thing on pins 9, 10, and 11; measure a high on the strobe line. Do not forget the reset on the SN74196 is inverted by one section of the hex inverter. An indication that all is not right is that one of the read outs has a preference for one digit. This can be caused by a bad connection from the SN7490 to the SN7475 or from the SN7475 to the SN7441. A bad connection from the SN7411 to the Nixie tube will show up as no digit on some count and a short will show one digit all the time. Trouble in the strobe line will cause it not to transfer or will flutter during counting. Trouble in the reset line can be detected by a continuous digit on the read out after the input has been removed and sufficient time has passed to reset. Be certain the connections between the SN7490, SN7475 and the SN7441 are correctly wired. An error in this department will cause the SN7441 to decode wrong. This could show up the same as a bad connection; that is, with no input and after the reset there would be some number on the read out. Depending on the cross and how it was decoded would determine what number would be on the read out.

With the grid dip oscillator couple some RF into the counter using about twenty turns of hook up wire around the GDO coil. Starting at the low end of the dial, slowly increase the frequency until you get up as high as you can, adjusting the wire positioned on the coil to get the best coupling to the counter. Don't be discouraged by dead spots or a frequency difference between the GDO and the counter. The hook up wire will load the GDO down and make the dial read high. Using a tube type Millen GDO, no trouble was experienced in getting the counter to work to 50 MHz. This system design will go up much higher in frequency, perhaps up to two meters.

LOW COST FREQUENCY COUNTER

There are still a few hams who have not built a digital counter, due in some cases to misconceptions regarding the complexity of digital devices. The counter to be described is about as basic as a counter can get, is very inexpensive to build, and will count to above 30 MHz. The readout is in decimal form, using 7 segment incandescent Minitron display.

Devices used for the digital section of the counter are all SN7400 series. Data sheets on these integrated circuits are available from many of the suppliers, such as Poly Paks, Solid State Systems, etc. A careful reading of the data sheet will reveal there is nothing mysterious or even hard to understand about the operation of these digital counting ICs.

These IC devices are analogous to building blocks. All that has to be done is to connect these blocks correctly to build a counter. There is no interface problem between the blocks. There is a slight interface problem where the signal to be counted enters the first 7400 device. For instance, the 7490 will operate with either sine or square wave, but if the input of the 7490 is left open, it will drift up; the device will lock up and refuse to switch if the signal is coupled through a capacitor. Since capacitor coupling is often desired, a resistor can be connected from the input pin to ground to keep the input from drifting up and locking.

A resistor of several hundred ohms will correct the lock up condition and allow switching, but it was experimentally determined that a biasing network was a better solution. If the input is held at about 1.2V, a signal which swings symmetrically about this point will be counted with a minimum of wasted voltage swing. This does lower input resistance, but this is a minor inconvenience. It was thought that the IC might be confused about whether it was at a logical 1 (up) or 0 (down). This has not been a problem.

The disadvantage of feeding the IC chain directly is the low input impedance inherent in the method. A better method is described later.

Frequency Measurement

Consider a 1 MHz sine wave, symmetric about a zero axis. Each cycle will have 1/2 T in the positive direction and 1/2 T in the negative direction, where T is the period and is 1/F. In exactly one second, 10^6 cycles will occur.

It is obvious that two things are required. The one second must be known to a high degree of accuracy and something must be available to detect (count) each of these micro-second intervals. The latter problem is simplified because each period is similar to the others and there is one positive excursion and one negative excursion during each period.

The 7490 will count this signal or any signal within its frequency range provided the negative part of the signal holds the 7490 input pin at less than .8V for at least 50 nano-seconds. The typical upper frequency limit of the 7490 is 18 MHz. Selection of the first 7490 will raise the frequency limit since some 7490s will switch faster than others.

Counter

A highly accurate timing signal is obtained by dividing down a 1 MHz signal from a frequency standard. The clock (Fig. 7-16) begins with a crystal

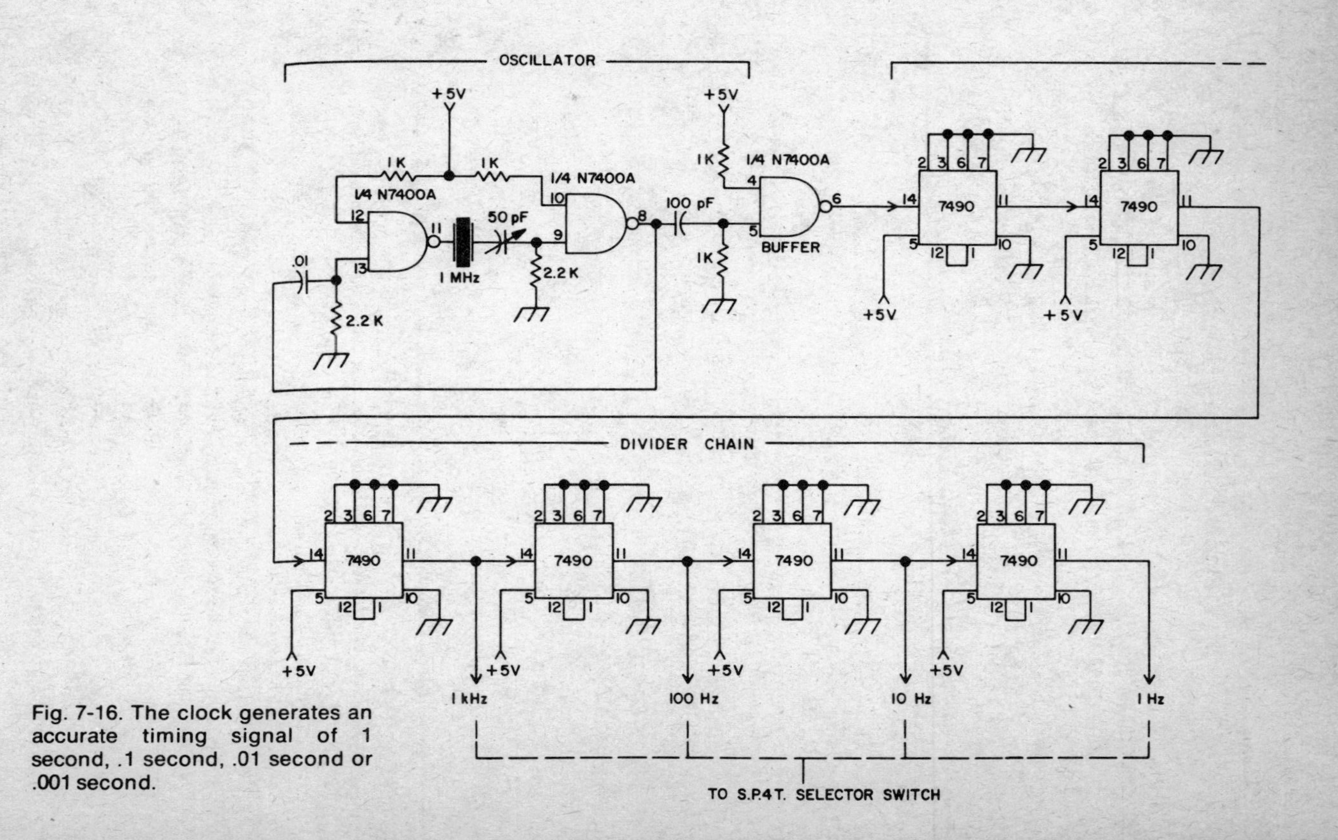

Fig. 7-16. The clock generates an accurate timing signal of 1 second, .1 second, .01 second or .001 second.

oscillator which uses two cross-connected gates as active elements. This signal is divided to whatever frequency is desired. The frequency is selected by a switch which feeds the gating circuit (Fig. 7-19). The signal coming from either of the clock outputs is a square wave, having an up or positive time of 1/2(1/F). (This assumes symmetric division, which is not necessary but is easier to understand.) This square wave goes up to the 7493 which counts to 16, overflows, and counts to 16, over and over. *The time this IC remains in any one count is 2(1/2) (1/F).*

When a count of 15 is reached a NAND gate detects the state and switches pins 2 and 3 of the first 7490 to ground enabling the 7490 to begin counting. The following 7490s are set up to count at this time and so the count chain (Fig. 7-17) operates. When the 7493 leaves count 15, the NAND gate switches pins 2 and 3 of the first 7490 high and the count stops.

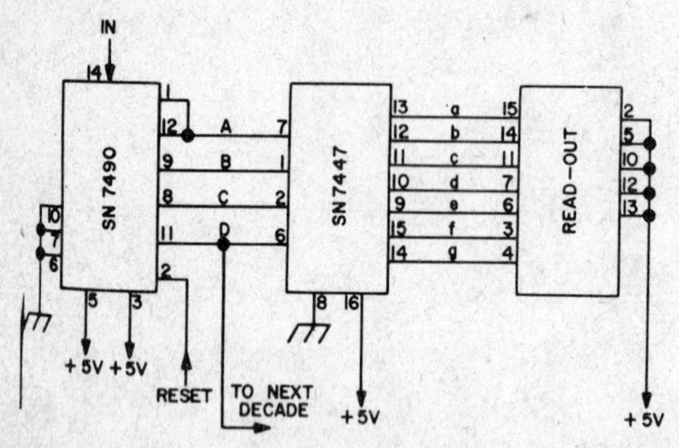

Fig. 7-17. This is the basic counting unit. It counts, decodes, and displays one digit. Four are required for 4 digit readout.

Since the four displayed decades have not reached a reset pulse and no pulses are coming in, the display stays up. The 7493 is still counting, having started over after 15 and when a count of 14 is reached the 3 other NAND gates reset the 4 displayed decades. At 15, the cycle repeats.

Since 1 MHz is a 7 digit number, seven 7490s, seven 7447 decoder/drivers, and seven readouts would be required to count and display all 7 digits. The last 7490 in the chain would contain the most significant digit (MSD), a 1. The input 7490 would contain the least significant digit (LSD), a 0.

However, if the counting period is divided by a factor of 10, the number to be displayed will also be divided by 10. The LSD has been eliminated, and 6 decades would be required to display all cycles counted.

In the same manner, if the count period is 1/1000 second, number of cycles will also be 1/1000. Frequency and the 4 digits displayed will be the most significant digits. Also, if only 4 digits are displayed and count time is 1 second and signal counted is 1 MHz, the counter will overflow and the 4 digits displayed will be the least significant.

By combining these operations, 4 digits can read out each digit of any signal within the range of the counter. Four digit readout (Fig. 7-18) appears to be a good compromise between cost and ease of operation. If the LSD is never desired, it is not necessary to provide a decoder/driver or readout for that digit. This saves the cost of the parts eliminated, and has the added advantage of increasing the upper frequency limit of the counter, since switching speed is partly a function of load on the IC.

For fast gate times the display appears to stand still, and will appear to 'track' a VFO. At the slowest gate time, the cycle takes 16 seconds. Since the primary use of the counter will be for MHz signals and resolution to the

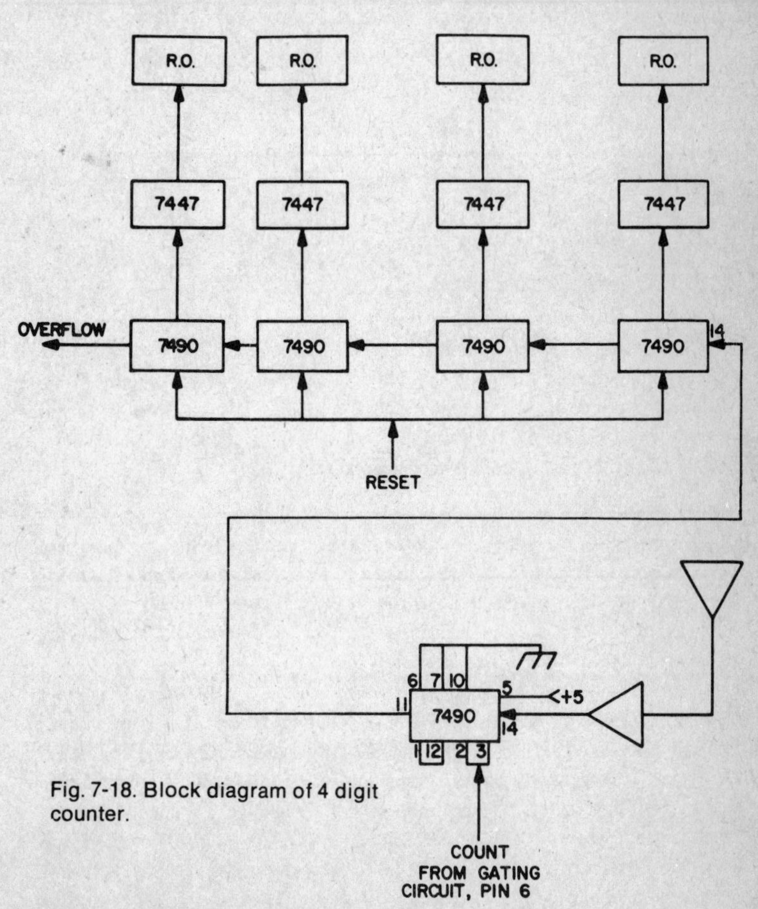

Fig. 7-18. Block diagram of 4 digit counter.

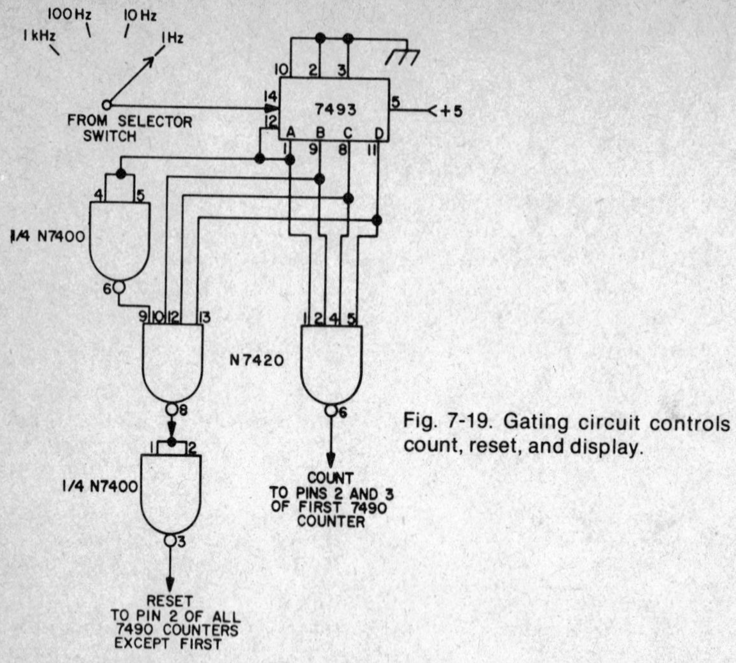

Fig. 7-19. Gating circuit controls count, reset, and display.

nearest kHz is usually sufficient, this is no problem. Another arrangement would be desired by a piano tuner.

The counter requires a good 4.75V to 5.25V supply. It is hard to damage the 7400 series if overvoltage is avoided, but at least one 7490 has been destroyed by a momentary supply voltage of 7V.

The supply described does a very good job (Fig. 7-20).

5V Power Supply

The 5V supply uses a Darlington connected emitter follower. Maximum current drain of the counter was calculated for worst case conditions and it was found to be approximately 1.3A. The 2N3055 will pass this easily.

Zero output resistance is desired, so that V_O will be constant with varying load current. Actual R_O is approximately .04Ω.

Output voltage will be Zener referenced voltage minus two base emitter drops. Zener was chosen to be 6V, 1W. Zener current = $(20\text{-}i_b)$mA. Zener dissipation will not exceed 6 (.02) W, which gives a good safety margin.

This is a reasonably small percentage of 20 mA, the maximum Zener current. Both transistors are operating well within power, voltage and current limits. Unregulated voltage comes from a center tapped full wave rectifier. The capacitive filter gives 8V from a 6.3V rms transformer secondary. The 5V supply is shown in Fig. 7-20.

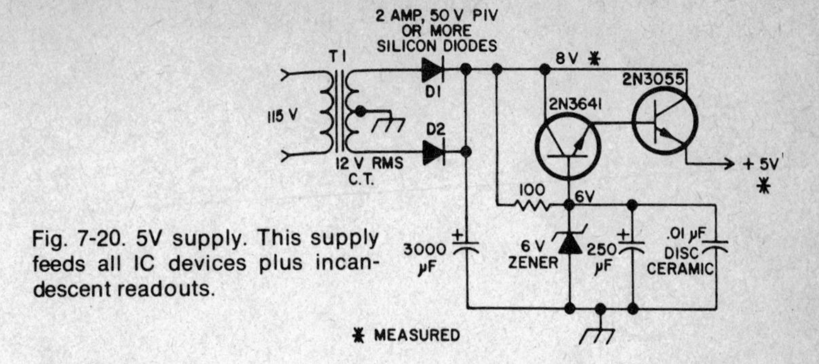

Fig. 7-20. 5V supply. This supply feeds all IC devices plus incandescent readouts.

Input Amplifier

The input amplifier is shown in Fig. 7-21. The MFE 3007 is a MOSFET having relatively high input resistance. It drives a direct coupled 2N3641 follower which gives power gain and impedance transformation.

From the data sheet, $R_L = 2.5\ K\Omega$, a resistor of 2.7 KΩ was selected. If drain current is chosen to be approximately 5 mA, about 13.5V will drop across R_L. It is desired to have about the same voltage drop across the transistor and this indicates a power supply voltage of at least 27V.

The bias network was determined experimentally, working from approximations obtained from the data sheet. The MOSFET is operating somewhat above voltage ratings on signal swings. Drain voltage maximum is 25V, while supply voltage is 35V, so bias was adjusted carefully to keep the device out of the high voltage region for extended periods. No ill effects have been observed due to the over-voltage. Output resistance of the MOSFET is taken to be approximately equal to the drain load resistor.

Output resistance of the emitter follower is about 50Ω. The 2N3641 is within all tolerances.

The resistor diode network at the output of the 2N3641 sets input conditions for the first 7490. The diodes conduct when signal input exceeds back bias plus

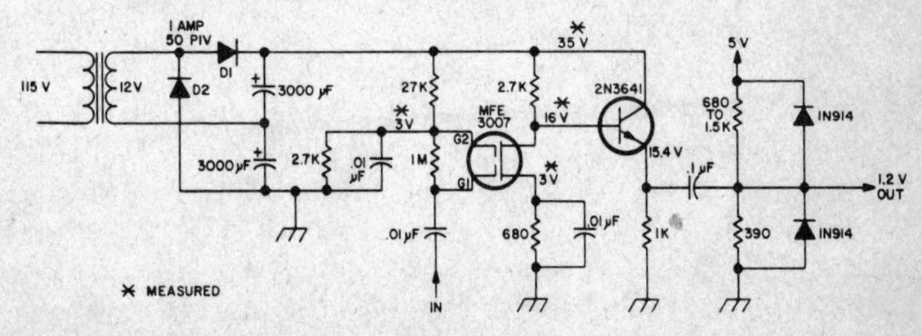

Fig. 7-21. Input circuit. Signal to be counted comes first to this circuit, where it is amplified, clipped, and fed to input of first 7490.

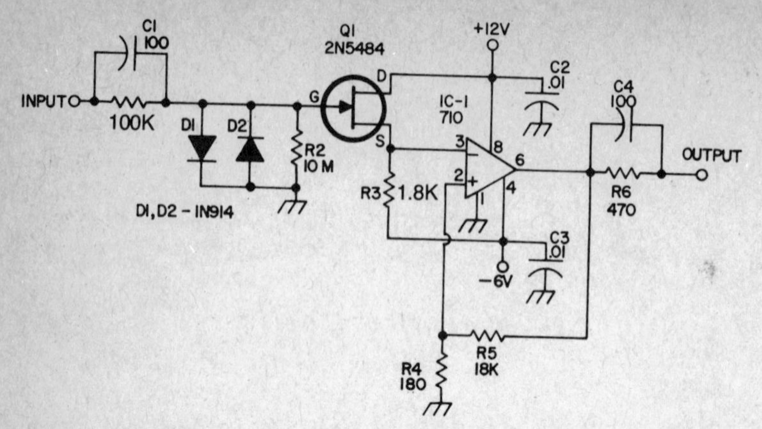

Fig. 7-22. Counter input circuit schematic diagram.

the forward silicon junction drop. This prevents burnout of the 7490 which otherwise might occur.

The input amplifier supply is a common voltage doubler. Ripple is about .1V.

The original model section, which was built from the schematics in this section will count reliably to above 37 MHz. The crystal oscillator can be zero beat to WWV to get a highly accurate timing signal. Accuracy of displayed count, of course, is directly related to accuracy of the clock. For all known frequencies measured, counter error has not been visible, that is, the undisplayed digit contained whatever error existed.

For instance, the color TV chrominance subcarrier frequency is phase locked to the network frequency standard which is highly accurate. This frequency is 3.579545 MHz. The counter displays 3.57954 with the LSD not displayed, as usual.

A counter is one of the most useful pieces of gear you can have around a ham shack, and at this price there's hardly an excuse not to have one.

FREQUENCY COUNTER INPUT CIRCUIT

There is no input level control in this circuit. The sensitivity is 10 dB better than the original circuit using a pair of inverters as a shaping circuit. The input is protected from high level input signals and the input impedance is much higher than the original.

The absence of an input level control is attributed to the components associated with the gate circuit of Q1. Resistor R1 and capacitor C1 couple the input signal to the gate of Q1. An input signal of approximately 43 mV is all that is required to drive the trigger circuit in a stable condition all the way up to 10

MHz. If the input signal becomes greater than 2 volts p-p the diodes, D1 and D2, will conduct and clip the input at this level. This protects the gate to the FET, Q1, and also eliminates the need for an input level control.

The IC is a Differential Voltage Comparator. This unit is a 10 lead flat pack so not as easy to work with as the TO-99, but in the end makes a very neat circuit. Figure 7-23 gives the outline details of the 710 flat pack and also of the MC170CG which is in a TO-99 package. Either will work well in this circuit and specifications are given for both units for reference purposes.

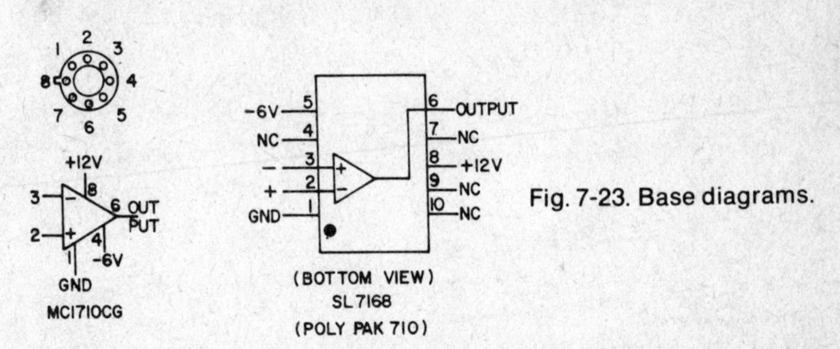

Fig. 7-23. Base diagrams.

IC-1 is hooked up as a Schmitt trigger. When the voltage on the source of the FET, Q1, reaches the upper trip point of about 7 mV the Schmitt trigger will latch on. The output will be driven negative very quickly by the feedback action to the non-inverting input (marked +).

The positive going input signal is applied to the "Inverting" (−) input. That is to say, the positive going signal will be amplified and inverted, thus becoming a negative going signal at the output. This 'Non-inverting' (+) input will amplify whatever signal is applied to it and be in the same phase at the output. Since this is the feedback point in our circuit the original input signal, inverted, is fed to this non-inverting input, amplified and comes from the output even more negative. The result is a very fast switching action and a beautiful square wave, as the IC is driven into saturation in each direction, to apply to counter circuits.

The reverse action is true also. As the input swings in the negative direction the output will switch into the positive direction. The voltage required to overcome the action in one direction and reverse it into the opposite direction is the hysteresis voltage and determines the minimum signal at the course of the FET which will trigger the IC. The hysteresis voltage in this circuit is about 20 to 30 mV. There are two cautions to take into consideration when building the circuit. One is that the bypass capacitors C2 and C3 should be mounted as close to the IC as possible. The other is that the value of R3 should be chosen to give a zero at the source of Q1 with no signal input. This value can be found by experiment or a 1k resistor used with a 1 or 2k trim control in series to set the correct value. The value will be between 1 and 2k and is almost exactly 1.8k on my unit.

There is a drawback to building such a circuit unless you already happen to have the voltages required. You need +12V and − 6V.

The best solution is given in Fig. 7-24. The transformer is 12.6V center tapped. The bridge rectifier is actually two full wave rectifiers now. One half supplies the +6V as previously required and the other half supplies the − 6V which is filtered and zener regulated for the IC. A voltage doubler is used to get +12V from one half of the transformer secondary, and since the oscillator in the counter uses +9V, this is also zener regulated from this doubler output. It is a good thing to use 1 μF tantalum capacitors on all supply outputs to prevent noise feedback from one supply voltage to the next.

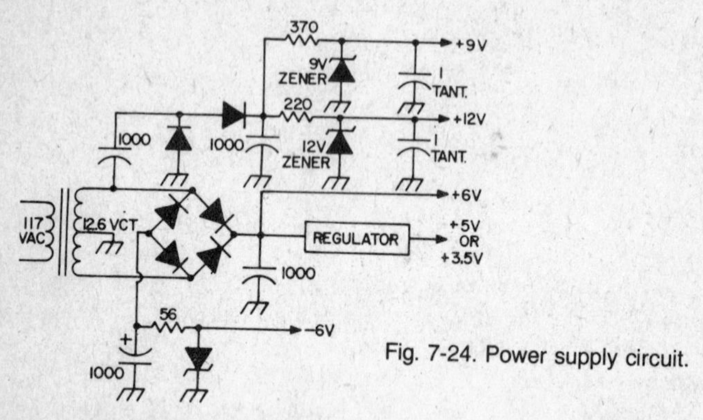

Fig. 7-24. Power supply circuit.

LOW COST FM DEVIATION METER

Most electronic experimenters have a lot in common. They like to own test gear, have limited funds, like to be original and have a hard time finishing one project before their mind wanders off to a new one. Another thing is that they may not really need the piece of gear they set out to build. Nevertheless, the junk box is usually raided (along with the bank account) and the curiosity is generally satisfied.

The search for a low cost FM deviation meter (Fig. 7-25) started with the idea that the low cost hand held Public Service Band receivers might offer an approach to the problem. The possibility of easily retuning those 30 to 50 MHz and 146 to 175 MHz receivers to 6 and 2 meters evoked promise. Another possibility was by tuning to the frequencies of the multiplier stages of 220 and 450 MHz transmitters, frequency deviation at their output could be determined by multiplying the value by the same amount the signal would be multiplied in the transmitter.

Investigation showed that the 10.7 MHz IF discriminator output voltage change was very small (as expected). In addition, the 1:1 receiver dial would make it hard to center the received signal at the crossover point of the ratio detector. This is important in order to get repetitive results. Even though the

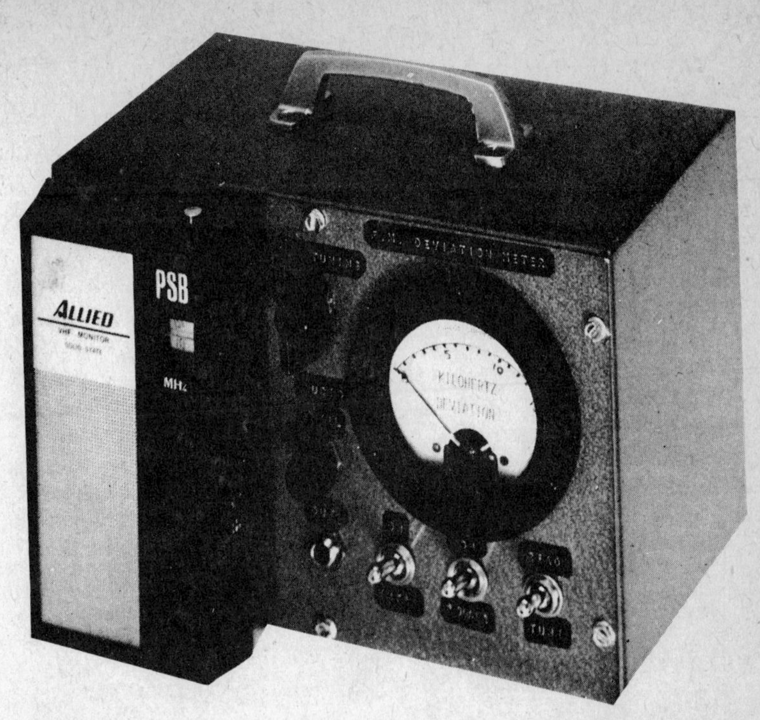

Fig. 7-25. The deviation meter makes a neat portable package that is an invaluable aid to anyone on FM.

transmitter to be checked would be close by, the possibility exists that another nearby transmitter could upset the output of the ratio detector and upset the reading. The decision to add another tuneable converter after the 10.7 MHz IF solved some problems with the following results:

1. Difficulty in tuning the signal was eliminated. Any signal heard through the receiver is heterodyned down to the 3.95 MHz range. The conversion oscillator tuning rate is much better.
2. Other signals which might appear in the 10.7 MHz range are attenuated.
3. By using an intermediate frequency inside an amateur band, the calibration is easier.
4. The output voltage per kHz deviation is much greater.

For the unit in Fig. 7-26, a 146 to 175 MHz P.S.B. receiver (usually found at hamfests) was used. The 10.7 MHz IF signal is tapped off the primary winding of the ratio detector transformer through a small coupling capacitor and shielded lead. Due to the very small coupling capacitor the de-tuning effect is

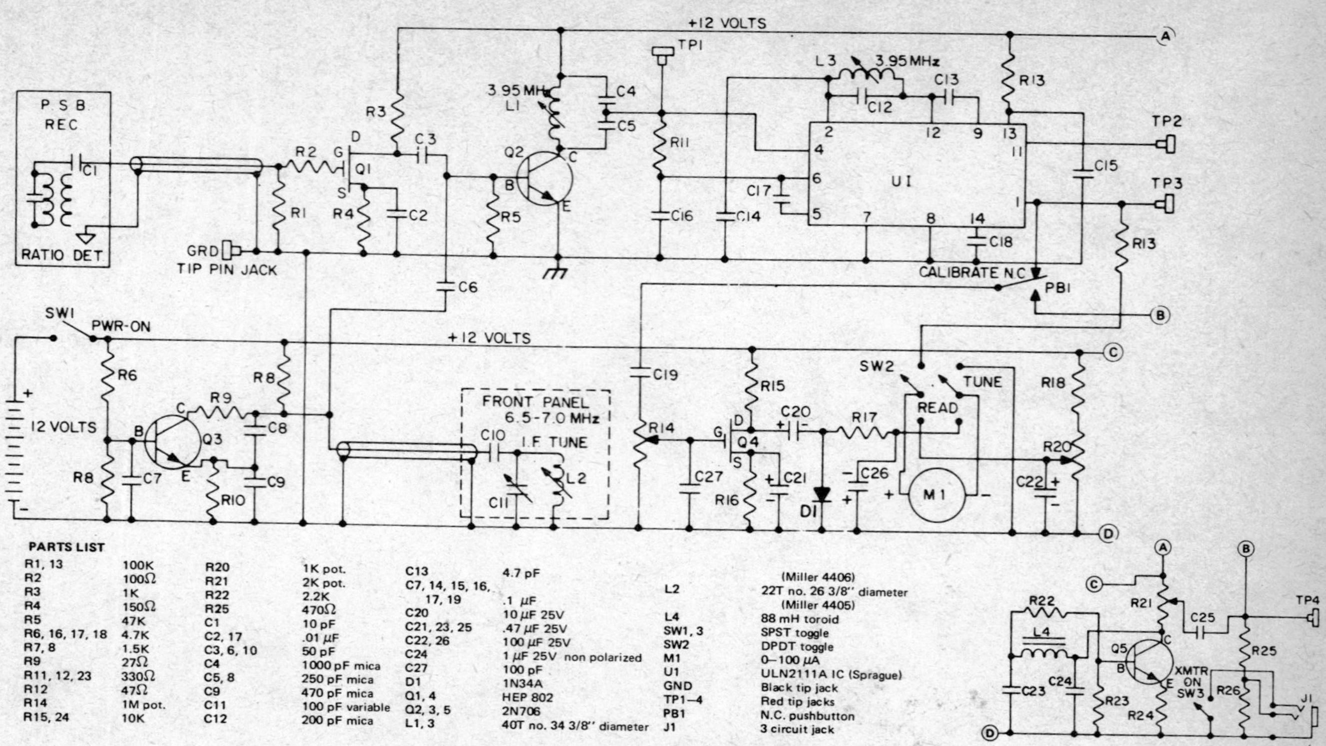

PARTS LIST

R1, 13	100K	R20	1K pot.	C13	4.7 pF		(Miller 4406)
R2	100Ω	R21	2K pot.	C7, 14, 15, 16, 17, 19	.1 μF	L2	22T no. 26 3/8" diameter
R3	1K	R22	2.2K				(Miller 4405)
R4	150Ω	R25	470Ω	C20	10 μF 25V	L4	88 mH toroid
R5	47K	C1	10 pF	C21, 23, 25	.47 μF 25V	SW1, 3	SPST toggle
R6, 16, 17, 18	4.7K	C2, 17	.01 μF	C22, 26	100 μF 25V	SW2	DPDT toggle
R7, 8	1.5K	C3, 6, 10	50 pF	C24	1 μF 25V non polarized	M1	0–100 μA
R9	27Ω	C4	1000 pF mica	C27	100 pF	U1	ULN2111A IC (Sprague)
R11, 12, 23	330Ω	C5, 8	250 pF mica	D1	1N34A	GND	Black tip jack
R12	47Ω	C9	470 pF mica	Q1, 4	HEP 802	TP1–4	Red tip jacks
R14	1M pot.	C11	100 pF variable	Q2, 3, 5	2N706	PB1	N.C. pushbutton
R15, 24	10K	C12	200 pF mica	L1, 3	40T no. 34 3/8" diameter	J1	3 circuit jack

Fig. 7-26. Schematic of the deviation meter.

minimal. The signal is amplified by transistor Q1. Its purpose is to regain some of the signal lost by the capacitive voltage divider action of C1 and the shielded lead capacitance. In addition, it helps to isloate the VFO from the receiver and furnishes a low impedance source for mixer stage Q2. The VFO tuning range is 6.5 to 7.0 MHz to heterodyne the incoming 10.7 MHz signals to 3.950 MHz. The ULN2111A integrated circuit is a combination limiter amplifier and quadrature detector operating at 3.95 MHz. Transistor Q4 functions as an AC amplifier for driving the rectifier and meter circuit. Control R20 is used to provide a small forward bias voltage for D1. This greatly improves the linearity of the meter at the low end of its scale. Transistor Q5 functions as a 1 kHz audio oscillator. Its output is used to modulate the transmitter requiring adjustment. It is also helpful for checking the AC voltmeter section or whenever a low level 1 kHz signal is needed. The whole unit is powered by 8 C cell flashlight batteries. The current drain is approximatley 30 mA. Smaller batteries may be used.

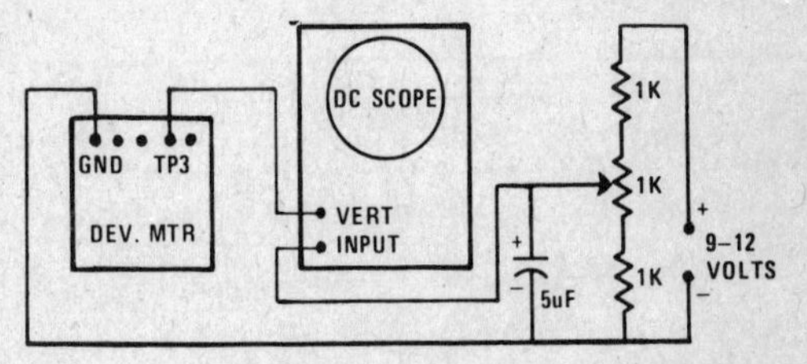

Fig. 7-27. Calibration circuit. Complete instructions are given in the text.

When using the meter it is only necessary to tune the signal on the receive until it is heard in the speaker. Next, vary C11 (IF tuning) and center the signal with meter M1 in the center of the "S" curve (see Fig. 7-28) with SW3 in the "tune" position. Switch SW3 to "read." By using the audio from jack J1 to modulate the transmitter, the clipping level may be read.

Construction

The unit is built in a 6 × 9 × 5 in. Bud utility cabinet fitted with rubber feet and a carrying handle. The receiver is bolted to the front panel. The receiver can be removed for servicing by unsnapping it from its case. Black plastic tape around the seam of the case prevents the receiver from accidentally coming apart. Prior to mounting the receiver to the front panel, the IF output lead should be installed. This consists of capacitor C1 connected at the top of the primary winding of the 10.7 Mhz ratio detector transformer. The opposite side of C1 connects to the center of a length of very small diameter shielded wire or coax. The shield is connected to the circuit ground in the receiver. Having a schematic of the receiver will help you locate the correct point at which to connect C1. Lacking this, it is possible to determine the correct point

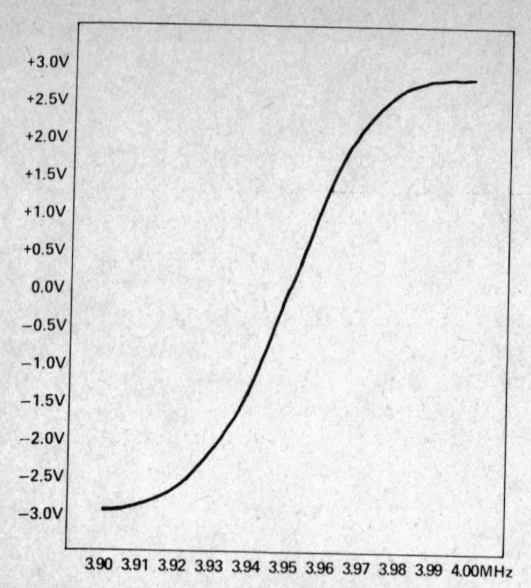

Fig. 7-28. Response curve of the quadrature detector without any resistance shunting coil L3. Zero voltage represents the quiescent state of the IC measured at TP3. This is the level to which all other points on the curve are referenced.

by using an RF probe or high frequency scope with a low capacity probe. Look for the point of highest signal level at the last IF transformer when tuned to a nearby transmitter. Check that the signal disappears when the transmitter is turned off. Also check for DC collector voltage at this point to make sure you have the primary and not the secondary or tertiary winding of the ratio detector transformer. Drill a hole through the rear cover of the receiver just large enough to clear the small shielded lead. Control R21 and capacitor C11 are mounted to the front panel. All other controls are of the screw driver adjusted type and mounted on the circuit board. The meter, jack J1 and switches SW1, 2, and 3 are the only other components fastened to the front panel. Not having the patience to design PC boards, the circuit was built on vector board. The only precautions are that shielded leads should be only as long as necessary and that L1 and L3 are separated to reduce any tendency for the IC to oscillate.

Testing and Alignment

Prior to applying power to the unit it is wise to check for shorts with the integrated circuit unplugged. Operate the power switch. Connect the common lead of an oscilloscope to the ground jack and TP4 to the vertical input lead. The output of the 1 kHz oscillator should be viewed when the audio control (R21) is advanced. Return the control to zero output.

In order to calibrate the meter it is necessary to make a new scale. At this point it must be decided what the maximum deviation will be. Should indications in excess of 15 kHz be desired, it will be necessary to check the linearity of the particular quadrature detector. This is good to do in any case. To increase the linearity it will be necessary to empirically shunt L3 with high values of resistance until the desired linearity is achieved. With the deviation being proportional to the output voltage shift of the IC, it is only necessary to calibrate the meter as an AC voltmeter. This is down by feeding a known signal into the Q4 AC amplifier with SW2 in the read position. Before introducing the signal, set control R20. (This is done by setting R14 toward its ground end. Slowly advance R20 until there is barely perceptible movement off of mechanical zero.). Monitor the input of Q4 with an oscilloscope (or other suitable means). Use TP3 with an external oscillator at 1 kHz or TP4 with the calibrate push button held down, to produce signal. Introduce a suitable signal (up to 2V peak to peak) as monitored on the scope. Set the scope gain so the signal occupies the same number of divisions on the scope as kHz you wish to indicate. Advance control R14 until the meter indicates full scale. By reducing the oscillator output and noting the scope readings, intermediate points on the scale may be calibrated. Next set the VFO to frequency by tuning a 40m receiver to 7.0 MHz. Set the oscillator capacitor C11 to approximately one quarter meshed. Tune inductor L2 until the VFO is heard in the receiver.

In subsequent tests a 75m transmitter connected to a dummy load may be used as a signal source. An unmodulated RF generator may also be used. In the later case, the introduction of signal to TP1 should be through a coupling capacitor. In either case the signal should always be maintained at the limiting level or above unless otherwise noted. When using the transmitter as a signal source, start with the drive control at minimum to avoid damage to the semiconductors. If you cannot get enough pickup with this method, attach a short radiator to the center lead of the coaxial cable feeding the dummy load. The 1 kHz oscillator output is adequate for setting the control R14 up to 20 kHz deviation on the mter. Should you desire the full-scale reading of the meter to read above this value, check at a lesser deviation (i.e., 5 kHz on the meter scale). If a wide band oscilloscope is to be used it may be necessary to bypass the input leads to keep RF patterns out.

Disable the VFO by shorting out capacitor C7. If the IC is not in its socket, temporarily remove power while it is installed. Set switch SW2 to the "tune" position. The meter should read approximately one-half scale. Connect a test lead or short piece of wire to TP1. Tune your transmitter up on 3.950 MHz in the CW mode so the output can be easily varied. Connect a VTVM with RF probe (or wideband oscilloscope) from the GND jack and TP2. Using the lowest range on the VTVM, couple just enough energy from the transmitter to get a reading. Inductor L3 may now be tuned for maximum indication while keeping the input level below the limiting level of the IC. It will be noted that the meter in the unit behaves like a zero-center meter monitoring a discriminator "S" tuning characteristics. L3 should be tuned for the center of the "S" curve (same meter reading as before signal was introduced).

Remove the power and connect the test circuit shown in Fig. 7-27. The scope must have a DC vertical amplifier. Before applying the power again, set the oscilloscope trace to the zero line with the DC amplifier on and with no signal applied (transmitter on standby) operate the "power" and "read" switches of the deviation meter and apply power to the test circuit. The oscilloscope trace will probably move from the zero line. Return it by using the 1K potentiometer of the test circuit. Turn the transmitter on and slowly raise the output level until the RF voltmeter connected to TP2 indicates the limiting level has been reached. It may be necessary to retune L3 if the scope trace is not on the zero line. Vary the transmitter frequency above and below the center frequency the amount corresponding to the maximum frequency deviation you desire to read. It is desirable to set the scope vertical gain so this deviation moves the scope trace an even number of squares. Note the amount the trace moved. Without moving the scope vertical gain adjustment and switching to its AC amplifier, the trace should again move to the zero line. Turn the transmitter off. Move the scope lead to TP4 and adjust the audio output control R21 so the negative and positive peaks coincide with trace excursions just noted. Depress the calibrate pushbutton and adjust R14 to read the deviation. Should the audio level be insufficient, adjust for a lesser deviation such as setting the audio level to one-half of what is required and calibrating the meter to one-half of full scale. Release the calibrate pushbutton and set switch SW3 to tune position. The meter should read somewhere in its mid-range. As a further check, the transmitter output can again be increased to the limiting level. The meter needle should move above and below the center as the transmitter is moved above and below 3.950 MHz.

All that is left to do now is to peak L1 to 3.95 MHz. With the output of the transmitter reduced to zero, connect one end of a test lead to the junction of R1 and R2. Slowly increase the transmitter output to get an indication of the RF voltmeter connected to TP2. Peak inductor L1. You should be operational.

The term "calibrate" is a misnomer to some extent as the instrument does not contain a standard. However, if the value of signal was checked at TP4 with an AC VTVM and recorded when R14 was set, it should be easy to recheck M1 using the same AC VTVM (connected to TP4) calibrate pushbutton.

AUDIO SIGNAL GENERATOR

An audio signal generator is a handy item when checking audio gear, and an AF oscillator is a lot more consistent than whistling into the mike. The oscillator described here is the answer. It's inexpensive, simple, and can operate from subaudio to RF.

The oscillator basically is a modified Wien bridge using complementary FET and bipolar transistors. Analysis of the Wien bridge of Fig. 7-29 shows the bridge is balanced when

$$F = \frac{1}{2\pi\sqrt{R1R2C1C2}}$$

a n d

$$C2/C1 = R3/r - R1/R2.$$

When C = C2 and R1 = R2 and the frequency is such that $R = X_c$, the impedance

$$Z2 = \frac{R}{\sqrt{2}}$$

$\angle -45°$ and $Z1 = R\sqrt{2} \angle -45°$.

Consequently, a third of the voltage applied to the bridge between points A and D appears across Z2. When R3 = 2r, one third of the voltage applied to the bridge appears across r, so there is no difference of potential between points B and C, and the bridge is balanced.

In the actual oscillator, R4 and R5 in parallel make the resistance r of Fig. 7-29 for AC. The reactance of C3 must be less than 10Ω at the frequency of oscillation. R5 is the resistance of a 6W 120V lamp whose positive temperature coefficient regulates the output of the oscillator and insures that signals within the oscillator are not clipped or limited.

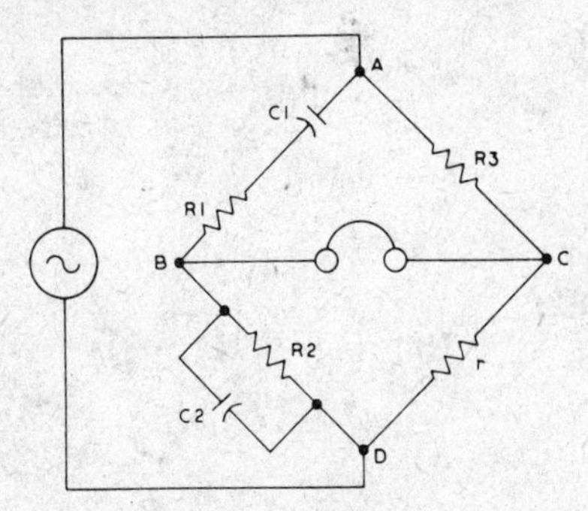

Fig. 7-29. The basic Wien bridge.

A typical 6W 120V lamp has a resistance of 200Ω when the voltage across it is 0.1V rms, and about 550Ω when the voltage is 3V rms. A good operating point is about 0.5V, or a resistance of about 300Ω. The actual operating point is determined by R3 and the particular lamp's characteristics. The larger R3, the higher the lamp operating voltage.

R3 is chosen to be slightly greater than twice the lamp's resistance when the voltage across the lamp is about 0.5V rms, and a value of 680 is about right. Initially, then, the bridge is unbalanced, and oscillations will start when power is applied. The AC voltage across the lamp increases the lamp resistance and brings the bridge toward balance. The bridge is brought to the balance point, which produces a bridge attenuation exactly equal to the amplifier's gain. The attenuation, the ratio of applied voltage (A to D) to output voltage (B to C), is infinite at true balance, and an infinity-gain amplifier would be required. Therefore, the oscillator must operate with some unbalance.

The match of R1 and R2 and C1 and C2 is very important and, although the circuit given can handle 5% tolerance components, a closer match is desired so that greater variations in lamp, amplifier gain, and loading can be tolerated. Loads heavier than 1 kΩ can best be accommodated by adding an emitter-follower buffer on the output of the oscillator.

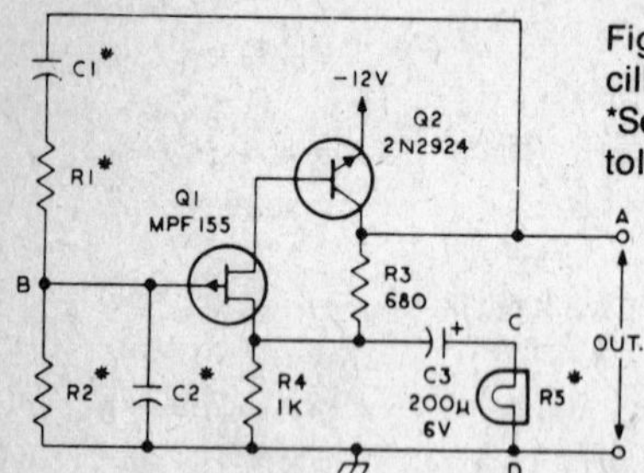

Fig. 7-30. Fixed frequency oscillator. All resistors 10% 1/4 W. *See text for values and tolerance.

Matching C1 and C2 can present an interesting problem if you don't have access to a good bridge. The technique used is slow, but it is a starting point. First, match a pair of resistors in the range of 270 kΩ with an ohmmeter and use them in the bridge shown in Fig. 7-31 to match the 0.1 µF capacitors. At the start, you won't know which of the capacitors is smaller, so shunt one of them and note whether the balance improves. Add to the one that improves the null, and try to balance within 2%. In matching the .001 µF, the bridge resistors should be in the order of 2.7 MΩ and a VTVM is essential for detecting the null.

Values of R1 and R2 between 2 kΩ and 1 MΩ and values of C1 and C2 above 470 pF are convenient. Within these bounds, these combinations are available:

Freq. kHz	C1, C2, µF	R1, R2, kΩ
.250	.01	63.4
.400	.01	39.2
1.0	.01	15.8
1.8	.001	88.7
2.5	.001	63.4
10	.001	15.8

The same general circuit approach can be used to make a wide-range variable frequency oscillator. In the variable oscillator, fixed resistors R1 and

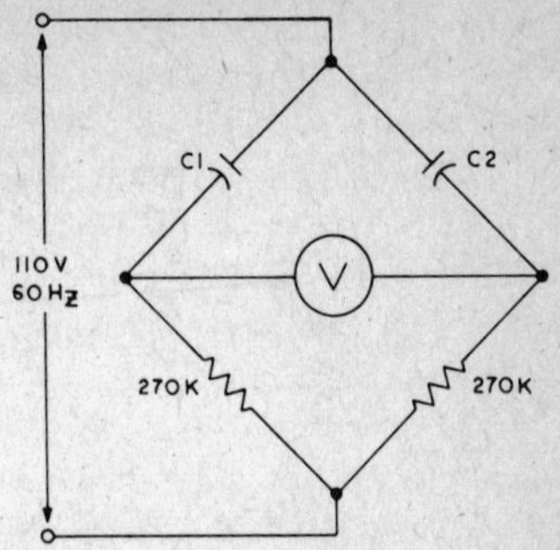

Fig. 7-31. Capacitor comparison bridge.

R2 are replaced with sections of a dual pot. If the tracking of the pot sections is within 5%, the circuit could be used directly, and a 100:1 tuning range could be achieved, but it isn't very likely that you will find such close tracking.

There are two possible solutions to the difficulty. One reduces the tuning range by adding fixed resistance in series with the variable resistances to reduce the percentage tracking error, and the other increases the gain of the amplifier so the unavoidable unbalance can be tolerated. In the oscillator shown in Fig. 7-32, both solutions are used. R_{ab} adds about 2.5 kΩ in series with each of the 200 kΩ pot sections and reduces the tuning range, and Q3 makes a higher voltage gain possible from Q2. The oscillator tunes from about 150 Hz to 6 kHz for the values given.

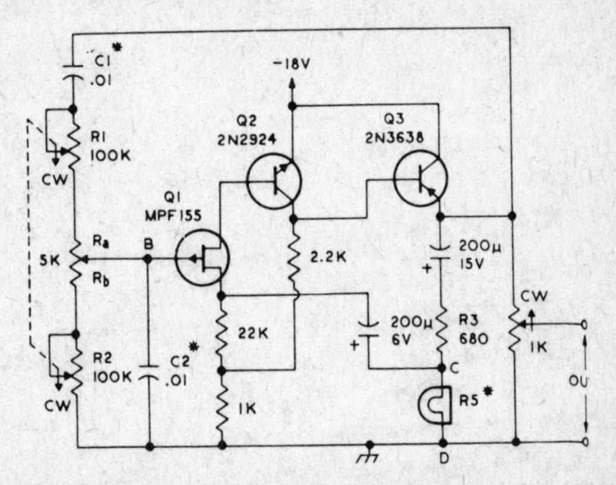

Fig. 7-32. Variable frequency oscillator. All resistors 10% 1/4W. *See text for tolerance.

The pot R_{ab} is adjusted to minimize the percentage difference between $R1 + R_a$ and $R2 + R_b$ as the dual pot is rotated fully clockwise. The power supply for the oscillator is not critical, and any voltage from 12–24V will do for the fixed-frequency oscillator, but the variable frequency oscillator should have a supply above 18V. The current required for the fixed oscillator is about 5 mA without a buffer, and the variable oscillator requires about 17mA.

The layout of the oscillator is not particularly critical for audio frequencies, but care should be taken to shield the high-impedance sections from power-line pickup. If an external power supply or batteries are used, the normal enclosure will be sufficient. The variable oscillator of Fig. 7-32 will fit in a minibox if an external power supply is used. The heft will be improved if you bolt a chunk of scrap steel into the bottom of the box.

If you prefer, you can change the polarity of the unit by making these direct substitutions: MPF 155 to MPF 104; 2N2925 to MPF6518; 2N3638 to 2N2923. Don't forget to change the polarity of the electrolytics. The heavy negative feedback makes the circuit very tolerant of component variations, but the following table of DC voltages may be comforting when you turn the oscillator on for the first time:

Fixed Osc.	**Emitter**	**Base**	**Collector**
	(Source)	(Gate)	(Drain)
A1	– 4.5	0	– 11.4
A2	– 12V	11.4	– 7.5
Var. Osc.	**Emitter**	**Base**	**Collector**
	(Source)	(Gate)	(Drain)
Q1	– 4.5	0	– 17.4
Q2	– 18.	– 17.4	– 14.4
Q3	– 18.8	– 14.4	– 18

An evening or two and a few dollars for parts are all that is needed to build the oscillator. A pair of these units operating at 400 Hz and 1.8 kHz is ideal for generating signals for adjusting the SSB rig.

A TWO-TONE TEST GENERATOR

In the testing of amplifiers and other devices associated with Single Sideband systems, the two-tone test has achieved a great degree of acceptance. Regardless of how the test is perfomed, whether looking at the output of the device under test with an oscilloscope (amplitude vs time) or with a spectrum analyzer (amplitude vs frequency), the basic input requirement is two good sine waves. In the laboratory, such a two-tone signal is usually obtained by using two good quality audio oscillators, like the Hewlett Packard 204C, and a resistive adding network.

In amateur testing, two-tone audio test signals are usually of much poorer quality, being derived from mike-preamps switched into oscillation by various types of frequency-dependent feedback networks. Even commercial "station monitor" systems often use only a pair of simple phase-shift oscillators (without *negative* feedback or automatic amplitude control). Although such simple methods of obtaining test tones are useful, they often yield test tones in which each sine wave contains appreciable distortion (contains harmonics), and this lack of purity can be incorrectly ascribed to distortion in the system under test. Basically, what it comes down to is that if your input tones aren't pure, you can't really tell how much distortion is from the test tone generator and how much is from the system under test. A really good two-tone test generator will help to "separate the sheep from the goats."

The two-tone generator described here uses two of the same basic Wien Bridge oscillator circuits as are used in most laboratory-type audio generators. However, by using modern Integrated Circuit (IC)Operational Amplifiers (op amps), each Wien Bridge can be built around a single semi-conductor package. Op amps are also used as active bandpass filters to further clean up harmonics of the two oscillators, and to sum the two pure tones. The block diagram of the generator is shown in Fig. 7-33.

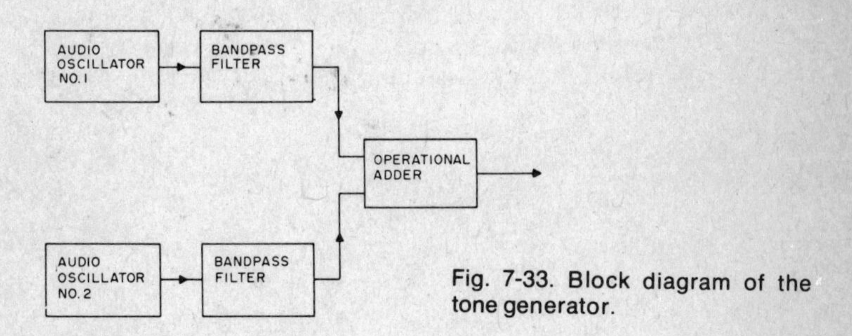

Fig. 7-33. Block diagram of the tone generator.

The particular oscillator circuit used is a form of Wien Bridge originally described by Bob Botos of Motorola. Its charm is that he uses a pair of back-to-back silicon diodes as the non-linear control element. Such a pair of silicon diodes is probably much more readily obtained, and less expensive, than a particular light bulb or thermistor—as used in most Wien Bridge circuits. The diodes prove to be very effective as non-linear elements; and they do not cause severe waveform clipping because of the 47K resistor in series with them. The 1K pot that is in the same arm of the bridge as the non-linear diode elements (R1 and R2) is used to set the oscillator level. This pot should be set to give a 10V peak-to-peak output at TP1 (and TP2).

An idea of how a filter composed of nothing much more than Rs, Cs, and an operational amplifier can have selectivity, can be gained as follows. The R-C network that controls the frequency is connected between the output of the op amp and its inverting input—that is, in the negative feedback path. The R-C network is reminiscent of the bridged-tee *null* network. At the network null frequency, the negative feedback is the lowest, and therefore the closed-loop amplifier gain is the highest. The frequency of the passband is adjusted by means of the 1K pot in the R-C network to match the oscillator frequency. (Adjust R3 for a maximum 2000 Hz output at TP3, and adjust R4 for a maximum 800 Hz output at TP4.)

The particular IC op amps used in the active bandpass filters are LM301As by National Semiconductor. These op amps are compensated in a way referred to as the "feed-forward" method of compensation. This way of compensating op amps allows them to have higher slew rates than with the normal single-capacitor compensation usually used with LM302AS. Because the "feed-forward" type of compensation is used in the active bandpass filter op amps, it would probably not be too good an idea to use other types of op amps here. The op amps used in the oscillators and operation adder, however, can be any of a variety of types. The MC1456 or Motorola or μA741 of Fairchild should serve well in these positions (with the 33 pF compensation capacitors omitted). Of course, there are a number of exact equivalents of the LM301A, MC1456, and μA741 made by a variety of companies other than the originators—these are not to be considered replacements but second sources. One could probably even use μA709s if one understands how to properly compensate them (and wants to go to all the bother); but if you are at all uneasy about substitutions, use LM301As throughout and the circuit of Fig. 7-34.

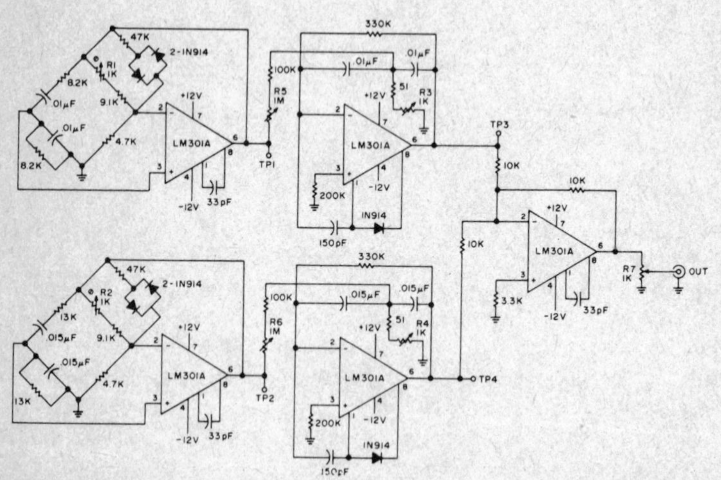

Fig. 7-34. Schematic diagram of the two-tone test generator utilizing Wien Bridge oscillators.

The last stage is the operational adder, or summing amplifier. The op amp is operated at a closed-loop gain of 1. The summing point of the two pure sine waves (2000 Hz and 800 Hz) is at the inverting input of the op amp. As connected, this port is a "virtual" ground; if you look at this point with a scope you will see nearly zero voltage. This is simply because the (high gain, 80 dB) op amp strives to keep the differential voltage between inverting and non-inverting input at zero, and so the non-inverting input is effectively grounded. So each sine wave "sees" the summing amplifier as 10K to ground; and the two sine wave generators cannot interact with each other to cause distortion. The summing amplifier is a true algebraic summer, which is why operational amplifiers thus connected were originally used in analog computers. If one sine wave is instantaneously at +5V and the other is at − 3V, the output will be +5 − 3 = +2V. And since there is no coherence between the 200 Hz and 800 Hz sine waves, we can expect to see plus and minus voltages as high as twice the peak value of each sine wave (− 10V peak, or 20V peak to peak). This should be more than enough level for most two-tone testing. The level of each sine wave can be controlled by the 1 Meg pot at the input of each active filter (R5 and R6).

Measurements of the output of the two-tone generator show that (for equal level tones) the harmonics and cross products are all down more than 70 dB from the desired tone. This sort of purity should be more than adequate for testing any amateur communication system.

A simple, but well-regulated power supply for the two-tone test generator is shown in Fig. 7-35. An integrated bridge rectifier is used with a center-tapped secondary transformer as a plus and minus full wave rectifier. Plus 12V is simply obtained by use of one of the new fixed-voltage three-terminal regulator ICs of Fairchild, the μA7812. This IC looks like a plastic power transistor (UGH 7812 393) and its common terminal is the heat sink tab. So screw it right to the chassis if you want to—with no mica washers, grease, etc.

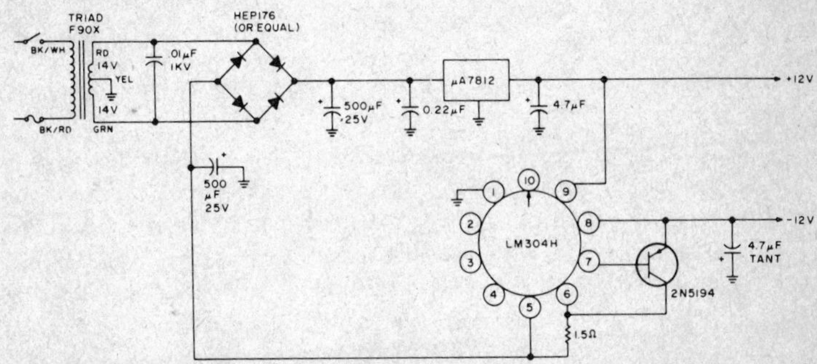

Fig. 7-35. Dual voltage regulated supply used to power the tone generator.

The negative regulator is slightly more complicated, but still simple. The National Semiconductor LM304H is used, with the regulated +12 serving as its reference voltage. In this way the plus and minus voltages *track*. A 2N5194 plastic power PNP transistor is used to increase the current capability of the LM304H. The 2N5194 must be heat-sinked in a conventional way, using insulating washer, etc., if the chassis is used as the heat sink.

SIMPLE NOISE GENERATOR

As anyone who has done any work at all on receivers knows, whether it is a conversion or simply substituting a "hotter" tube in the front end, we get to the point where we begin to wonder if the adaptation was worth while or have we been fooling ourselves.

A noise generator using one of the noise diodes (IN21 or IN23) can give an indication if any improvement has been made.

The circuit (Fig. 7-36) is straight forward, but with one addition that others do not have. The voltage is regulated by a Zener diode.

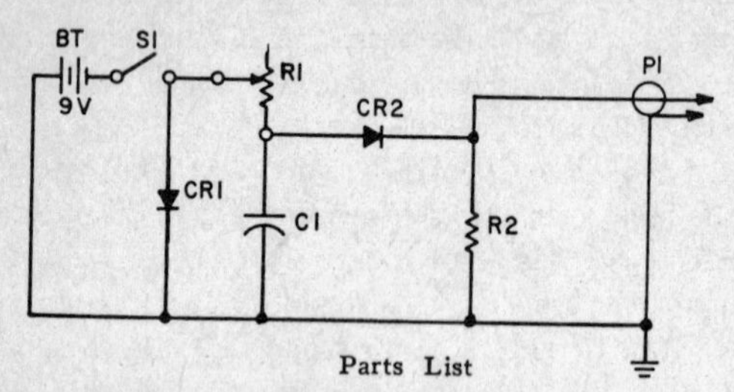

Parts List

2—¼ x 2¼ x 4 Minibox
Bt—9 volt battery
Cr1—6 Volt Zener Diode
Cr2—IN 21 or IN 23 Silicon Diode
R1—10M-50M Variable
R2—51 ohm or 75 ohm (according to your line)
C1—.001 to .005 disk ceramic
S1—S.P.ST. this may be on your variable resistance
P1—So-239

Fig. 7-36. To avoid excess wear and tear on the zener diode and the battery a 200 ohm resistor should be inserted between the 9V battery and S1.

The reason is obvious to anyone who has worked with the simple type of noise generator. The results are not always consistent from measurement to measurement and from day to day. The voltage and current vary with the setting of the variable resistance and due to the normal aging of the battery.

The Zener Diode eliminates this by maintaining a constant line voltage. In this particular instrument it is six volts. Of course we must use a battery in excess of six volts. Nine volts is a good value. Used transistor radio batteries still have enough life in most instances to last for many tests.

One of the main requisites of a noise generator is that it must be shielded throughout. Therefore we must give some thought as to the placement of the various components.

A Mini-box 2 1/4 × 2 1/4 × 4 is an ideal size. As for a connector, use the SO-239 coaxial. If a direct connection to the receiver is desired merely attach it through the double connector type DKF-2 made by Dow Key. On the other hand if it is desired to have the controls of the noise generator close at hand merely connect a length of coax of eighteen inches or so.

In the construction of this noise generator just remember a few basic rules. Keep all connections as short as possible. The noise diode and bypass condenser and resistor (50 or 75 ohms as the case may be) as close as possible to the output plug. Remember to use pliers to absorb the heat when soldering the leads of the diodes.

To mount the silicon diode, which has one large end and one small, we must improvise to a certain extent. For the small end, a lug from one of the old tube sockets will do. For the large end use a small fuse clip.

Don't be too fussy about the variable resistor. For most purposes any value from 10M up to 50M can be used.

The battery you choose will determine the manner of mounting.

POOR MAN'S UNIVERSAL FREQUENCY GENERATOR

As precise frequency control and measurement becomes more and more a part of the amateur radio game, the need develops for test instruments that deliver a wide range of both RF and AF signals of high accuracy. It would be ideal if everyone could have a frequency counter and a synthesizer type RF and AF generator but that is hardly the case. Most amateurs must utilize their basic station gear along with selected accessory items to test out and adjust equipment. Here is a very useful accessory item that for a modest cost goes a long way toward having some of the expensive test equipment just mentioned. The item to be described is somewhat like a grid-dip meter in that it is basically a simple type of oscillator but as one gets to know and use it, new uses for it are found and its versatility constantly expands.

Circuit Description

Figure 7-37 shows the circuit diagram of the test generator. Basically, it consists of a string of SN7490 decade counters which are used to divide down a selected input signal by a factor of 10 or 2. The input signal can come from a 1 MHz master oscillator, a special crystal oscillator for externally used crystals or from any external sine-wave source. The special crystal oscillator which uses an SN7400 will operate with almost any basic or overtone crystal in the HF range. It can be used for crystals in the low frequency and lower VHF range also by a simple modification. One gate of the SN7400 crystal oscillator is used to drive an LED which will indicate that the crystal is oscillating so it serves as a crystal activity indicator as well. When an external sine-wave source is used, it is first coupled through a SN74121 multivibrator. This stage squares off the sine wave so it can better drive the subsequent frequency divider chain.

The frequency divider chain is fixed, although one could easily switch the individual SN7490 units to divide by different ratios when desired. This should

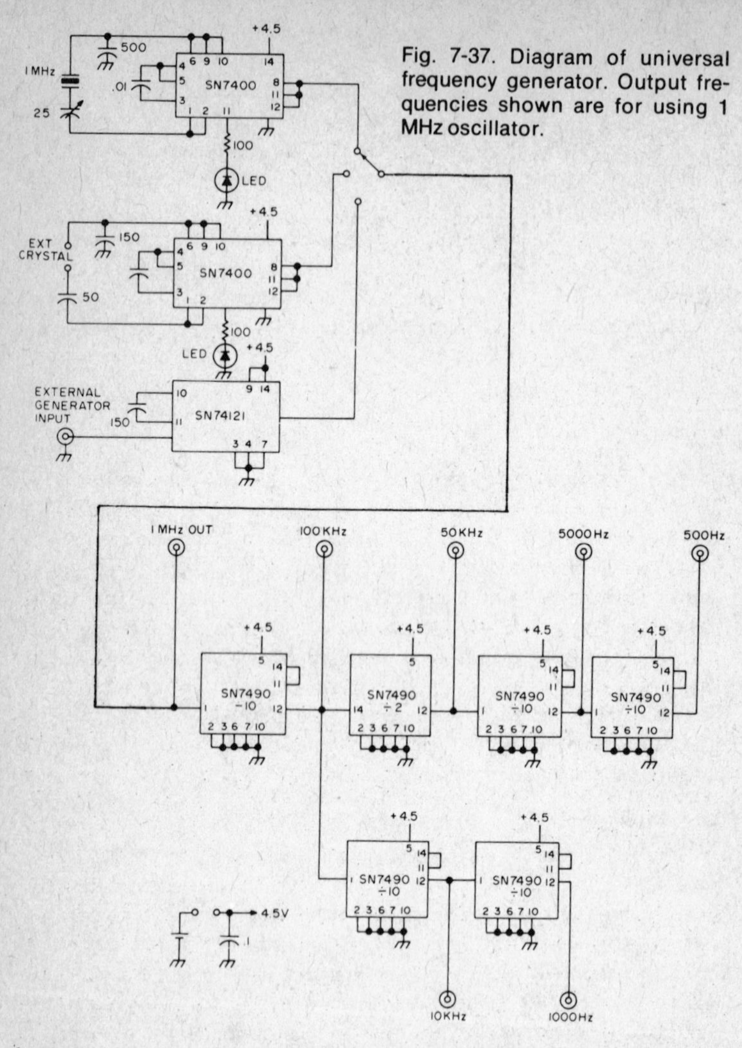

Fig. 7-37. Diagram of universal frequency generator. Output frequencies shown are for using 1 MHz oscillator.

be obvious by noting the wiring of the divide by 2 SN7490 with that of the divide by 10 units. However, the variety of frequencies which can be generated then with different input sources becomes confusing and more than would normally be needed.

The fixed divider chain follows the sequence: divide by 10, divide by 2, divide by 10, divide by 10. A separate branch after the first divide by 10 unit goes through two other divide by 10 stages. In the case of the divider chain being driven by the 1 MHz master oscillator, this results in the following output frequencies being simultaneously present: 1 MHz (basic oscillator output), 100 kHz, 50 kHz, 10 kHz, 5000 Hz, 1000 Hz and 500 Hz. With any other frequency input source you can easily calculate what frequency outputs the divider chain will bring in both the RF and AF regions. Many surplus crystals will produce interesting frequencies of high stability in the AF region that can be used for test purposes.

When using the special crystal oscillator, the LED will glow to indicate that oscillation is taking place. As shown with a 150 pF capacitor from one side of the crystal oscillator circuit to ground, the oscillator will work satisfactorily with HF crystals. Its range of oscillation can be extended to lF as well as high frequency overtone crystals by changing this capacitor. The value of capacitor required in picofarads is 500 divided by the frequency of the crystal in MHz. This value need, however, be only approximate unless you require an absolutely square wave output from the unit.

When using the multivibrator input about a 1 1/2 to 2V peak input, either sine-wave or approximate square wave is required.

Construction

The whole unit can be constructed on a piece of perforated board about 3 × 2 in. and made completely portable if powered by a 4 1/2V battery (Burgess No. 532) or just three D cells in series. This arrangement does not provide the absolutely best stability for the 1 MHz master oscillator but unless you intend to use the unit for marker frequency generation in the VHF range, it

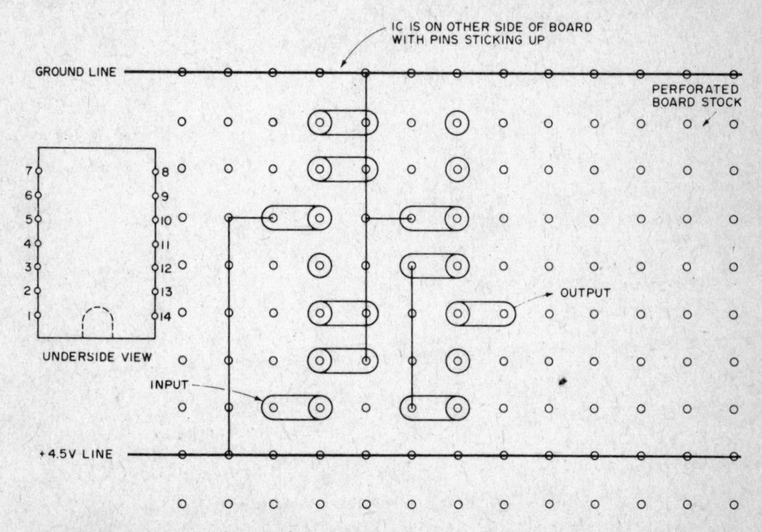

Fig. 7-38. Perforated board wiring of ICs. One SN 7490 divide by 10 units is shown wired.

is a perfectly satisfactory arrangement. Alternatively, one could power the ICs from any standard 5.5V regulated supply used for IC digital circuitry.

One simple way to wire the relatively small number of ICs involved is to purchase perforated board which has hole spacing to fit standard DIP and preferably with a copper pad still left around each hole. The ICs are then placed on the board and the appropriate pins which either go to ground or to the 4.5V line bent in different directions. The ground line is run along one side of the IC and the 4.5V line along the other side and bare wire used to connect the appropriate pins to either line. Figure 7-38 illustrates the wiring for one of the divide by 10 ICs. When one starts this process on the board, it will be surprising how fast the wiring is completed. Individual insulated wire jumpers are used to make the input/ output connections between ICs. The wiring is not critical and using a receiver to hear the markers, or an audio amplifier for the lower frequency outputs, one should be able to determine quickly if the circuit is working. The frequency of the 1 MHz master oscillator may be brought exactly on frequency using the 25 pF trimmer in the circuit and checking against WWV with a harmonic of the oscillator or by using a counter.

Applications

As mentioned before, the applications that you can find for the generator really begin to unfold only after you have had it around the shack for awhile. Some of the applications would be:

1. A frequency marker generator for receiver calibration. The markers are usable up into the VHF range.
2. To extend the range of present RF or AF signal generators into lower frequency ranges than they presently cover.
3. To perform stability checks on high frequency variable oscillators. The divider chain will always perform precisely and you can monitor the change in frequency of a higher frequency oscillator with a stable low frequency receiver.
4. A frequency generator to generate precise RF or AF square wave signals at any frequency desired by choosing the proper crystal.
5. A crystal activity checker.
6. By taking two or more of the simultaneous outputs together via mixing diodes and a series tuned circuit resonant at the desired frequency, you can also mix the divider outputs to generate a variety of intermediate frequency outputs.

PRECISION WAVEFORM GENERATOR

Here is a compact signal generator that simultaneously outputs square, triangular, plus, unlike others, sine waves in the frequency range of 0.05 Hz to 1 MHz. Although the author's unit is a signal generator, it can be used as a frequency modulator or voltage controlled oscillator with only minor circuit modifications.

The signal generator utilizes the advanced Intersil 8038 monolithic chip which features:

- simultaneous sine, square and triangular wave outputs
- low distortion (1%)
- high linearity (0.1%)
- wide frequency variation (.001 Hz to 1 MHz)
- variable duty cycle (2% to 98%)

Typical amatuer applications for this waveform generator are:

- RTTY AFSK Keyer
- FM modulator
- voltage controlled oscillator
- signal generator

Signal Generator Circuit

The heart of the signal generator is the Intersill 8038 waveform generator. The circuit consists of the wave form generator, timing capacitors and potentiomenters, and a DC coupled buffer amplifier. This amplifier is switched to the desired wave shape output. Three dedicated buffer amplifiers and output terminals may be used for additional flexibility.

Components

The timing capacitors C1 to C8 should be high Q, low tolerance components where possible. These capacitors determine the frequency decades of the signal generator. R1 serves for frequency tuning within each range; R2 determines the frequency coverage of R1; R3 and R4 limit the upper frequency of the ranges. A good quality linear tape potentiometer with at least 270° tape function should be used for R1. Using a linear potentiometer for R1 the dial scale will be semi-logarithmic.

With the resistor and capacitor values as listed, the following frequency ranges are covered (R1, R2, R3 and R4 are adjusted to give a 600 to 6000 Ohm tuning range):

Capacitor	Frequency Coverage
C1, 500 μF	0.5Hz–.5Hz
C2, 50 μF	.5 Hz–5 Hz
C3, 55 μF	5 Hz–50 Hz
C4, .5 μF	50 Hz–500 Hz
C5, .05 μF	500 Hz–5 kHz
C6, .005 μF	5 kHz–50 kHz
C7, 500 pF	50 kHz–500 kHz
C8, 250 pF	100 kHz–1 MHz

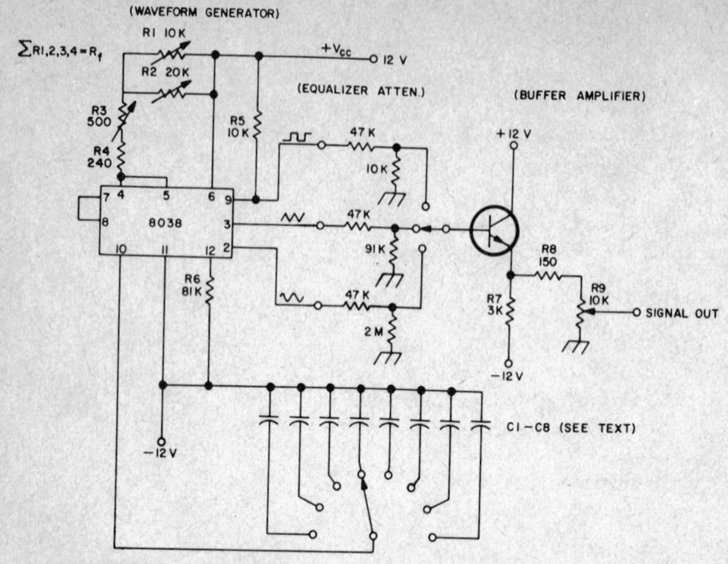

Fig. 7-39. Schematic of the triple-wave output signal generator.

If different values of capacitors or resistors are more convenient, the resulting frequency for different RC values may be calculated from the following formula:

$$F = \frac{150{,}000}{R \times C}$$

C is the timing capacitor in μF and R is the total resistance between +VCC and terminal 4 + 5 in ohms.

The permissible resistance values between terminals 4, 5 and +VCC range from 250Ω to 500kΩ. The permissible supply voltage may vary from ±5V to ±15V and preferably should be regulated. A single supply of +10 to +30V may be used, but it is then advisable to decouple the output with a capacitor because of the large DC offset voltage. This will likely cut down on square wave and low frequency response.

The sine, square and triangular wave outputs at chip pins 2, 9 and 3 have different output levels. These levels are (with 100kΩ load resistor) 0.9 × Vs for the square wave, and 0.2xVs for the wine wave signal. (Vs=total supply voltage). Thus, with 24V supply voltage, the available output levels are 21V peak to peak (square wave), 7.5V peak to peak (triangular wave) and 5V peak to peak (sine wave). The square wave and triangular wave outputs are therefore attenuated to the same level as the sine wave output before going into the buffer amplifier.

Buffer Amplifier

A simple buffer amplifier is shown in Fig. 7-39. The resistor dividers provide equal signal levels to T1 and reduce the loading effect of the low base input impedance. T1 is a 2N3709 high gain amplifier transistor which is wired as an emitter follower to provide a lower output impedance than the 8038 chip. The signal level for all waveshapes at ±12V supply voltage is approximately 5V peak to peak.

Circuit Alternative AFSK Generator

Because of its high stability and low distortion the sine wave output is perfect for RTTY audio shift frequency keying. Two methods of frequency shifting are possible with this unit. The value of the frequency determining resistor may be switched, or a frequency shift voltage may be introduced to terminal 8 as indicated in Fig. 7-40. If the first of the two methods is used, typical values of Cf and Rf are (with Cf = .033 μF):

Rf = 3135 Ω
f = 1450 Hz mark frequency
Rf = 2806 Ω
f = 1620 Hz (170 Hz shift)
Rf = 1976 Ω
f = 2300 Hz (850 Hz shift)

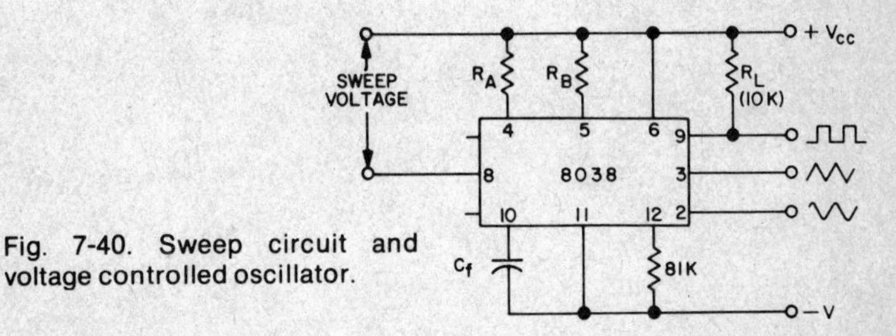

Fig. 7-40. Sweep circuit and voltage controlled oscillator.

Frequency Modulator

Figure 7-41 shows a typical schematic for narrow band frequency modulation. The frequency of the waveform generator is a direct function of the DC voltage at terminal 8, measured from +VCC. By altering this voltage, frequency modulation is performed.

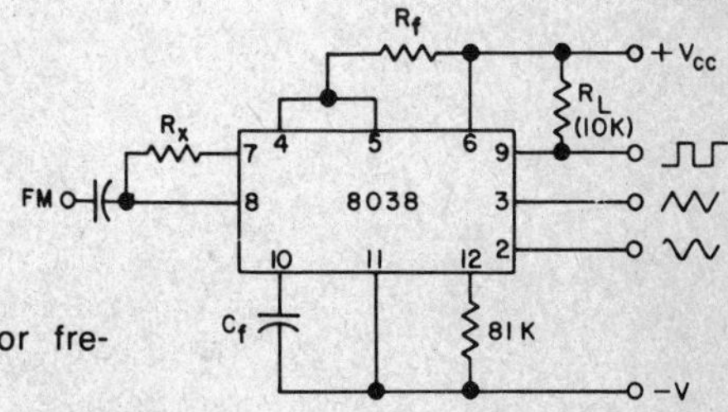

Fig. 7-41. Connections for frequency modulation.

For small deviations of about 10%, the modulating signal can be supplied to pin 8 through a decoupling capacitor. An external resistor between pin 7 and 8 is not necessary but can be used to increase the input impedance, which then increases from 8kΩ to 8kΩ + R.

For larger FM deviations or frequency sweeping, the modulating voltage is applied between the positive supply voltage and pin 8. A 1000:1 sweep range can be achieved with a change of f = 0 at V sweep = OV. The potential at pin 8 may not exceed 2/3 of +VCC.

A typical ham application would be to produce a frequency modulated (e.g., 455 kHz IF) signal for mixing purposes. Another use for this circuit is the determination of filter bandpass curves by frequency wobbling.

Variable Duty Cycle Oscillator

If the timing resistor circuit (R1 through R4, Fig. 7-39) is changed as outlined in Fig. 7-42, the duty cycle of the output signals can be adjusted from 2% to 98%. Thus, a variable mark-space ratio square wave or a sawtooth shaped triangular wave can be generated. The frequency of a (360°) wave stays constant regardless of the position of R_{AB}. Rf permits about one decade of frequency adjustment without changing Cf.

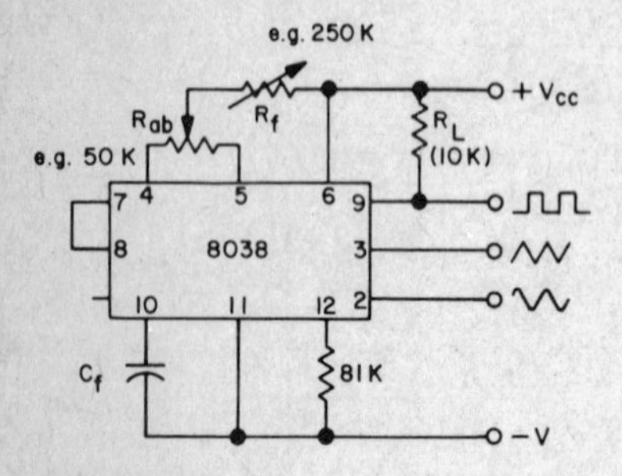

Fig. 7-42. Variable duty cycle oscillator.

Comments

After you build this signal generator the dial has to be calibrated. This work requires a frequency counter. The high value timing capacitors (50 μF and 500 μF) are electrolytics, and dial calibration may be somewhat "out" on these 2 low frequency ranges. The wave outputs of the 8038 deteriorate slightly above 500 kHz.

DIPPER THING

Have you ever wished your grid dip meter were more accurate at higher frequencies or had the capability of going lower in frequency than the 3 MHz range that seems to be so popular on most units?

Solution: Use a simple detector (Fig. 7-43) lead in conjunction with an RF generator for grid dip purposes. The probe is inexpensive and easy to construct and use. It utilizes an inexpensive movement which is available from most mail order houses under the title of "light meter movement." The cost is usually about a buck and a half and is typically a 45 μA (basic sensitivity) meter. The sensitivity is not critical; however, something in this range is necessary for an indication which is easily seen. A VOM or other similar indicator can also be substituted at some sacrifice in compactness and operating convenience.

In place of the standard RF generator lead, which usually incorporates a 50Ω or 75Ω terminating resistor, connect the detector lead shown below.

The probe has a BNC connector on one end and miniature alligator clips on the other. Components are mounted on the rear of the meter and this assembly is secured to the clip end of the lead with a good cement. Direct connections may be made to the circuit or a two-turn loop may be connected at the clips for inductive coupling to the circuits to be dipped.

You may now dip circuits whose low frequency limitation is governed by the output of the particular RF generator and the meter sensitivity. No problem has been encountered using the third harmonic of 48.3 to tune circuits in the 145 MHz range.

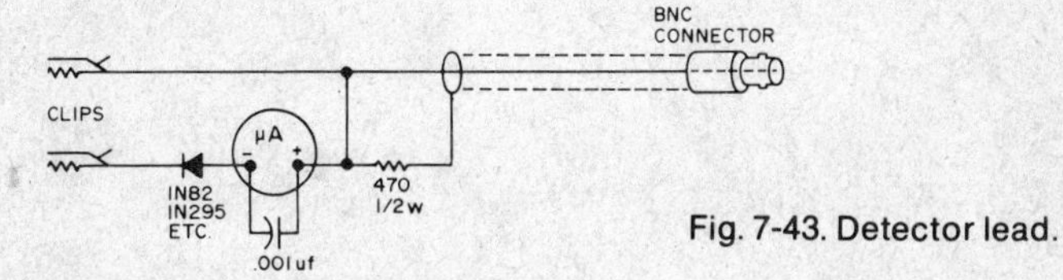

Fig. 7-43. Detector lead.

GENERAL PURPOSE GOOD-BAD TRANSISTOR CHECKER

A good-bad test on a transistor (Fig. 7-44) can be devised on the basis of simple checks which can reveal the existence of a short or open condition between elements.

An AM ohmmeter reads resistance of a substance by applying a small voltage across the material. If the resistance is slow, the applied voltage can force a relatively high amount of current through the unit. The meter indicates the amount of current on the scale which is calibrated in ohms.

The instrument will bias the transistor in both directions. If forward bias occurs, large current flows indicating the resistance is low. Reverse bias will indicate at least a ratio of 20 to 1 if the transistor is normal. If a transistor does not respond as described, it may be considered defective. However, weak units, or those with leakage between elements, might respond satisfactorily. This test is basic and only intended to indicate whether the transistor is completely inoperative.

Make sure all checks with the tester described are made with the ohmmeter set to the highest resistance range which gives a convenient reading. Certain types of transistors that are not classed as general purpose may be damaged by the voltage from the ohmmeter.

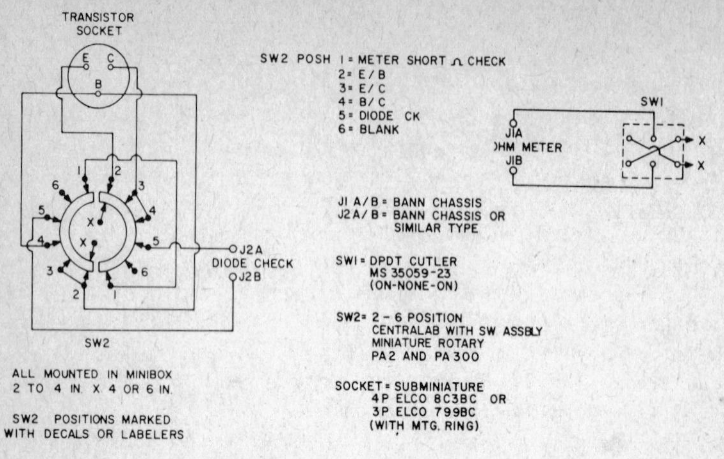

Fig. 7-44. Schematic diagram of the general purpose, Mark 1 transistor checker.

The purpose of the tester described is to simplify as well as speed the test in a comfortable and stable position. Whether the transistor is PNP or NPN, simply plug in the unit, connect the ohmmeter to the banana jacks input, set the meter to the highest range that will give the convenient reading, set sw2 to the test position (position 1) to zero the meter. Then make transistor checks for E-B in No. 2 position, E-C in 3 position, B-C 4 position and No. 5 for diodes, with the diode in the diode jack, of course.

In the diode position, it is also possible to check condensers for open-short. If it is desired to check components in circuit and if conditions warrant it (one end of component disconnected), simply run a pair of test leads with alligator clips at the diode jacks, leave switch 2 in diode position and use reversing switch 1 to test forward and reverse currents.

If desirable, the No. 6 position can be used to run a pair of test leads directly, or perhaps another set of jacks to hook up the test leads.

A TRANSISTOR PARAMETER TRACER

Most transistor manufacturers include on their spec sheets such data as maximum allowable voltage measured from collector to emitter (VCE), maximum allowable collector current (IC), maximum voltage collector to base (VCB) or base to emitter (VBE), maximum power dissipation at 25° Centigrade (PD) with derating factor for other operating temperatures, and typical beta (HFE) or forward current transfer ratio measured at a given set of parameters. This amounts to the minimum amount of data necessary for basic design purposes, but once the transistor is purchased and a circuit must be constructed, a number of problems suddenly arise. For instance, if the specifications for a gain of 100 are given as VCE = 5 V, IC = 1 mA, F = 1 kHz, what change in the value of beta can be expected when VCE = 12 V, IC = 2 mA, and f = 7 MHz? Just as important is the missing value of base current (IB). Since the transistor is a current amplifying device this parameter is as important for optimum gain as the grid bias in voltage amplifying tube circuitry.

This test unit (Fig. 7-47) is essentially a variable DC transistor power supply with three meters which allow the parameters VCE, IC, and IB to be monitored simultaneously. Thus, the designer may vary one or more DC parameters and, by plotting the resulting values on linear graph paper, have a permanent record of the DC curves for each individual transistor. This is important, due to the fact that most transistors are produced in "batch" quantities and no two of the same type have exactly the same parameters. In addition, the value of beta at different bias points may be computed by using the set of DC curves. The value of such design data should be apparent. With the DC parameter information obtained from the Transistor Tracer, circuitry can be designed and constructed without the usual trial and error method. By selecting a suitable bias point on the DC curves for the required value of beta and computing the necessary resistances by Ohm's Law, there will be no question of possible saturation, cut-off, or low gain. An additional feature of the Transistor Parameter Tracer is its ability to be utilized as a power supply with variable voltage control and a safe output of 12 V at 400 mA or 15 V at 300 mA.

The unit is constructed using a Bud 5 × 4 × 3 inch Minibox. No extra panel space is available and the SPST power switch is mounted on the left side of the box. The size of the meters used determines the space left for the controls when using standard-size miniboxes.

The transformer is mounted on the right side of the minibox with the wires toward the center of the top panel, and just high enough to allow clearance for the bottom of the box. The four diodes, 2-watt resistors, and pilot light dropping resistor are mounted on a piece of Vectorboard which is attached to the top of the transformer frame with epoxy glue. Not much space is needed for mounting these components, but overall board dimensions will be determined by the depth of the meters.

Due to the change in parameters which occurs when a transistor is subject to temperature variation, the transistor socket should be mounted on the opposite side of the box from the transformer. A number of holes, 1/4 inch in diameter, should be drilled above the transformer on the side of the minibox and on both sides of the other section of the enclosure, just below the transformer, to allow air circulation. In cases where the transformer produces excess heat, it will be necessary to place a heat shield around the socket. The first filter capacitor is soldered across the bridge and is supported by its own leads. Two electrolytics are not necessary, but the added filter capacity is desirable.

The value of the bias control is not critical, with the exception that too low a value will draw unnecessary current, requiring a larger rheostat. Also, since no series meter resistor is used, the medium setting of this control should provide meter protection in the event that the transistor under test is shorted. In order to be able to monitor voltage and current when the Transistor Parameter Tracer is utilized as a power supply, the voltmeter is connected across the circuit at all times. When the function switch is in the 12 V/400 mA position, the 0–50 mA meter is shunted with constant resistance wire (R1) of a value which allows the meter to measure 0–500 mA, or ten times its normal scale. The value of this shunt resistance is determined by the formula R = Rm/n-1, where R is the necessary shunt resistance, Rm is the internal resistance of the meter, and n is the factor by which the original scale is to be multiplied. Thus, with a meter resistance of 2 ohms, the necessary shunt resistance is R = 2/10-1 = .22 ohms. If constant resistance wire having a resistance of 2 ohms/foot were used, 2/12 = .22/length = 2.64 inches of this wire would be needed for the shunt. It is best to cut the length slightly longer to allow for soldering, then adjust by cutting about 1/8 inch at a time from the length while monitoring the output with a variable 5-watt load connected in series with a reference milliamp meter. With a variable load between 24 and 120 ohms, the meter reading from 100 to 400 mA can be checked. An alternate method, not requiring a reference meter, uses composition resistors, arranged in series or parallel, to give a value of 30 ohms with a power dissipation of 5 watts. With this load connected across the output, a 12 V reading on the voltmeter should give a reading of 400 mA on the milliamp meter when the shunt is the correct length. Keep in mind that the power rating of the supply is 5 watts, limited by the 25 k ohm control, and any reading over 12V at 400 mA will be close to exceeding this value, causing possible damage to the rheostat. All meters should be calibrated with the help of a reference meter and the zero-set screws made fast with a drop of clear nail polish or glue.

Operation

A volt-ohm meter should be used to check out wiring and to assure that the function switch has been connected with the correct polarities (NPN, PNP) in reference to the transistor socket pins. To preclude applying too high a voltage to the transistor under test and to protect the meter in case of an inner transistor short, the VCE/IC control should be set to maximum resistance before applying power. Potentiometer IB should be set at mid-range for similar reasons. In addition, since transistor breakdown from possible transients can occur when inserting a transistor into the socket with the power on, the line switch should remain off until the transistor to be tested is inserted and the controls are set as above.

After setting the function switch on the polarity of the transistor to be tested, inserting the transistor into the test socket, and adjusting the controls as outlined above, power may be applied. The VCE/IC and IB controls are interactive; that is, changing the setting of one will affect the meter reading controlled by the other. Therefore, it is possible to obtain a DC curve within the limits of the meters and the transistor under test. By advancing the VCE/IC control until a reading is shown on the milliamp meter, the first set of values is obtained. Varying the IB control will show whether the base is drawing current or if it is cut off. Continual advancing of the BCE/IC control in convenient steps will give additional sets of values, as advancing the IB control will give different levels of base current. By readjusting both controls for each set of values, a constant base current curve at different values of VCE and IC can be noted.

These values of VCE, IC, and IB for each setting can be transformed into a DC curve for the transistor under test by utilizing linear graph paper. Figure 7-45 shows such a DC curve for a 2N918 transistor, plotted from values obtained with the Transistor Parameter Tracer. Figure 7-46 shows the DC curves included on the manufacturer's spec sheet for this transistor. The value of beta, or the gain which can be expected from the transistor under test, can be computed at any point along the DC curves. Since beta = IC(ua)/IB(ua) for any constant value of VCE, beta may be determined for any corresponding values of IC, IB along the chosen vertical VCE axis. Table 7-1 shows the

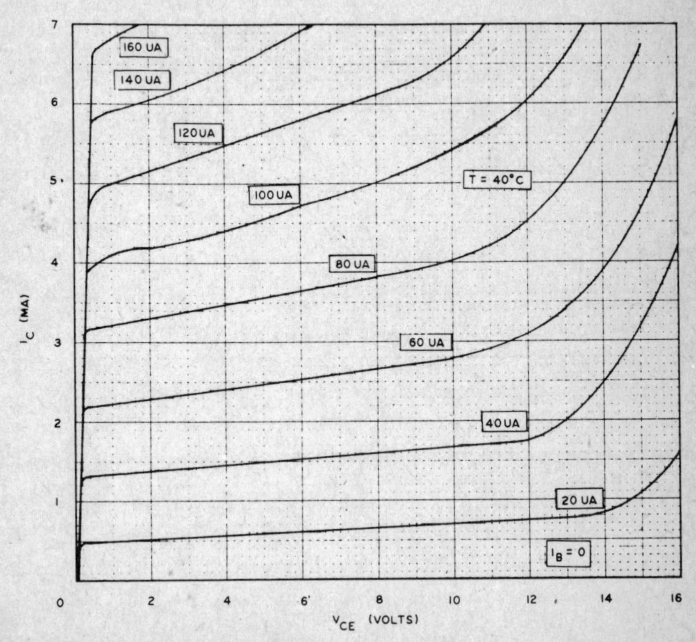

Fig. 7-45. DC curves of 2N918 transistor obtained with the Transistor Parameter Tracer.

Table 7-1. Values of Beta Taken from Figure 7-45.

$$\text{beta} = \frac{I_C\ (\text{uA})}{I_B\ (\text{uA})};\ V_{CE} = 12\ \text{V}$$

I_C	I_B	beta
1.4 mA	20 uA	70
2.3 mA	40 uA	58
3.5 mA	60 uA	58
4.8 mA	80 uA	60
5.8 mA	100 uA	58
7.2 mA	120 uA	60

computed values of beta taken along the VCE = 12V axis for the 2N918 transistor. The values of beta obtained with this Transistor Parameter Tracer are valid only for the common emitter configuration. Naturally, the available gain will not be as great when the transistor is used in the common base configuration, and cannot be expected to provide any gain when utilized in the common collector configuration.

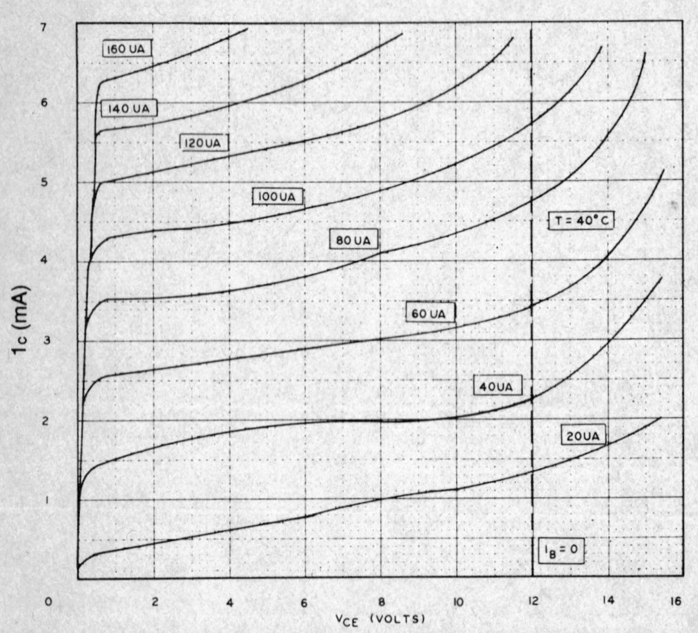

Fig. 7-46. DC curves of 2N918 transistor from manufacturer's specification sheet.

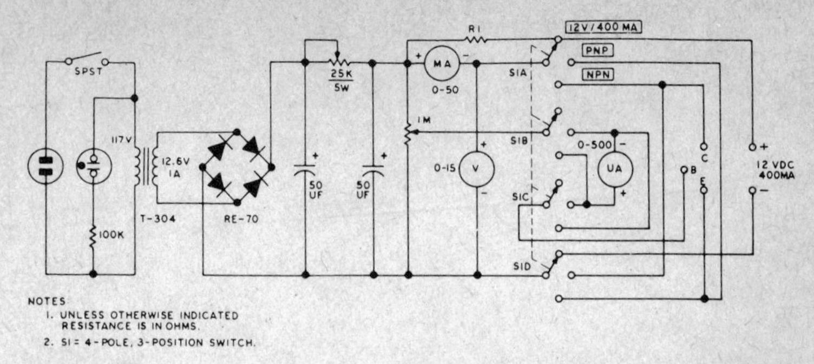

Fig. 7-47. Schematic of the Transistor Parameter Tracer. See text for component description.

A SOLID-STATE PRESELECTOR

There are many receivers now on the market and a great many used receivers that can have their performance improved significantly by the addition of a preselector. Such receivers are the ones to which a beginner is attracted by consideration of price.

These receivers usually share such characteristics as lack of stability, lack of selectivity, and lack of ability to reject images. (Sometimes this last-named characteristic is linked to a high inherent noise level.)

Stability usually is a matter of basic electrical and mechanical design and is not often amenable to corrective measures. Selectivity sometimes can be enhanced by the addition of a Q-multiplier or a crystal receiver. By using solid-state active components, a preselector can be constructed as a single, self-contained unit. Such a unit can be added to a receiver without modifying or in any way lessening the resale value of the receiver, an important consideration for one who intends to upgrade his station equipment by swapping or trading in.

What is desired for a preselector? Here are 10 points:

1. Selectivity for rejection of images (not necessarily for rejecting adjacent-channel signals).
2. No gain. Even a small loss would be acceptable. Too much gain could degrade the cross modulation of the intermodulation characteristics of the receiver.
3. No increase in cross modulation, intermodulation, or noise.
4. No instability.
5. High tolerance of impedance match, both for the input and the output circuits.
6. One control; single tuned circuit.
7. Simple construction.
8. Low construction cost.

9. Parts readily available.
10. Capable of operation for a self-contained power supply.

This preselector is designed to meet the conditions of these 10 points.

The unit is for use over the whole high frequency spectrum; that's why tapped coils are used in both the antenna and the tuned circuits (Fig. 7-48). High input impedance (for best selectivity) and freedom from cross modulation and intermodulation indicated the use of a field-effect transistor; using a junction FET as a source follower further enhances these desirable characteristics. Isolation from reactive effect requires the use of a low-gain, ultrastable stage coupling the source follower to the output.

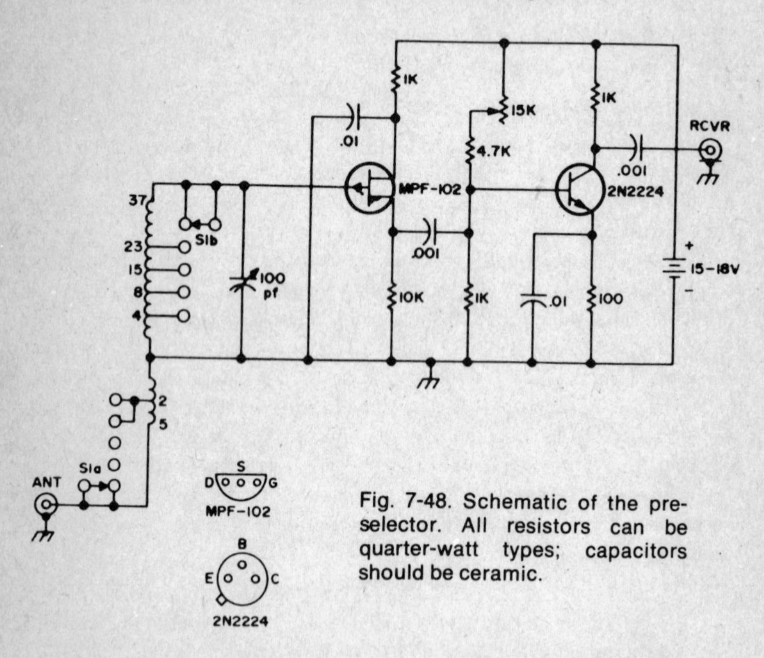

Fig. 7-48. Schematic of the preselector. All resistors can be quarter-watt types; capacitors should be ceramic.

You can wind your own coil quite easily, using one of the popular plastic pill boxes for a form and a piece of Air Dux 832 T coil stock. It's 1 in. in diameter and has 32 turns per inch—also, it just happens to be the right size to make a firm pressfit into the top of one of those pill containers, a very convenient way of mounting the coil upright. Five turns up from the bottom, cut the wire and leave a one-turn gap before the start point of the secondary portion. The primary is tapped two turns from the cold end (the end nearest the secondary); this tap is used for the two highest tuning ranges. The secondary is tapped (from the cold end) at 4, 8, 15, and 23 turns; it has a total of 37 turns. Don't take these figures as sacrosanct; stray values of inductance and capacitance can affect them markedly.

The band switch is a two-pole, five-position rotary switch, mounted for short leads to the coil and to the variable capacitor. Note that two of the primary switch points are left unused. Nothing is served by altering the primary turns in small steps.

The tuning capacitor is a two-bearing unit, quite sturdy. It measures 100 pF fully meshed, and minimum capacitance is 10–12 pF.

Nothing is unusual about the MPG-102 junction FET. There's a 1 kΩ decoupling resistor in the drain circuit to encourage RF to use the 0.001 μF capacitor as a path to ground. Coupling from the source to the NPN bipolar transistor is through a 0.001 μF capacitor.

You may be astonished at the low values of the base-to-ground resistor and the collector load resistor. These were selected to keep gain low and stability high. The gain can be controlled (at the expense of stability) by the 15Ω variable resistor in series with the 4.7Ω resistor used for applying positive bias to the base. It's best once set and then forgotten. In the emitter circuit, there's a bypassed 100Ω current- stabilizing resistor.

RF output is taken off the collector through a 0.002 μF blocking capacitor to a short piece of 50Ω coax.

A word about that bipolar resistor. You'll not find the 2N2224 listed in many supply houses, yet it shows up on surplus printed-circuit boards. It's good, although several other NPN RF transistors will work quite well, too. Don't be concerned about getting any particular type of transistor. If you can't get an MPF-102 easily, the Motorola HEP-802 is a good substitute.

The preselector performs amazingly well. The tuning is sharp, so much so that a slow-motion dial might be a good investment. It's quite astonishing to note just how much of the "crud" that you thought was inherent to a band disappears when the preselector deletes images.

THE 73-A-PHONE

Ever need to shout louder than you can? Perhaps on a DX'pedition when you were on top of the hill and your buddy was just starting up—without the chow?

Here's a device which can help you out in such a situation. It's called the 73-A-Phone because (if you use the specified speaker or one of similar characteristics) it will produce a 73 dB sound level at a distance of some 90 feet, or the same 73 dB sound level inside a room 10 feet high by 15 feet wide by 20 feet deep. This 73 dB is twice as much power as average conversational speech levels, so should be clearly heard.

As you can see by the schematic (Fig. 7-49), there's not much to the 73-A-Phone. A carbon mike, a 10 ohm resistor, the 2N234A transistor, and the speaker are all there is to it except the batteries. For portability these can be flashlight cells, but the larger lantern types are preferred for longer life.

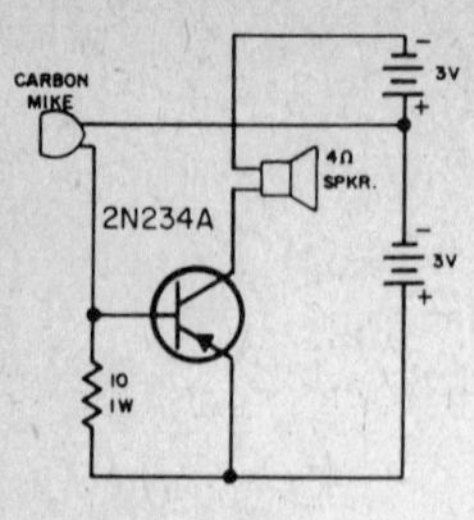

Fig. 7-49. 73-A-Phone, schematic.

While the University MIL paging speaker is recommended and is the only one tried, any 4 ohm speaker will work with this circuit. The MIL has a self-contained horn and is a high-efficiency unit, which is why it was chosen, but is admittedly a trifle expensive for such an otherwise simple device.

The transistor delivers approximately 1/3 watt in this hookup. This may not sound like very much, but if it's driving an efficient speaker it's a pretty potent package.

Many variations of this simple circuit are possible, but we don't recommend any drastic increase in supply voltages. Under no circumstances should you increase the base supply voltage above 3 volts; you're free to increase the collector supply as high as 9 volts, though, at the risk of incurring possible distortion. This, naturally, would increase power output somewhat.

With conventional speakers, this can make a most usable portable PA system for hamfest use. The batteries, transistor, and resistor can be placed inside the speaker box, leaving only the wire running to the mike button. Take care to avoid feedback in such use, though—this gadget will surprise you with its sensitivity!

ECONOMY TR SWITCH

Here (Fig. 7-50) is an electronic device that permits instantaneous "switching" of a single antenna from transmitter to receiver. It is simple to

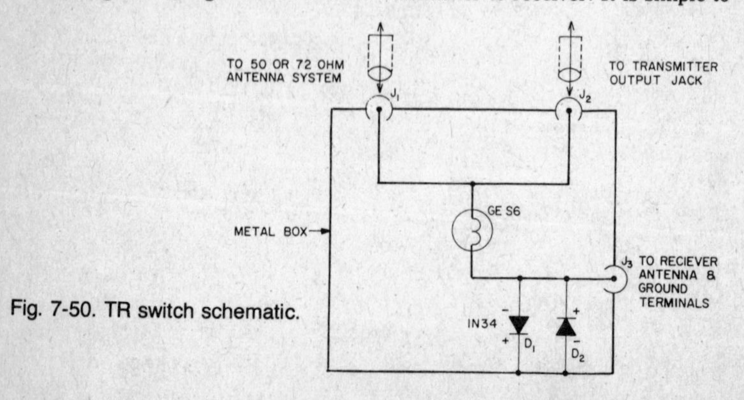

Fig. 7-50. TR switch schematic.

build and requires no external power supply. It will handle output up to 500 watts and will also operate on "flea" power when an outboard converter is used. The only disadvantage found in this unit is that it does attenuate receiver signal slightly.

The unit can be built in a small metal box where all parts can be easily mounted. It should be remembered to use a heat sink when soldering the 1N34.

This unit works very well on both CW and SSB and a single relay can be added and used on AM.

LIGHT BULB AMATEUR WATTMETER

This amateur wattmeter is a very useful gadget for determining the RF power output of solid-state VHF—UHF transmitters in the difficult range to measure, from abut 10 mW up to 5 watts. It does not read watts directly; but by a simple comparison of calibrated pilot light brilliance, it will tell you how many watts you are putting out, to within less than 5%. It allows you to check power increases and estimate your efficiency quite closely.

Principle Involved

While this unit is not by any means a "trick," it does not read RF directly. You first light a pilot light as a good dummy load, matching it into the RF tank circuit of your transmitter by the normal means, also noted here.

You then switch on a second bulb of the same type by means of a battery, controlling the light output with a $1.30 wirewound potentiometer in series, as shown in Fig. 7-51. This pot must be previously calibrated in milliwatts, as by the method of "volts times milliamperes equals milliwatts." You then match the brilliance of the bulb lit up with RF or its dull glow at some 18 to 25 milliwatts if you're just getting your transmitter going, and read the watts on the wattmeter dial. It's astonishing how well it works, how repeatable it is, and how you wouldn't be without it once you build and calibrate it.

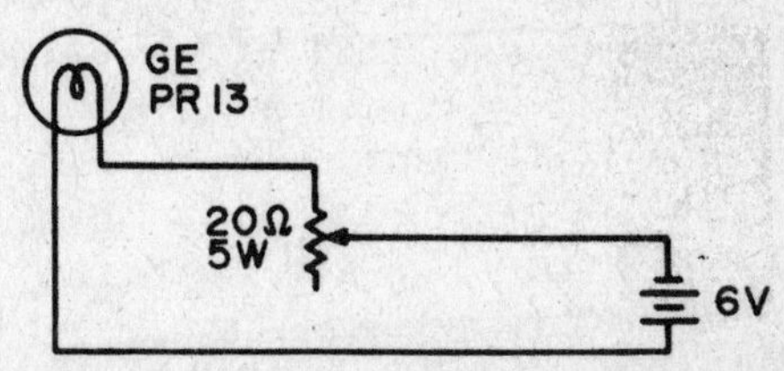

Fig. 7-51. The circuit supplies the brilliance "standard" for comparison. When the "standard" lamp is mounted adjacent to the dummy load, the pot permits variation of the standard to match the load. If the resistance is panel-marked in watts, a good power indication is achieved.

Brilliance Standard

Figure 7-52 tells almost the whole story at a glance. You can, of course, put as much calibration on the dial as you have time for. It is quite important to orient the bulb filaments in the same relation to your eyes for best matching. There isn't much in back of the panel except one 6V battery which can be obtained in any hardware store.

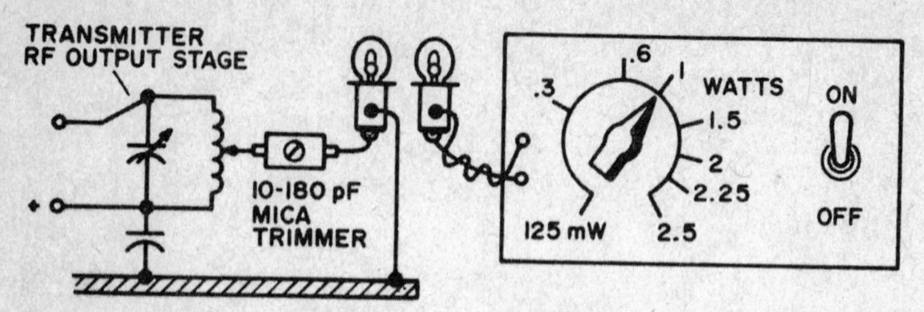

Fig. 7-52. A series capacitance loads the RF indicator for comparison. The capacitance value will decrease inversely with frequency increases.

RF Matching

Not that it is particularly critical, but be sure and note the need for a large range of series capacitors for the RF pilot lights as you go up in frequency. This can be seen clearly in Fig. 7-52 and 7-53. The block diagram, Fig. 7-52, shows a 6 meter setup. As you go up in frequency the series capacitance drops. A good matched load on 432 MHz can be obtained as shown in Fig. 7-53. It is important to vary the amount of coupling, and thus the series capacity, by spacing the tab capacitor closer or further away from the ground plane, as detailed in Fig. 7-53. You also can use as many bulbs as you can solder onto a tuned RF inductor, even though they don't all light up with the same brilliance. You can match them all up, but you don't have to. Just check the wattage, or milliwattage, of each one and add them up for the total.

The number 48 and 49 bulb, listed at 2V and 60 mA, is rated at 120 mW, and glows dim at about 12 to 15 mW; so it can be used for low-power receiver oscillators, etc. With two other bulbs found in hardware stores connected and matched to the RF inductor, such as the PR13 (5V at 500 mA), you can read correctly up to 5W. From there on up you're on your own, although a good variable 115V DC supply can be made up to work around 50 to 100W.

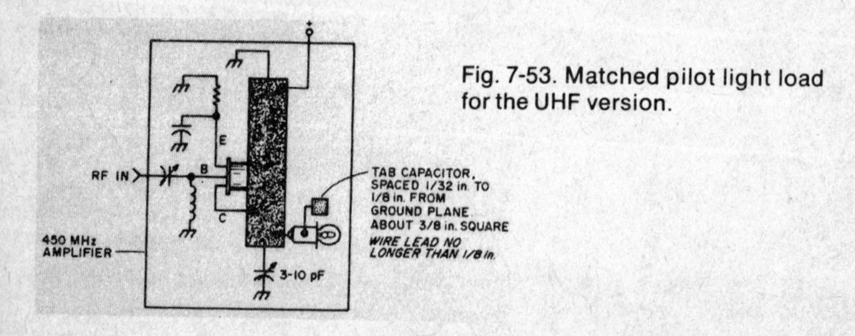

Fig. 7-53. Matched pilot light load for the UHF version.

AN IMPEDANCE MULTIPLIER FOR THE VOM

A voltohmmeter or similar device can be found in almost every shack. The typical VOM is inexpensive, rugged, and independent of the AC mains. However, it has a disadvantage in that it can't be used to measure voltages in relatively high-impedance circuits without some loss in accuracy. Consider the circuit in Fig. 7-54. A VOM is being used to measure the voltage drop across R2. A typical VOM has an input impedance of 20 kΩ/V (or less). This means that the input impedance of the VOM, in ohms, is 20,000 times the full-scale voltage.

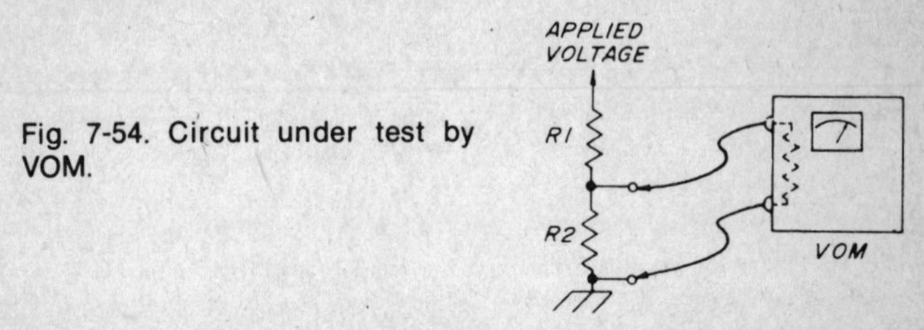

Fig. 7-54. Circuit under test by VOM.

On the 10V scale, for example, the input impedance is 200 kΩ. If R1 and R2 are relatively small, say 2 kΩ, then the effect of shunting R2 with the VOM will be small. In Fig. 7-54, if the applied voltage is 20V, then the VOM will read approximately 9.9V. On the other hand, if R1 and R2 are 2 MΩ, then the VOM will read approximately 2V! Clearly, one solution to the problem is to buy a VTVM with an input impedance of 22 MΩ or so. A less expensive solution is to build the impedance multipliler.

The schematic of the unit (Fig. 7-55) consists of a voltage divider which presents a 25 MΩ impedance to the voltage being measured, and an operational-amplifier unity-gain voltage follower.

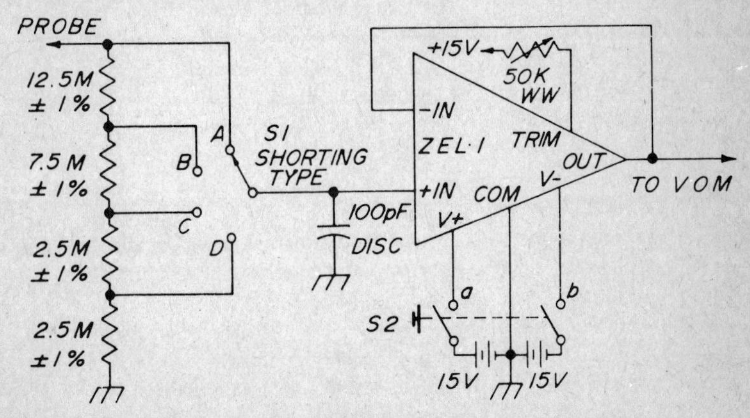

Fig. 7-55. Schematic diagram of the impedance multiplier.

The voltage follower has characteristics which make it ideal for the task at hand. Its input impedance is on the order of several hundred megohms so that it does not degrade the accuracy of the voltage divider. Its output impedance is less than an ohm and it can supply up to 5 mA (or more, depending on the amplifier used), so is virtually unaffected by the VOM. Finally, as its name implies, its output voltage is (for all practical purposes) the same as its input voltage.

For the ham who is looking at his first operational amplifier circuit, the following may be helpful.

An operational amplifier is a linear, high-gain, direct-coupled amplifier, usually provided with a differential input, but with its output referenced to ground. If a small voltage (a millivolt or more) is applied, the output voltage will swing to one or the other of its limits (typically ±10V). Only for voltages less than abut 0.1 mV (for the ZEL-1) will the output voltage not be at one of its limits.

The polarity of the output voltage depends directly on the polarity of the input voltage. That is, with the positive lead of the source connected to the positive input of the amplifer, the output voltage will be positive, and vice versa. Further, instead of connecting one voltage source across both input terminals, a separate source can be connected to each input, with the second lead of each source connected to ground. In this case it is the difference of the two input voltages which appears across the amplifier input. The key to the operation of the follower is that one of these voltages is the one we wish to measure while the other is provided by the amplifier.

For the ham who is starting from scratch, two amplifiers can be recommended. First is the ZEL-1, the type specified in Fig. 7-55, which is available for $11 from Zeltex, Inc,. Concord, Calif. While the ZEL-1 is expensive, it does tend to be forgiving about having its output shorted to ground and the like. For the less cautious ham, the μA709 is available from Poly-Paks, Lynnfield, Mass. 01940. It is not nearly so forgiving. A circuit which accommodates the μA709 is shown in Fig. 7-56.

Other than the fixed-value components, the main difference in the circuit requirements of the two amplifiers is the 50 kΩ trim pot which is used with the ZEL-1. The pot is used to adjust the follower output to exactly zero when the probe to the voltage divider is shorted to ground. In some applications, the absence of such a control in the μA709 circuit would be a handicap. However, an error of only a millivot or two is likely to result in this application. If a ZEL-1 amplifier is used, the pot may be replaced by a fixed resistor once the correct value has been experimentally determined.

The accuracy of readings made with the impedance multiplier in use will depend in part on the accuracy of the VOM involved and in part on the accuracy of the resistors used in the voltage divider. Since the garden-variety VOM has an accuracy of no better than ±2%, use of better than 1% tolerance resistors seems unwarrented.

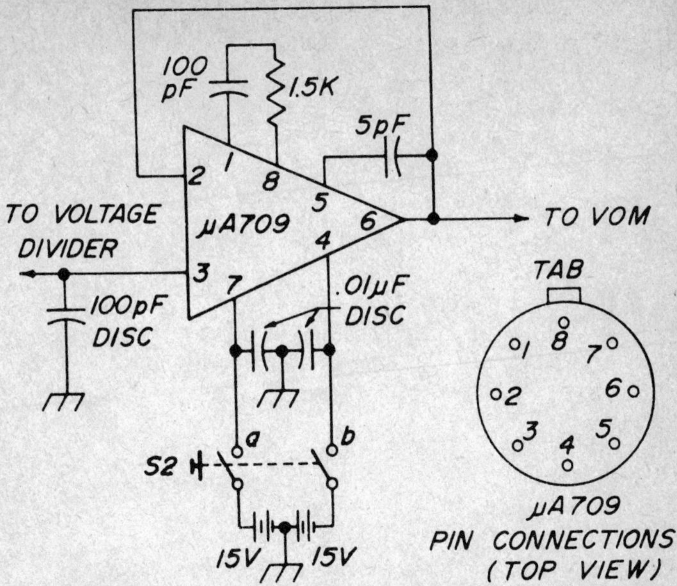

Fig. 7-56. Circuit modifications for μA709 amplifier.

The DPST switch should be provided as shown, in the interest of long battery life. Also, battery voltage should be checked frequently, since the output voltage swing capability of the amplifier depends markedly on supply voltage.

Physical layout is not at all important. The device will fit into a small minibox which can be provided with male plugs so that it can be plugged into the VOM in place of the probe leads.

Operation of the impedance multiplier is straightforward. With the VOM switched to its 10V range, a full-scale range of 10, 20, 50, or 100V is provided when the range switch (S1) is in position A, B, C, or D, respectively. The unit may also be used with the VOM switched to a more sensitive range. As before, the voltage which appears at the output of the amplifier is the same as that at the input of the amplifier, but the full-scale capability of the combination is reduced. For example, with the VOM switched to a 2V range and the voltage divider switched to position D, the voltage to be measured must be less than 20V or the VOM will go off scale.

The impedance multiplier may be used to make either DC or low frequency AC (audio) measurements. The only restriction is that not more than 10V may be applied to the input of the voltage follower.

SIMPLE LIGHTNING DETECTOR

Not everyone *needs* a lightning detector, but it is very handy for those stations that normally operate round the clock and unattended—repeaters, private repeaters, and remote control systems such as autopatches. In that case the lightning detector can temporarily disable the station and ground the antennas when a storm is in the vicinity.

Figure 7-57 shows the basic circuit. A small, sensitive SCR is in series with a relay coil. When a positive pulse appears on the SCR gate, the SCR turns on and energizes the relay. A 20 foot pickup wire attached to the gate and strung about the house works quite well to couple lightning-caused pulses into the diodes. Once the SCR latches, the circuit remains on until the power is removed.

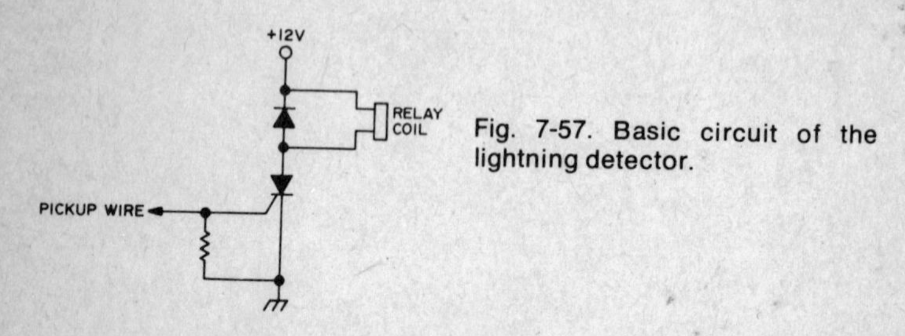

Fig. 7-57. Basic circuit of the lightning detector.

Figure 7-58 shows an improved version which automatically resets itself after a while. The SCR works as before, but when the relay closes, +12V is applied to the top of the capacitor, which then slowly discharges through the 10K resistor and the base of the transistor. The transistor is across the SCR and lets the SCR release immediately after the lightning stroke is over. The relay stays pulled in for a short time interval, depending on the gain of the transistor and the quality of the 10 μF capacitor, and then opens.

In operation, make sure the pickup wire is far enough from your transmitter and antenna so that it does not pick up RF energy from there. If any RF does get into the pickup wires, use a loop of wire whose far end is grounded. If the sensitivity is too high, simply use a shorter pickup wire.

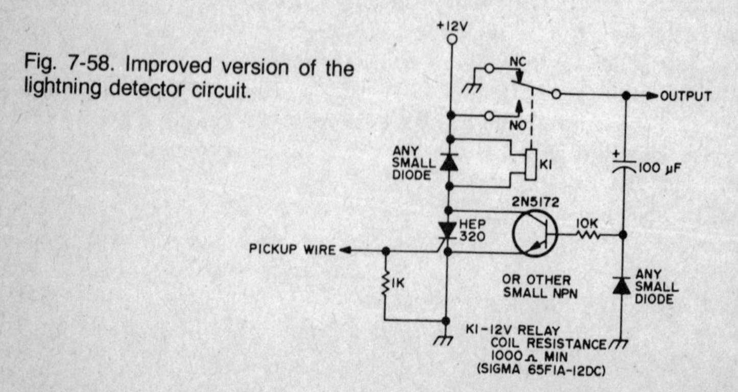

Fig. 7-58. Improved version of the lightning detector circuit.

ALL PURPOSE METERING CIRCUIT

The builder of more complex equipment is often faced with a choice of three evils, whether to meter all circuits which should be metered with individual meters, with a single switched meter or forget the whole thing and trust the equipment to perform in a satisfactory manner after initial adjustment. Individual metering is rather expensive and space-consuming and must in most instances be confined to "price is no object" projects. Obviously, lack of metering is an invitation to trouble, expense, and work, repairing and replacing ruined components. Switched metering too has its shortcomings, but when used in conjunction with full-time metering of the most important circuit, it serves well.

As a typical example (Fig. 7-59), a linear amplifier might well have a plate current meter plus a second switchable meter which could be used to monitor filament, screen, plate, and bias voltage, screen current, relative power or any desired combination. It will be noted that these voltages and currents are of various polarities and magnitudes and may (as in tetrode screen current) be both positive and negative depending upon operating conditions. It would not be easy to design a conventional switching circuit capable of measuring all these parameters.

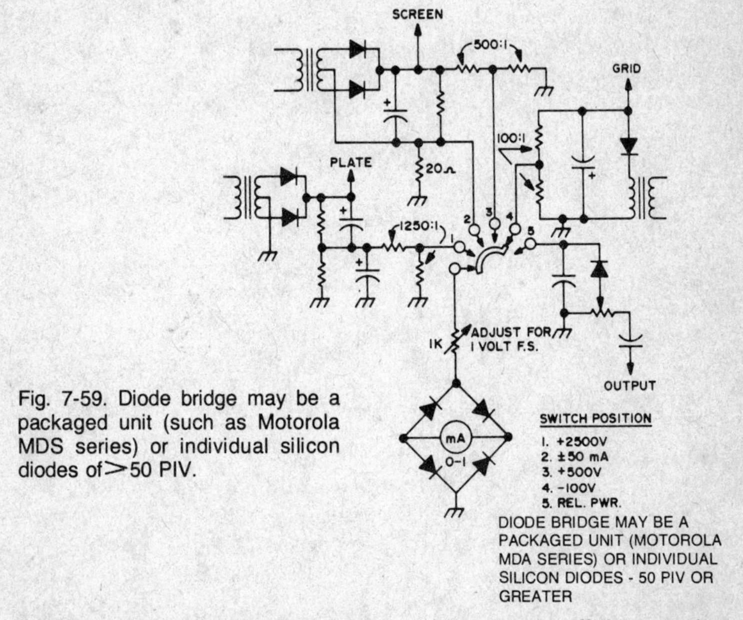

Fig. 7-59. Diode bridge may be a packaged unit (such as Motorola MDS series) or individual silicon diodes of > 50 PIV.

The easiest method of attacking this problem is to install the meter in a diode bridge in such a way as to cause an upscale deflection no matter which polarity is applied. By proper selection of component values it is possible to use the basic meter scale times a factor for each function. The meter/bridge assembly is made to read a set value (one volt is handy) and all inputs to the switch are arranged to produce this voltage under full scale conditions of the range desired. In the example, this system is used in conjunction with a 0–1 mA meter movement to read ±50 mA screen current, + 500 V screen voltage, + 2500 high voltage, and – 100V bias, not to mention relative power. The switch is a simple single pole, five-position type. In your design let Ohms Law be your guide and be sure to use a voltage divider where required to keep surges out of the meter. Remember too to use a non-shorting switch to prevent connecting adjacent circuitry together while switching.

SUPER SIMPLE BFO

Have you ever wanted a small BFO to receive CW or SSB on one of the many multi-band transistor portables which are available on the market today?

Many articles are available which state "Simply modify an old IF transformer as follows, etc., etc." Attempts to remove the built-in capacitor or simply cut the unused and unnecessary winding from a miniature transformer is enough to drive a good man to drink. By the way, that drink doesn't help a large unsteady hand cut that elusive #42 wire from a small transformer with any less effort. Should you be fortunate enough to remove the lead without damage to the remainder of the transformer you usually find the tuning rate to be too fast to be practical if not just impossible.

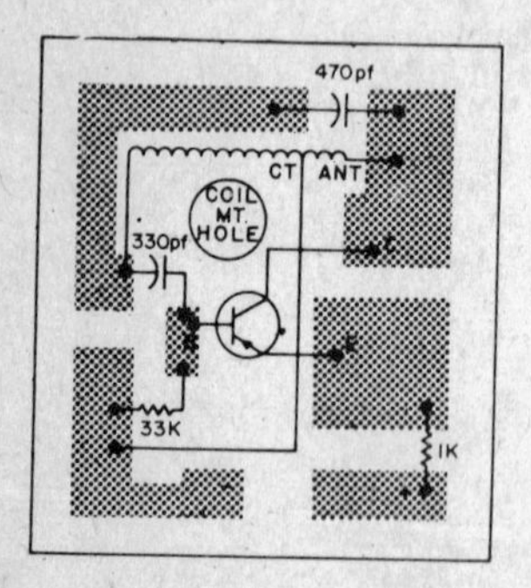

Fig. 7-60. Actual size PC board.

The circuit described herein (Fig. 7-60 and 7-61) utilizes an inexpensive Vari-loopstick. There are four additional components and any PNP transistor gives sufficient signal that no direct coupling to the receiver is needed. Many replacement loopsticks are supplied with a small chrome knob and in this circuit two revolutions are needed to go through zero beat at 455 kHz. Therefore, adjustment of SSB or CW signals by changing the oscillator frequency is not at all critical in the event the tuning rate of your receiver is too fast.

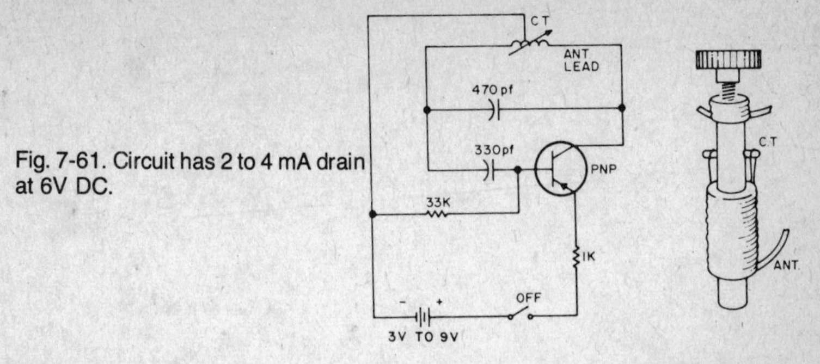

Fig. 7-61. Circuit has 2 to 4 mA drain at 6V DC.

Wiring is not at all critical; however, short rigid leads are desirable and should you elect to use the circuit board or a metal mounting plate the effects of hand capacitance when tuning will be eliminated. As a final precaution against instability due to the construction of the loopstick, coat the coil with paint or epoxy glue for rigidity. The PC board may be mounted to the bottom of the loopstick and the coil then inserted through a hole in the receiver case. Outboard operation is also possible by merely placing the BFO close to the receiver being used.

LED READOUT CRYSTAL SWITCH

The search for something distinctive is nowhere more apparent than among ham builders. The latest state-of-the-art circuitry is constantly employed to produce electronics far surpassing the commercial variety. An even larger group of hams is perennially modifying their commercial equipment to improve its performance or convenience.

A dial readout is described here that was designed for displaying FM crystal channels, but which can be made applicable to many detented switch schemes. It uses diode matrixing and Light Emitting Diode (LED) readout for a minimum of current draw and a maximum of flexibility and reliability.

The readout unit itself is constructed on a one-sided glass epoxy printed circuit board. The copper is etched into the seven segment pattern using standard printed circuit techniques. Either thin tape or liquid resist may be used for the pattern. Size is dependent on the space availability in the individual rig. Compartments are then constructed using flashing copper or metal reclaimed from a "tin" can. The can must be tin plated steel and not aluminum, as it is too difficult to solder to aluminum. A child's magnet will easily separate the steel from the non-ferrous metal if you are not sure. An LED is placed in each compartment and secured with a glob of silicon rubber sealant, such as Silastic or RTV. The chambers are then sealed by soldering on a metal top. This shields the LED from the effects of RF, and due to the long

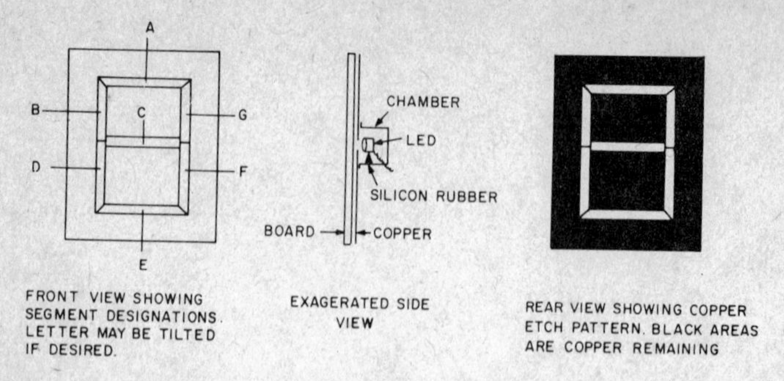

Fig. 7-62. Construction details of the readout from single sided glass epoxy board.

life of an LED there should be no need to open the compartment, once sealed. The use of the glass epoxy board is recommended due to the translucency and stability of the material.

Alternatively, slots could be cut in the panel of the rig with a saw or nibbling tool, and thin plastic frosted with steel wool placed behind the panel Fig. 7-63. Such plastic may be a piece of acetate used for wallet photo compartments or notebook picture protectors. To frost, rub a fine grade of steel wool over the plastic until it becomes uniformly etched with fine scratches. Mount the plastic (with the frosted side in) on the outside of the unit. The slots will then be visible only when illuminated, as is the case with the printed circuit board, and a more distinctive, less homebrew look will result. Compartments here can be made with either metal or plastic, glued to the rear of the panel. If available, of course, a standard seven segment commercial LED readout could be used.

The matrix is a seven diode maximum per character circuit that uses from two to seven diodes to form each digit. The matrix desired is selected by an additional wafer on the transmit and receive crystal switch. A single-pole

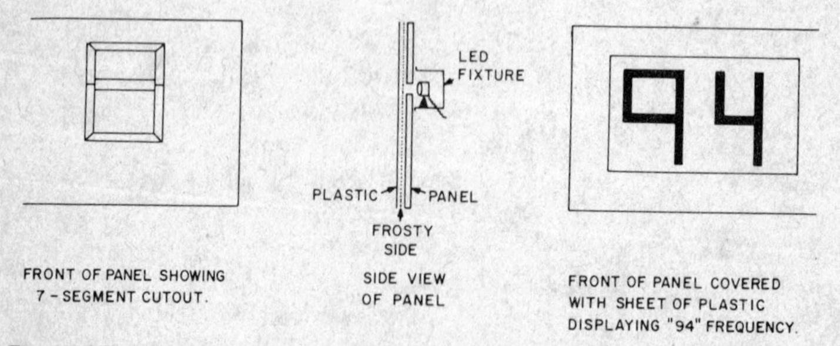

Fig. 7-63. Alternate construction technique using panel cutouts and frosted plastic.

switch is used, with as many positions as needed for the number of channels to be covered. Each switch position is connected to two matrix busses in order to provide two digits. Figure 7-64 is a sample two digit readout set up for 76 and 94. Diodes are placed in the matrix in accordance with the scheme shown in Fig. 7-65 to provide any of the desired digits. If a decimal point is needed, another LED with its resistor could be connected directly to the supply to provide that function.

A sample schematic is shown in Fig. 7-66. Here three transmit and receive frequencies are provided for: transmit frequencies are 16, 28 and 94; receive frequencies are 76, 88 and 94. Double ended boards are used for each pair of readouts. A large board with four matrices could easily be used.

While printed circuit technique with double sided boards is undoubtedly the most convenient way to make the matrix, perforated board can be used with flea clips and bus wires. A hybrid board, using one set of etched conductors and one set of bus wires, may be the most satisfactory solution to many builders. Figure 7-67 illustrates the various types of construction.

The entire unit is powered from a 12 volt supply and is directly applicable to mobile rigs. For base station use, a simple voltage divider from B+, or a tap onto a 12 volt supply can be used. Voltage is not critical, and anything from 10—15V DC will work.

Parts for this setup are not the most critical in the world. The LEDs can be bought from several companies, among them Poly-Paks, for around one dollar

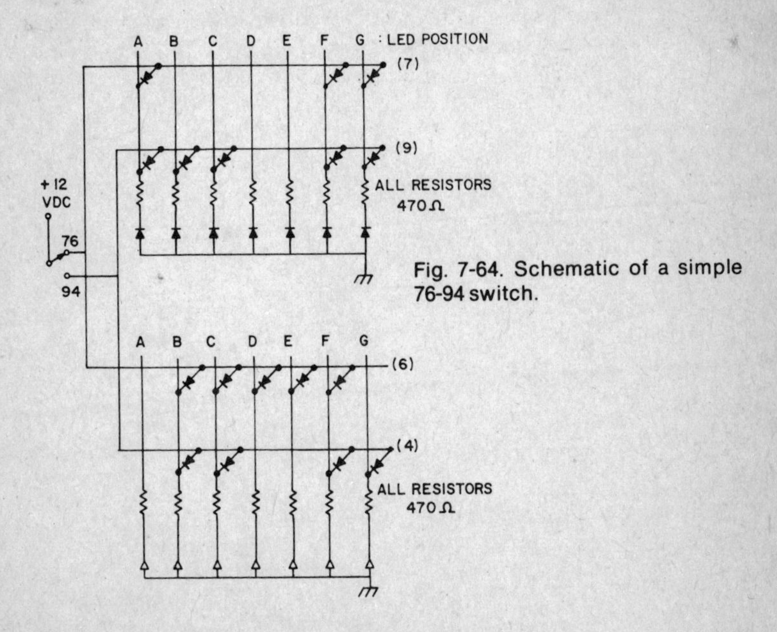

Fig. 7-64. Schematic of a simple 76-94 switch.

SEGMENT

DIGIT	A	B	C	D	E	F	G
1						X	X
2	X		X	X	X		X
3	X		X		X	X	X
4		X	X			X	X
5	X	X	X		X	X	
6		X	X	X	X	X	
7	X					X	X
8	X	X	X	X	X	X	X
9	X	X	X			X	X
0	X	X		X	X	X	X

X = DIODE NEEDED

Fig. 7-65. Encoding scheme for the matrix. Referring back to Fig. 7-64: To display 76, diodes are connected from the 76 position on the switch to the A, F, and G buses for the number 7, and the B, C, D, E, and F buses for the number 6.

each. The diodes are the common type that can be had from many sources for around ten for a dollar. Boards, resistors, etc., are stock items.

Although this is a simple scheme, it works as well as much more complicated ones. Seven segment decoder/drivers are available as integrated

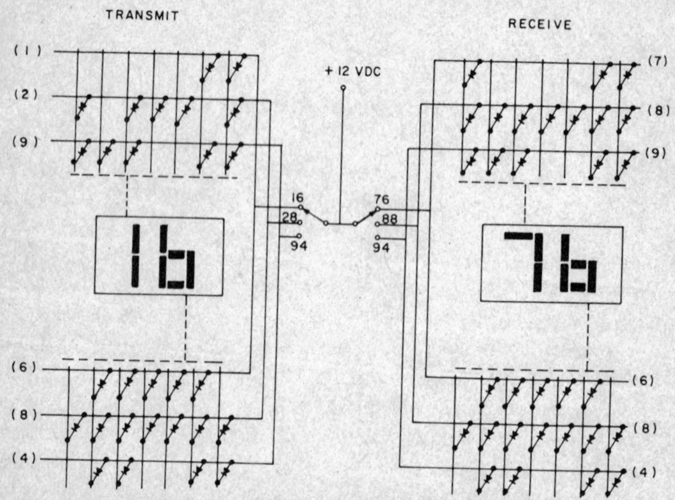

Fig. 7-66. Typical circuit that will display 16-76, 28-88, and 94-94.

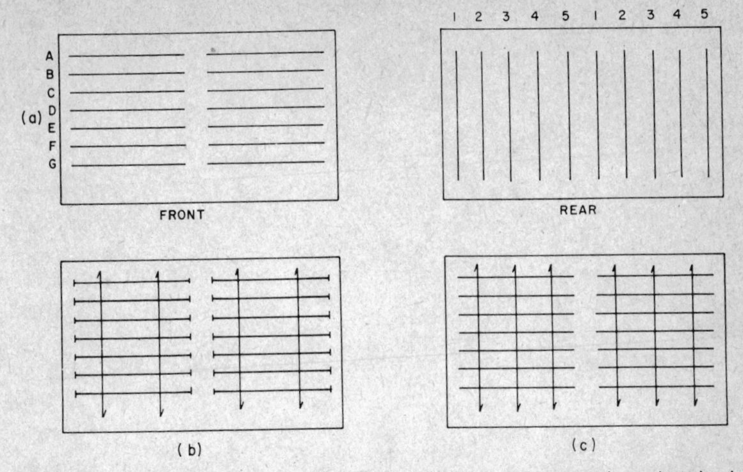

Fig. 7-67. Types of matrix construction. (a) PC double sided board etched for 5 frequencies, 2 digit readout. (b) Perf board with flea clips for 2 frequencies. (c) Perforated PC board with etched segment busses and wire individual digit busses.

circuits that will take the place of the matrix used here, but their cost is prohibitive. It is felt that the techniques used here allow a distinctive look to be imparted to many projects at a level most builders are able to afford. The seven segment readout and matrix can be applied to items besides FM gear: receiver band-switches, clocks, and anywhere else a highly legible data display is needed at a minimum of cost.

WIDE-RANGE RF MILLIWATTMETERS USING HCDs

The present low price of about one dollar for hot carrier diodes HP2811 or 2800 helped greatly in working out the two RF milliwattmeters shown in the circuits and photographs. They cover the range from audio up to 450 MHz.

A milliwattmeter is extremely useful in checking the output of any transistor or tube oscillator such as those used in transmitters, VFO units, receivers, and VHF converters—providing the oscillator has a 50Ω output connection. A temporary 50 Ω output connection can be made to any oscillator, doubler, or tripler tuned circuit by one or more turns or wire around the coil and running the RF output to the milliwattmeter through a short lead of coaxial line. If a measurement at any frequency shows a milliwatt or two into a 50Ω load, one can be reasonably sure of enough RF injection even into a high-impedance mixer. Some FET mixers require about 5 mW injection, so if the measurement into a 50Ω RF milliwattmeter indicates this amount is available, most of the tedious work is done in designing or checking this part of a receiver or transmitter. A VFO frequency control unit may need to have constant output over its whole range. Measurements of RF power over the whole range is needed in this case which may only take a minute or two. The time to iron out VFO irregularities is something else!

The parts needed in a simple RF milliwattmeter are relatively few in number and moderate in cost. The microammeter is used only as a reference indicator so no scale calibration is required. A black line of the meter face cover at the desired 5 or 10 or 15 μA is all that is needed. The RF power calibration is made on the dial or scale of the high resistance variable resistor or potentiometer such as used for AF gain controls.

The HP2811 (or 2800) hot carrier diode is remarkably uniform in characteristics from unit to unit. It is also usable with a forward bias DC voltage which greatly increases its sensitivity as a detector. All diodes have a minimum voltage below which the RF current flow is too low to be useful in a microammeter which is a current indicating device. If an ordinary diode has forward bias it may become a good noise generator, or erratic in operation. The diodes used in these RF meters work very efficiently with some forward bias from a small 1.5V battery. The current through the diode is limited by a series resistor of some value between 180 and 200 kΩ. The DC path is completed through the diode and the RF terminating resistor of 50—51Ω. This bias voltage also causes a current to flow through the microammeter, which preferably should be balanced out for low RF power measurements. This current is greatest when the power indicator variable resistor is set to minimum resistance.

By balancing out this current by means of a screwdriver-adjustable potentiometer so the meter reads zero with no RF power input, one reference line on the meter face can be used for any RF power input at any frequency. The DC drain on the battery cell is a value of less than 1 mA when making measurements, but the life of the battery can be extended greatly by having a switch in this circuit.

The 51Ω 1W resistor Fig. 7-68 should be a noninductive, carbon or metallic film type suitable for RF service. Actually a 1/2W type has a little better RF

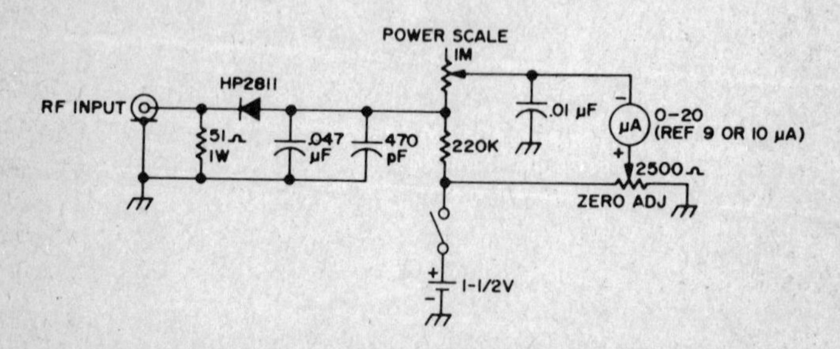

Fig. 7-68. Schematic diagram of 0.05—500 mW wattmeter.

characteristic but no resistor should be used at full rating. This terminating resistor is soldered across the BNC RF input fitting with as short leads as possible. The diode and its parallel bypass capacitors are also mounted at this input jack. All other components can be mounted anywhere on the 4 × 8 × 2 inch chassis panel (or other sized case if desired).

In the other higher range RF meter (Fig. 7-69), the RF resistor consisted of two resistors in series with short leads across the BNC input jack. A 39Ω 1W and an 11Ω 1/2W resistor in series make up the 50Ω RF load. The diode is tapped into this resistor in order to keep within the 15 or 20 PIV rating of the HP2900 or 2811 diode. (An HP2800 with its 75 PIV rating could be used across a 50Ω resistor.)

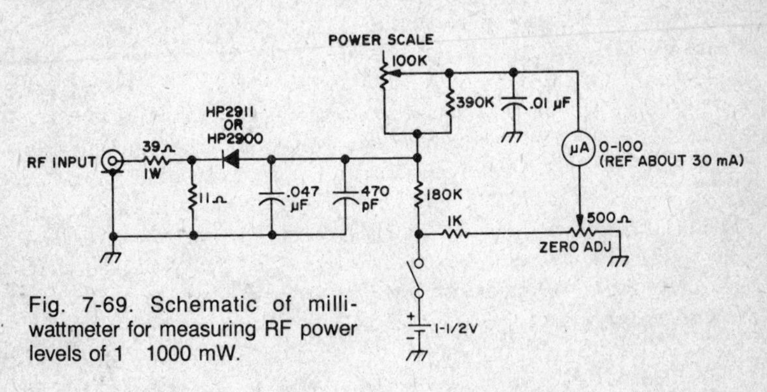

Fig. 7-69. Schematic of milliwattmeter for measuring RF power levels of 1 1000 mW.

Parallel bypass capacitors were used from diode to ground of the input jack in order to use the instrument over the whole range of 50 kHz to 450 MHz. With even larger bypass values in parallel, the milliwattmeters could be used throughout the audio range as well; and calibration of the power scale resistor could be made more easily. A sensitive AF or low RF voltmeter and an oscillator within the voltmeter frequency range can be used to calibrate the devices. The oscillator needs to have an attenuator, a low impedance output, and a power output of up to 1W for connection to the milliwattmeters through a piece of 50Ω coaxial line. Calibration measurements can be made using the low-power stages of a 450 and 144 MHz transmitter limited to about 500 mW. A series of 3, 6, 10, and 20 dB resistor pads (1W maximum ratings) with 50Ω impedance values are then cut into the coax line to check calibration points at 144 and at 432 MHz. The maximum errors should be less than 15% and at most frequencies, much less.

The component values shown in Fig. 7-68 resulted in an instrument having a range of .05 up to 500 mW. The low end of the power range will depend on meter resistance to some extent.

The unit shown in Fig. 7-69 used a small square microammeter without microampere calibration but the full-scale deflection point seemed to be a little over 100 μA. The reference line of RF power indication was simply a black ink line on the face of the meter. The popular imported 50 μA meters may be used in this same circuit though the higher meter resistance may prevent getting down to less than 0.25 mW readings in the circuit of Fig. 7-68. The minimum reading of 1 mW up to 1W or the unit in Fig. 7-69 may be easily obtained with nearly any range of microammeter and "power indicating" resistor. This variable resistor or potentiometer can be nearly any size of AF gain control with its maximum resistance either limited by a shunt resistor or by the pot value itself, depending on the range or reference value of the microammeter.

This resistor scale is hand calibrated on an aluminum panel in the higher-range instrument and on a brown Bakelite copper-plated board in the low-range unit.

The zero adjustment circuit of Fig. 7-69 with a 500Ω pot and a fixed 1000 or 1200Ω resistor makes it an easy matter to zero the meter. Less than 500 mV of bias is needed for the meter circuit. A few microamperes of forward bias current through the diode is enough to enable measurements down to a small fraction of 1 mW.

About 30 different germanium and silicon diodes were tried in these circuits. Even the "fastest" computer diodes were not very good at 450 MHz. These function, but at lower efficiency than at HF or VHF—so the power scale error becomes objectionable. The hot carrier diodes are much better.

SENSITIVE RF VOLTMETER

This RF voltmeter has full-scale ranges from 0.03V to 10V and frequency response flat from 40 kHz to over 200 MHz, making it useful for most solid-state work. It is portable and battery operated.

The performance and accuracy of the finished meter depends greatly on how much care is taken with the matching of the semiconductors in it, and the adjustment and calibration. These procedures will therefore be described in considerable detail.

A two-stage amplifier is used. FETs were chosen for low pinch-off voltage, therefore low current drain. This meant the voltage drop in the biasing resistor and the drop from drain to source could also be low. The stage would work with supply voltage from a 9V battery, even a fairly old battery which had fallen off to 7V or so. The gain of the stage was low, too, but the following high-gain pair of transistors in the second stage took care of that. The complete circuit in Fig. 7-70 would easily drive a rugged 1 mA meter. At the same time, it would not deliver enough overload current to hurt the meter. The total battery drain was about 5 mA from a 9V battery.

The RF probe is a voltage doubler circuit, using silicon diodes which are forward-biased for maximum sensitivity. Germanium diodes were tried first, but the silicon diodes proved to be much less affected by temperature. With

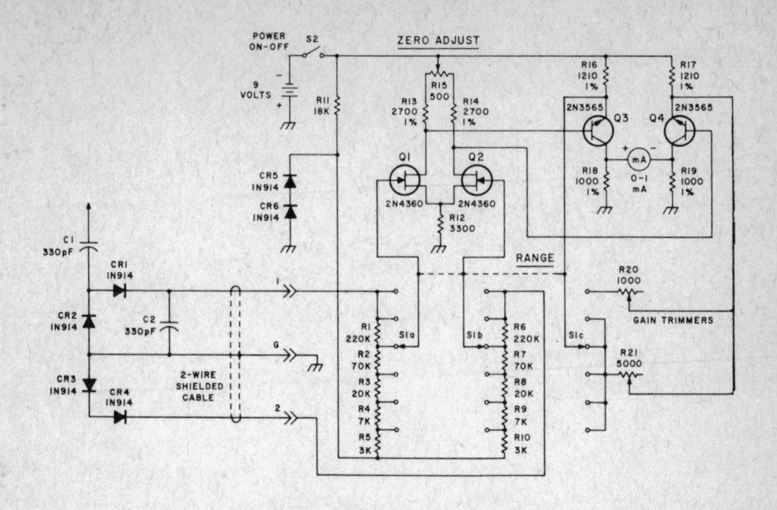

Fig. 7-70. Circuit diagram of the WA6NIL RF voltmeter.

proper forward bias, there is very little difference in small-signal performance between germanium and silicon types. A second pair of diodes with no RF input signal is mounted in the probe and connected to the other side of the DC amplifier's balanced input circuit, thus compensating for any temperature drift in the rectifier diodes. The capacitors C1 and C2 in the probe are deliberately kept small to restrict the low frequency response of the meter for several reasons: to make the meter insensitive to hum pickup, to make it read carrier level on an audio-modulated signal regardless of the modulation percentage, and to minimize the strain on the diodes if the probe is accidentally touched to a point of high DC potential.

Referring to Fig. 7-70, diodes CR1, CR2 and capacitors C1, C2 are the voltage doubler. The compensating diodes CR3, CR4 are mounted in the probe also, so they will stay at the same temperature as CR1 and CR2. The diodes are type 1N914, which have excellent high frequency performance, reasonably high voltage rating and low cost. These parts are mounted in an old lipstick case. The business end of the probe is a piece of No. 14 tinned bus wire, straightened out and filed to a point. The removable 50Ω 2W load is built into the lipstick cap as shown in Fig. 7-72. At the 10V maximum input to the instrument, this load just reaches full dissipation. The reverse voltage rating of the diodes is high enough so there is a comfortable margin of safety at 10V RF input. The probe connects through two-conductor shielded microphone cable to a two-pin microphone plug mating with the two-pin connector on the cabinet. Any small three-wire or two-wire-plus-ground connector would do.

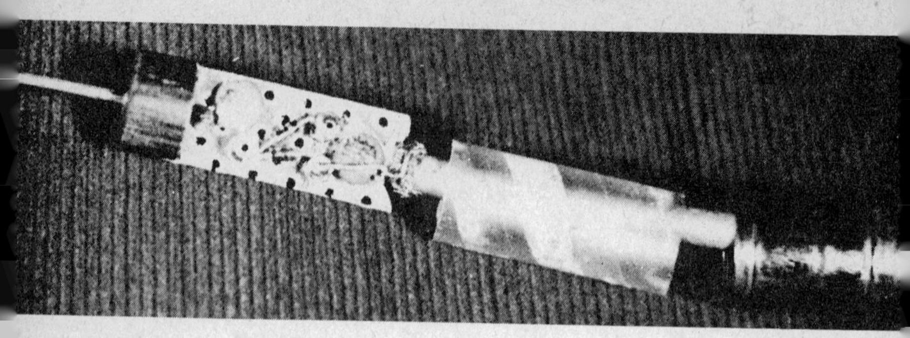

Fig. 7-71. The RF probe is built on a small piece of perf-board and then tucked into your favorite-brand lipstick case.

The Amplified Circuit

The DC signal and reference outputs from the probe are each connected to a voltage divider associated with the range switch S1. The cold ends of the two voltage-divider strings are connected to the bias source, diodes CR5 and CR6, which are supplied about 0.4 mA current through resistor R11. (This is far more current than will ever flow in the voltage dividers, so there is always current flowing through CR5 and CR6.) The negative voltage across CR5 and CR6 forward biases the signal diodes CR1 and CR2 for best sensitivity. Even with no RF input to the probe, a tiny forward current flows from the bias source through resistors R1, 2, 3, 4, 5 and the signal diodes. An equal current flows through R6, 7, 8, 9, 10 and the reference diodes CR3, CR4, so that if the diodes and resistors have been well matched (as described later), equal

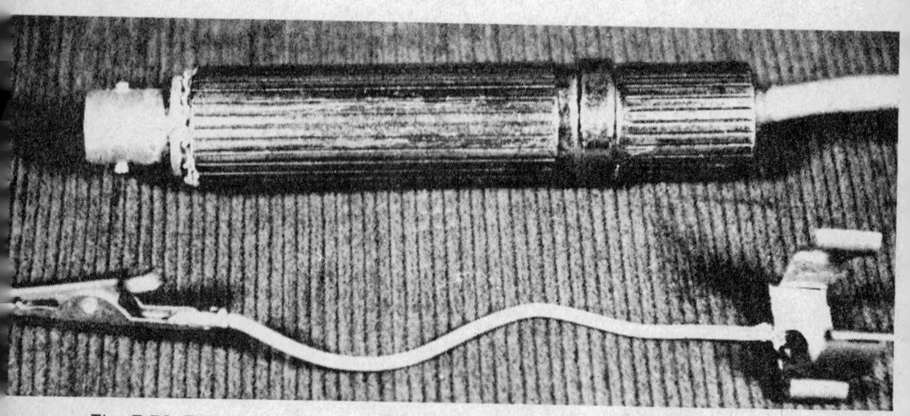

Fig. 7-72. The top of the case contains a 50Ω load and plugs onto the RF probe. The construction is detailed in Fig. 7-71. Ground connection is accomplished via the short piece of wire and clip.

voltages will appear at the output contacts of switch sections S1a and S1b when no RF signal is applied to the probe, and the meter will read zero. As RF signal is applied, the DC voltage at S1a (feeding the gate of FET amplifier Q1) becomes more positive, while the voltage at S1b feeding the gate of Q2 stays put.

Field-effect transistors Q1 and Q2 are matched and the Zero Adjust control R15 is set so that the drain voltages of Q1 and Q2 are equal when their gate voltages are equal, that is, when no RF input voltage is applied to the probe. The large common-source resistor R12 causes the total drain current of Q1 and Q2 to stay about constant, so that while Q1 drain voltage goes negative as RF signal is applied, Q2 drain voltage goes positive by a nearly equal amount.

The second-stage amplifiers Q3 and Q4 are high-gain NPN transistors, selected for highest gain and matched so that the meter between their collectors reads zero or very near it when the voltages at their bases (from Q1 and Q2 drains) are equal. A slight adjustment of the Zero Adjust control will then make the meter read exactly zero at no signal input.

The large resistors R16, R17 in the emitter circuits of Q3 and Q4 would, without the gain trimmer resistors, make this stage very low gain and stable because of the large inverse feedback. One of the gain trimmer variable resistors, R20 or R21, as selected by the range switch S1c, is connected between the two emitters; when this resistance is small, the inverse feedback is cut down and the stage gain is high, while when the gain trimmer resistance is high, the stage gain drops to a low value. It was originally intended that this method of changing gain would be used for switching between the three most sensitive ranges, leaving the other ranges to be switched by the voltage divider R1-R5. However, the efficiency of the diodes in the probe is very low on the 0.03V range, so there is a large difference in gain between this and the 0.1V range in the DC amplifier. The Q3-Q4 stage must run at nearly full gain on the 0.03V range, and almost at minimum gain on the 0.1V range. The voltage divider then provides switching to all the higher ranges. A third gain-trimming resistor was originally provided and can be seen in the photo, but will not be needed if the voltage divider resistors are adjusted as described later.

The 1 mA meter is a surplus item. It is a large rectangular type; a large size is desirable since four scales must be put on it. The meter originally had an odd-ball scale, furlongs per fortnight or something, but this did not matter as new scales had to be drawn and hand-calibrated anyway. The original scale was used to determine the proper size and location of the new scales. The dial plate was then turned over and the new scale glued on its back with rubber cement. The meter is connected between Q3 and Q4 collectors to read the current unbalance between them.

Construction

The layout and wiring of the instrument is not critical. All the high-frequency circuitry is in the probe, and the rest can be laid out in any

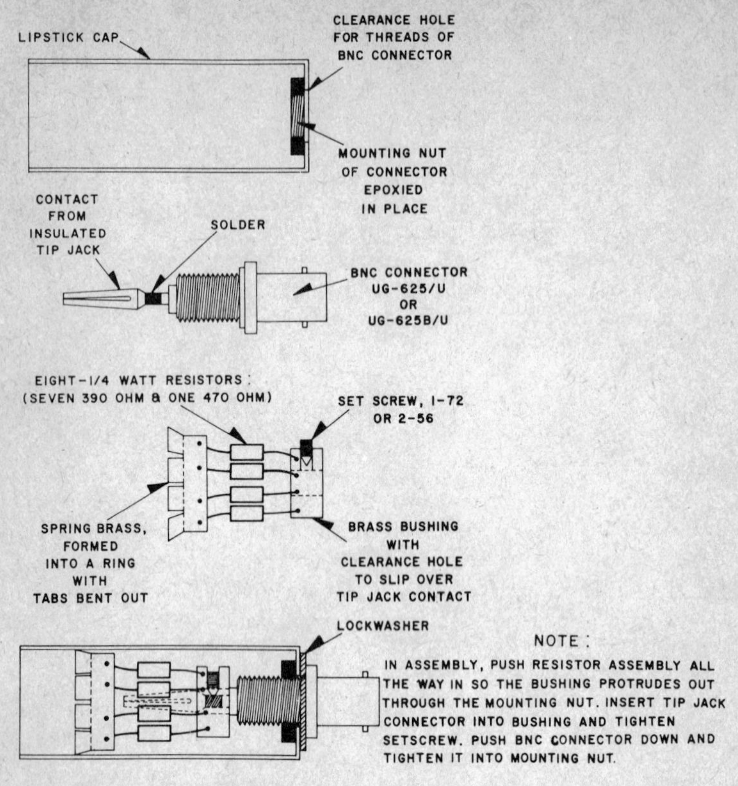

Fig. 7-73. Construction of the 50Ω probe adapter.

convenient fashion. The only front panel controls are the range switch, the battery on-off switch and the zero-adjust potentiometer.

Mounting the voltage dividers R1-R5 and R6-R10 on the range switch will make the job a lot simpler, since not so many wires will run between the switch and the circuit-board. The FETs and transistors are mounted in sockets rather than soldered in, to make matching easier.

The first thing to do is to lay out the parts in the cabinet or chassis and panel, and mount the meter, zero adjust control, on-off switch, battery holder and the connector for the RF probe. The range switch with its associated resistors should be mounted temporarily where it will be accessible, as the values of the resistors will have to be adjusted later. Then the circuit board is laid out and built up with the sockets for Q1, 2, 3, 4 and the rest of the parts. Most of the semiconductors in the circuit must be selected and matched, but the bias source diodes CR5 and CR6 are not critical and may be mounted at this time. When installing diodes, it is good insurance to make a small loop in the lead wires between the body of the diode and the solder terminal, to relieve the strain. If you can, use close-tolerance resistors where they are called for. Buying 1 percent resistors in small quantities is an expensive sport unless you can find them in surplus. Matching of pairs of resistors in the two sides of the circuit, such as R13 with R14, or R16 with R17, is more important than their actual value. In the case of the voltage divider resistors, their values will be trimmed later anyway, so the main reason for using precision resistors here is their stability. If you are forced to use ordinary composition resistors, pick matched pairs with a good ohmmeter and use the same precautions you would when soldering in transistors or diodes—leave little length of lead between the body of the resistor and the solder joint, and put an alligator clip or other heat sink on the lead while soldering.

When this much of the circuit is built up and checked out visually and with the ohmmeter, you are ready to select the FETs and transistors.

Selection of Semiconductors

This circuit does not use any of the fancier techniques such as chopper stabilization to hold down unbalance and drift. It is therefore very necessary to select and match the semiconductors for best results, especially since the inexpensive devices used in this circuit have a much wider spread in characteristics than do vacuum tubes. It is true that carefully matched pairs of FETs can be had, but these could easily cost more than the whole instrument. It seems much better for amateur purposes to buy a good quantity of each type and select out pairs, leaving the rest available for other uses. It is a great help if you have access to a stock of them. In any case, you should have available at least half a dozen each of the 2N4360s and 2N3565s and a dozen 1N914s or 1N914As. It won't hurt to have more, as they are all very useful devices.

The first step is to select the FET pair, Q1 and Q2. Set up the test circuit of Fig. 7-74, remembering that the FET pin connections are different from transistors. Plug in the 2N4360s one after the other and select all those which draw between 0.4 and 0.5 mA. Pick the two which draw most nearly the same current and put them in Q1 and Q2 sockets.

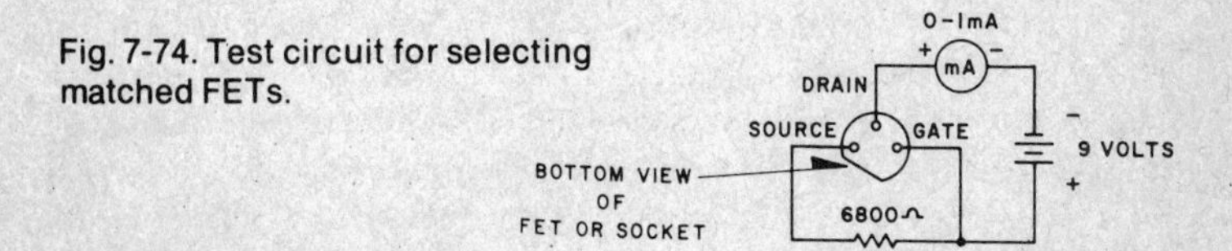

Fig. 7-74. Test circuit for selecting matched FETs.

Now to match up the second stage transistors. Switch to the 0.03V range and turn gain trimmer R20 to zero resistance. Connect the circuit to the 9V battery. Temporarily connect the base pins of Q3 and Q4 sockets to each other with a clip lead. or jumper. Now try various pairs of 2N3565s in the sockets, looking for pairs that make the meter read nearest zero. If you have a transistor DC beta (HFE) tester available, pick the pair that shows the highest beta. If not, remove the temporary jumper from Q3-Q4 sockets. Plug each selected pair of 2N3565s in turn into Q3 and Q4 sockets, and choose the pair which produces the greatest meter deflection for a given small amount of the zero adjust control. Leave the chosen pair in Q3 and Q4 sockets.

This much of the circuit is a nice sensitive DC voltmeter of about 10 mV DC full scale deflection at maximum gain. It is now ready to be used for matching up the diodes. Set the range switch to its 0.1V range and adjust the gain trimmer R21 to maximum resistance. Connect one diode between input pin 1 of the probe connector and ground as shown in Fig. 7-75; be sure to connect it the right way round. Temporarily jumper input pin 1 to pin 2 and turn the zero adjust control to put the meter needle on some mark near the center of the scale. Now remove the jumper between the input terminals. Try the other 1N914 diodes between input pin 2 and ground, sorting them into groups by the meter deflection they cause.

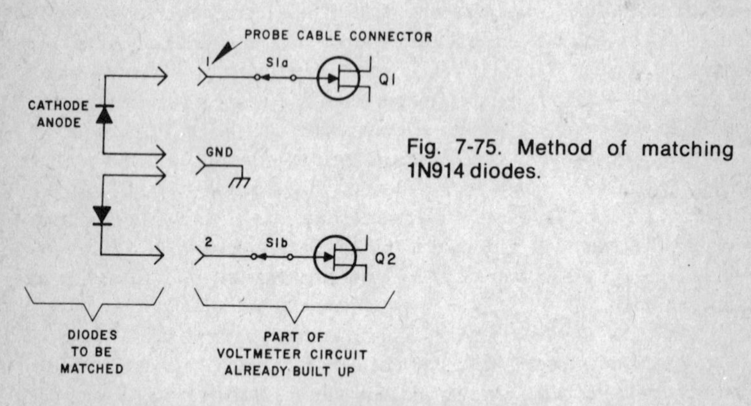

Fig. 7-75. Method of matching 1N914 diodes.

Switch to the 0.03V range, set R20 to zero resistance, and again jumper the input pins together. Turn the Zero Adjust control to put the meter needle on the mark. Remove the jumper, and try pairs of diodes from the various groups for match by connecting one diode from pin 1 to ground and the other from pin 2 to ground. The idea is to pick pairs of diodes which bring the meter needle closest to the mark, thus having exactly the same voltage drop as closely as possible.

Use one of the best-matched pairs of diodes for CR1 and CR3, and another well-matched pair for CR2 and CR4. It is not necessary that CR1 and CR2 be alike, or CR3 and CR4, though it will do no harm. These diodes may now be installed in the RF probe, observing the usual soldering precautions, and the probe assembly finished up.

The meter is now ready to be calibrated.

Range Adjustment and Calibration

Calibration of the voltmeter requires a radio-frequency signal of known and adjustable level. If you can borrow a good signal generator with 2 or 3V RMS maximum output, it will do nicely. An uncalibrated signal source can also be used if you have some way of measuring its output, such as a calibrated oscilloscope or another RF voltmeter. If a signal generator is used, be sure to consult its instruction manual to find out whether it delivers its rated voltage output into a matched load or an open circuit. Most generators built in the U.S. deliver rated voltage into a 50Ω load. Assuming you have such a generator, attach the 50Ω load to the RF probe and connect it to the signal generator output.

Switch the range switch to 0.03V. With no signal input, turn the Zero Adjust control to zero the meter. Put in the 0.03V RMS signal at a frequency of a few megahertz and adjust gain trimmer R20 for full scale meter deflection. (If full scale deflection cannot be obtained, you will have to pick a higher gain pair of FETs Q1 and Q2, or higher gain transistors Q3 and Q4.) Now switch to the 0.1V range and, with no signal input, again zero the meter. Put 0.1V RMS into the probe and adjust trimmer R21 for full scale deflection. On each range, recheck the zero after adjusting the gain trimmer; readjust both the zero and the gain as necessary.

Switch to 0.3V range and re-zero the meter with no input signal. Put in 0.3V RMS and note the meter reading. If it is over full scale, R2 is too high. Connect a variable resistor of several megohms maximum resistance across R2 and adjust it for full scale deflection. Remove the variable resistor, check its resistance with an ohmmeter and connect a fixed resistor of this value across R2 with the usual soldering precautions. (If the meter reading was below full scale, R1 is too high and must be shunted down in the same way, but this is not likely.)

Now switch to 1V range, re-zero the meter, and put 1V rms in. The meter will probably read over full scale again, indicating that R3 is too high. Shunt it down in the same way. If the change in R3 is more than 5% or so, go back and check the 0.3V range again.

Switch to the 3V range, re-zero the meter and put 3V RMS in. Shunt down R4 to make the meter read full scale.

Switch to the 10V range and re-zero the meter. Put in 10V RMS (if available) and shunt down R5 as necessary for full scale reading. If your RF source will not put out this much, you can assume that the meter scale is linear on this range, and adjust to make the meter read at the right point on the scale for the voltage you do have.

The meter may now be hand calibrated. Starting with the 0.03V range, recheck the zero and full scale readings of the meter as above, then put in levels of 5, 10, 15, etc., up to 30 mV, and write down the deflection on the original meter scale for each input voltage. Calibrate each higher range in the

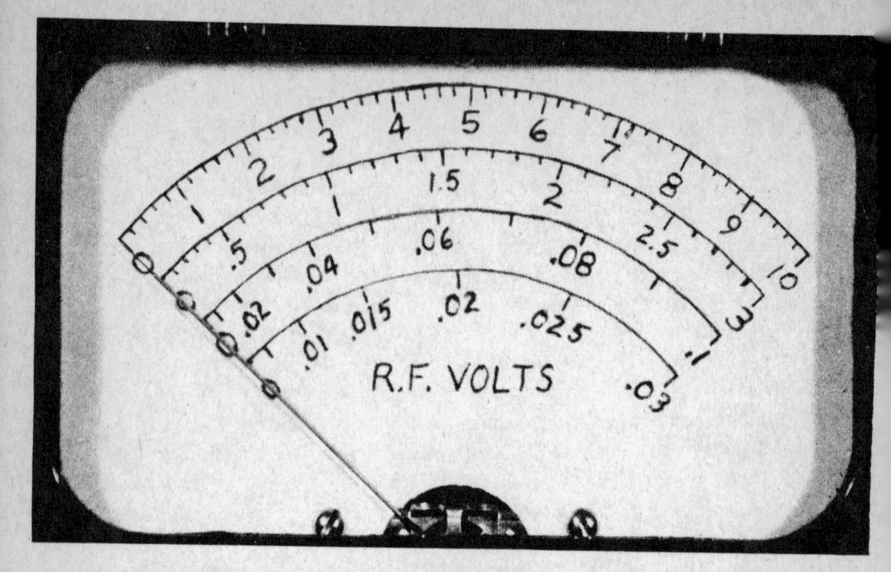

Fig. 7-76. Close-up of the recalibrated meter face. Because four separate scales are required, a large meter allows direct calibration without the scales crowding together.

same way, using appropriate steps of input voltage. The meter scale may be drawn up by any of the usual methods, such as making an enlarged layout and photographing it. The finished scale is shown in the photograph. Since the scales get very non-linear at small voltages, separate scales are required for the 0.03V and 0.1V ranges. One scale can be made to do for the 0.3 and 3V ranges, by compromising a little, and one scale for the 1 and 10V ranges.

As an added refinement, you can tinker with the circuit to reduce the shift in meter zero when changing ranges. If there is much shift when switching between 0.3V and 0.1V ranges, try interchanging transistors Q3 and Q4. If the shift reverses when this is done, a better matched pair of transistors is needed. If it stays the same, a tiny adjustment of resistors R16 and R17 for better match is needed.

If the zero shifts when going to higher ranges, adjust the resistors R7 through R10 as necessary. The zero shift should be very small if they are made equal to the corresponding resistors R2 through R5.

The range switch may now be installed in its permanent location and the assembly of the instrument completed.

Use of Other Types of Semiconductors

The diodes in the probe should be silicon high speed types. As mentioned, the 1N914 is quite satisfactory. It is made by several manufacturers and is quite inexpensive, and there seems to be little reason to use anything else. If you have some others on hand you want to try, they should have something like the 4 nanosecond recovery time of the 1N914. For the 10V maximum range of this meter, a peak inverse rating of 40 or 50V would do; the 100V of the 1N914 gives a desirable safety margin. The bias diodes CR5 and CR6 should be the same type used in the probe.

The 2N4360 FET is a P-channel junction type. It is made by Fairchild and may not be as widely available as those of some other manufacturers. Other P-channel units may be used; the main thing is to get or select some that have a fairly low pinch-off voltage, so that they will work at low enough supply voltage in the given circuit. N-channel types could be used also, but this would mean that the following transistors Q3 and Q4 would have to be changed from NPN to PNP types, and the polarity of all the diodes, the meter and the battery would have to be reversed.

The 2N3565 is a very high gain type; the ones in my meter have an HFE (beta) of 275. This again is a Fairchild device. Other NPN transistors which look suitable are the 2N3117, 2N3692, 2N3711, 2N4124 and the 2N5131. A browse through the manufacturers's catalogs will turn up others. If you should change the circuit to use PNP types as mentioned above, the 2N3965, 2N4062, 2N4126 and 2N4250 are some of the possibilities. You could get by with less beta by using a more sensitive meter than 1 mA, or by doing without the 0.03V range; this would allow many more silicon types to be used. (Dropping that lowest range would simplify the circuit considerably as well.)

Conclusion

The frequency response of the completed meter was measured to be 1 dB down on the low end at 40 kHz. At the upper end, it was still flat at 216 MHz, the upper limit of the available signal generator. The input impedance of the probe was measured with an R-X meter; it proved to be 4 pF shunted by a resistance of 30,000Ω at 1 MHz. At higher frequencies, the capacitance remained 4 pF, but the shunt resistance dropped gradually, being 8,000Ω at 30 MHz and 1000Ω at 200 MHz.

With the 50Ω load attached to the probe, the meter will indicate RF power from a few microwatts up to 2W maximum. (Power equals E^2/R.) Without the 50Ω load, the probe may be used directly on sensitive circuitry, provided its 4 pF capacitance can be allowed for. Thus the meter will be found very handy for working on nearly all solid-state and other low power circuits.

Chapter 8

Power Supplies

A VOLTAGE SEXTUPLER POWER SUPPLY

This circuit is an extension of modified quadrupler. The resultant output voltage is an extension of a modified quadrupler. The resultant output voltage is more than six times the input voltage, about 7.5, because the capacitors charge up to the peak of the AC voltage imposed on them, and after rectification more DC results than the average value of the AC input. Output DC is 900 volts with low voltage and bias.

Figs. 8-1, 8-2, and 8-3 show how the circuit was developed. In fact, this voltage multiplication process could be extended an infinite number of times, but electrolytic capacitors have to be used in series, and when they do the effective capacitance is reduced. So there is a practical limit to this multiplication business, as the regulation begins to suffer as the effective capacity is decreased.

Voltage regulation is good, but not as good as with a quadrupler, due to the fact that the output capacitance is made up with C2, C4 and C7, three 300 μF capacitors in series, giving an effective 100 μF. C5 and C6 in series make up the capacitor for the fifth multiplication, two 200 μF at 450 volt units giving an effective 100 μF. All capacitors are of the twist-lock, can type, except C5 and C6 which are tubular type. The 900 volts high voltage drops about 80 volts under 250 mA voice peaks, which is still under 10% regulation. Not bad for this type of power supply, or any other HV supply. The low voltage tap only drops about 5 volts under voice peaks.

A 5 × 9 × 2 inch aluminum chassis is used to mount the components on and a front panel used to mount a 4 in. speaker. The lone transformer is for the filaments, and is surrounded by the forest of capacitors. The schematic of the complete supply is shown in Fig. 8-4.

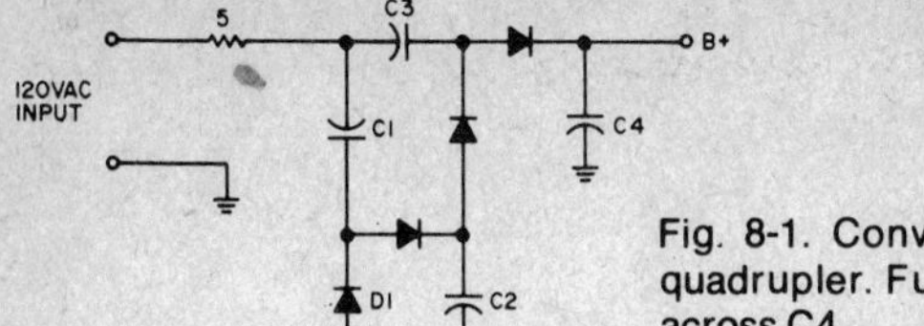

Fig. 8-1. Conventional half-wave quadrupler. Full output voltage is across C4.

Mention should be made here of the surge resistor arrangement. It is an absolute necessity to protect the diodes, when the power supply is first turned on and the capacitors look like a dead short; however, after thirty seconds the resistor is shorted out by the time delay relay K1, as the resistor serves no other purpose, and if left in the circuit causes a voltage drop as current is drawn through it. A one volt drop in the input causes approximately a 7 1/2 volt drop in the output voltage at the high voltage tap.

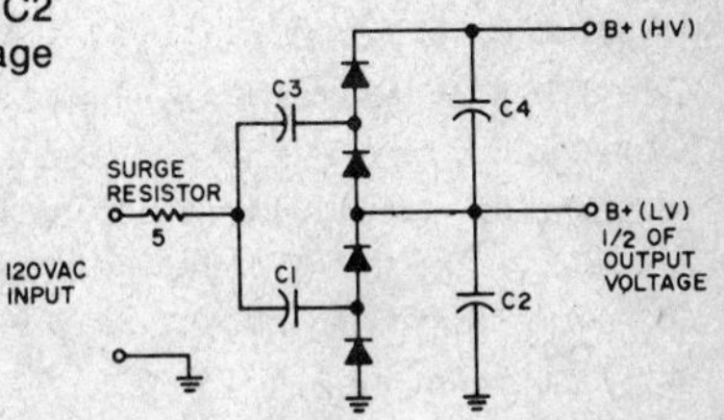

Fig. 8-2. Modified and redrawn quadrupler. By putting C4 and C2 in series, a lower voltage capacitor can be used for C4.

Each diode has an .01 disc ceramic capacitor across it for transient protection.

All of the precautions should be used on this power supply since it operates with one side of the AC line grounded. However, nothing is to be feared if the AC line plug is inserted correctly. This can be determined by two different methods. The first is to run a ground lead from the power supply chassis to a

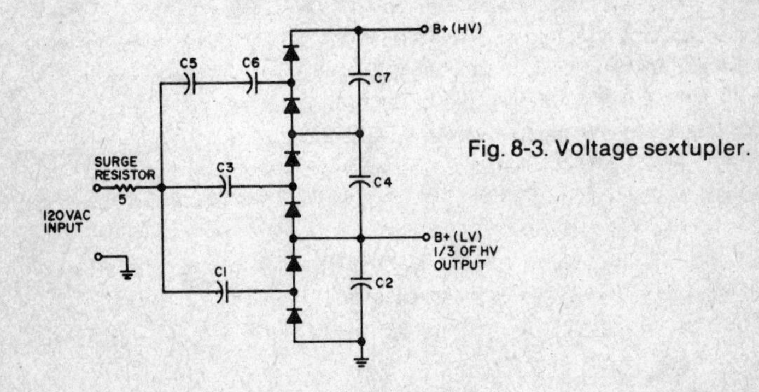

Fig. 8-3. Voltage sextupler.

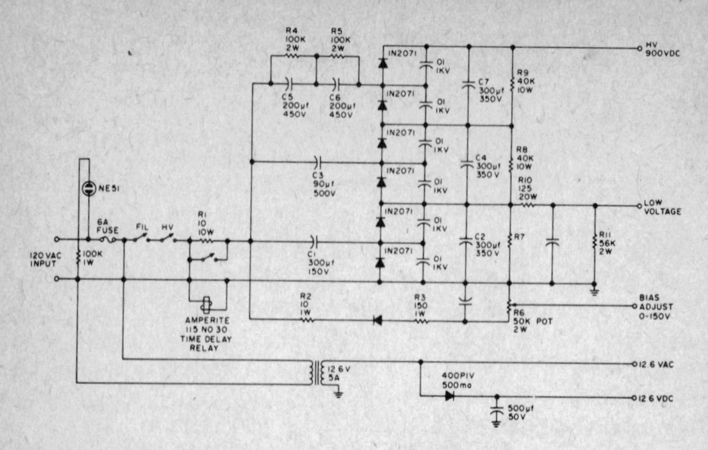

Fig. 8-4. Complete schematic of sextupler power supply.

good ground, cold water pipe, or a driven ground rod. With the switches in the power supply turned off, insert the AC plug into the wall outlet. If the neon bulb ignites, the plug is in correctly. If the neon bulb doesn't light, just turn the plug over, and you are in business.

The other method is to determine which one of the sockets in the AC wall outlet is the neutral or grounded leg of the incoming power. City electrical wiring codes specify that the larger of the two sockets be the grounded or neutral leg, but be sure and play it safe and find out for sure. After you determine which is the hot and neutral, use a polarized AC plug. This is the preferred method, and can be determined with an AC voltmeter by hooking one lead to a ground and plugging the other lead into either socket which gives you 120 volts. The one that doesn't give a reading is the neutral leg, therefore ground.

This power supply has been used with all of the commercial transceivers on the market requiring an external power supply, with excellent results. The value of R10 may have to be changed to drop the voltage on the low voltage tap to give the correct amount for your particular brand of transceiver. The light weight, 5 1/2 pounds, and good voltage regulation make this type of supply an excellent choice for the modern day, compact SSB transceivers on the ham market today. And the cost isn't too bad.

After first building this power supply, it was used to power a linear amplifier using 4 TV horizontal sweep tubes in a grounded grid circuit. With this arrangement, 900 watts PEP input was run without a power transformer in the whole station.

ELECTRONIC VARIAC

In the realm of electronic experimentation, it becomes necessary to use a variable voltage source as a means of precision voltage control.

The Electronic Variac circuit outlined in Fig. 8-5 will provide a full range of voltage control for the primary of any 120 VAC 60 Hz transformer requiring less than 10 amperes of primary current.

The Electronic Variac may be constructed in a small 3″ × 5″ utility box, or mounted in the front panel of the controlled source. The only modification to the equipment in which the Variac is added is a single 3/8″ mounting hole for the 150K pot. If an outboard system is more versatile, then a line cord and socket will be needed for ease of connection.

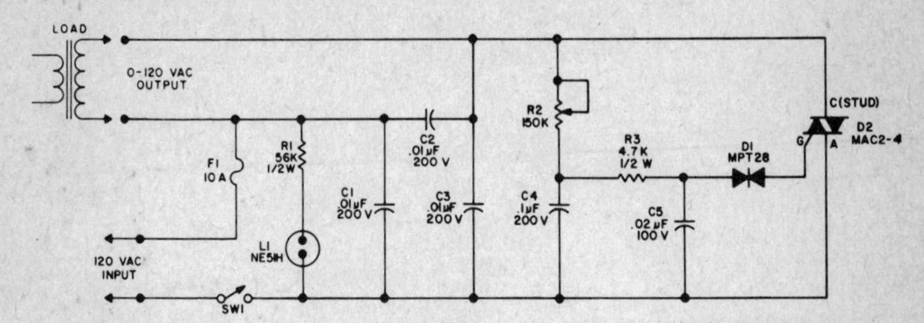

Fig. 8-5. Electronic Variac schematic. C1 to 3—.01 μF 200V Capacitor; C4—.1μF 200V Capacitor; C5—.02 μF 100V Capacitor; D1—MPT28, 3 Layer Diode (Motorola); D2—MAC2-4, 200V Triac (Motorola); F1—10 Amp Fuse; L1—NE51H Neon Lamp and Socket; R1—56K, 1/2 Watt ±10% Resistor; R2—150K Pot. Lin. Taper, 1/2 Watt; R3—4.7K, 1/2 Watt ±10%; SW1—SPST Switch, 10A; Misc.—Line Cord, Terminal Strip and Chassis.

Construction

Choose either of the two above mentioned mounting arrangements and begin by mounting the MAC2-4 triac to a good heat sink surface. Since this is a stud mounted device, a single 1/4″ hole will be needed. Apply silicon grease or IRC heat sink compound to both insulating washers and the base of the device before mounting. Check for electrical shorts between the stud and chassis after mounting the triac, for the stud carries full line potential. When connecting the MPT-28 trigger diode, be certain to heat sink the leads before soldering. The leads should be left their full length for heat dissipation. Both leads of the MPT-28 are identical and no polarity need be observed. Although the anode and cathode of the MAC2-4 are identical, the gate is the shorter of the two devices and it must be connected to the trigger diode MPT-28.

Addition of the on and off switch, fuse, neon lamp and capacitors C1, C2 and C3 are all optional. The capacitors are installed to eliminate the RF or noise generated by the system. Experimentation with the requirements for these capacitors is left up to the individual.

Applications

The control of many types of AC loads is within the capabilities of this simple circuit. One example is the addition of the Electronic Variac to the circuit shown in Fig. 8-6. A battery charger or high current bench supply that is variable from 0-30 VDC is very useful in checking the mobile rig on the bench or powering bread board circuits. Fig. 8-6 shows a straightforward power supply using a full wave bridge if needed. A well stocked junk box should yield most of the parts required for the supply.

Other applications include a variable supply for hobby use such as electric trains or slot cars, plating of metals and light control. The Variac can be used to replace that worn out hunk of Variac in the plate supply of your high power rig. The Electronic Variac is small, produces little heat, and for less than $8.00 worth of parts, does the job of a large variable transformer costing more than three times as much.

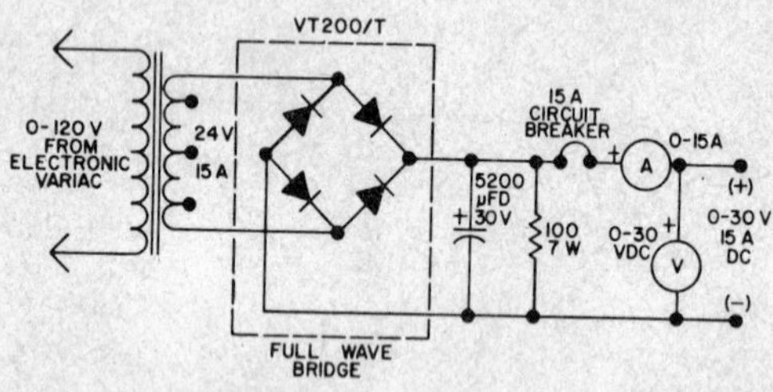

Fig. 8-6. DC battery eliminator/charger/power supply. C1—5200 μF, 30V, (GE, 86F147M); CB1-15A Circuit Breaker (375-215-010 Wood Electric); D1—Full Wave Bridge; (VARO VT200/T), (or: 2-MR1120, and 1-MR1120R, Motorola); M1—0-15 Amp Ammeter (EMICO); M2—0-30 Volts DC, Voltmeter (EMICO); T1—24V 15A Filament Transformer (Knight 54F2335) or Equivalent; Misc.—Chassis, Terminal Strip, Binding Posts; R1—100Ω, 7W Wire Wound Resistor.

LOW COST TRANSISTOR POWER SUPPLY

This circuit is a novel one. It prevents any switching transients from occurring, thereby removing this cause of transistor failure (Fig. 8-7). The reason for this is the fact that the feedback transformer (T1) secondary is equally loaded through all parts of the switching cycle. Reverse voltage on the base–emitter junction of the "off" transistor is limited to the diode (D1 or D2) voltage drop across it. This allows use of inexpensive 2N3055s, which have emitter-to-base breakdown voltage ratings of 7V. These transistors, which are available very reasonable have collector current ratings of 15A maximum, so no large heat sinks are required.

The main power transformer (T2) is a dual-winding 6.3V filament transformer. The secondary windings are used as the primary and vice versa.

After much experimentation, it was found that a small 115-to-24V 250 mA unit made a good feedback transformer (T1). When connected as shown, it furnishes more than enough base current to insure driving the 2N3055s into saturation, providing low switching losses and good efficiency. Measurements indicated 315 mA of base drive per transistor. According to specifications, this

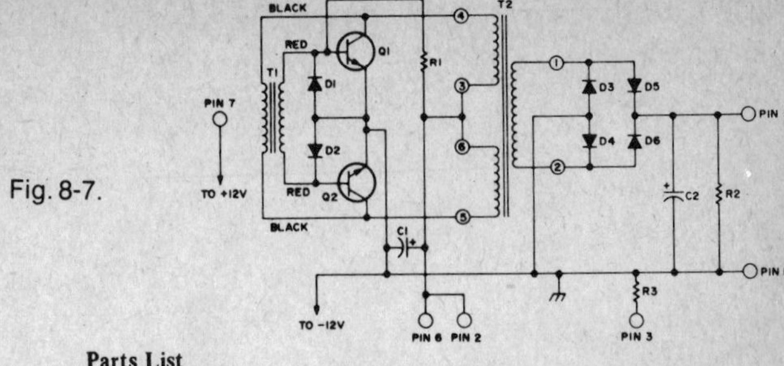

Fig. 8-7.

Parts List

T1	=	115V to 24V 250 mA	D5, D6,	=	silicon diode, 1A and
		(Fair Radio MW#4528)			400 PIV or more (Poly Paks)
T2	=	115V to 2/6.3V 1.2A	C1	=	100 µF 25V electrolytic
		(Fair Radio #7629809	C2	=	20 µF 350V
D1, D2	=	silicon diode, 1A 50 PIV	R1	=	1000Ω
		or more (Poly Paks)	R2	=	470kΩ
Q1, Q2	=	2N3055 (Poly Paks)	R3	=	150Ω
D3, D4,					

is sufficient for switching 9.45 amperes collector current, assuming a minimum beta of 30.

Resistor R1 provides a small base–emitter forward bias. As the secondary DC resistance of T1 is only 8Ω and small by comparison to R1, it is not necessary to bias each transistor with a separate resistor. Imbalance is immeasurable.

Capacitor C1 helps filter out any transient spikes appearing on the 12V line. The secondary circuit is a conventional bridge rectifier and filter setup which furnishes about 210V under transmit conditions.

This same circuit can be used with regular 12V vibrator transformers, making cannibalization of an old car radio worthwhile to obtain a suitable power transformer.

Owners of earlier FM gear using vibrator supplies can transistorize them inexpensively, gaining efficiency and reliability in the process. Just remember to provide adequate transistor heat sinking, if you plan on making a 100 or 200W supply, for example. The circuit will also provide an ideal receiver supply for 450 MHz FMers who want to "duplex" their surplus mobiles.

About the only difficulty which might be encountered in the construction of this supply is its failure to oscillate. Should this happen, merely reverse *one* set of leads on the feedback transformer.

Don't worry about overheating *either* transformer.

VACUUM TUBE LOAD BOX

How do you test a power supply? By asking it to supply power. Put it to work, and check its output voltage for various load conditions, its hum characteristics, and its overload performance. These simple, informative tests should be easy, but usually they are not. How about something which will make them nearly as straightforward as testing resistors?

The very simple load box described here uses a pair of 6GE5s as variable resistors to dissipate the supply's power (Fig. 8-8). Its minimum resistance is roughly 1000 ohms and it will load supplies at as low as 30 or 40 volts. The rated dissipation of 35 watts may be exceeded by a factor of two or three for brief tests.

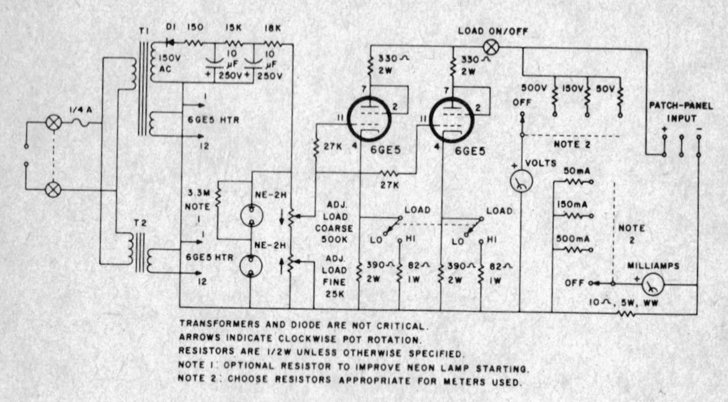

Fig. 8-8. Load box schematic.

A small part of the power dissipated by the load box goes to heat resistors in the load circuit. Most of it appears at the 6GE5 anodes and is thrown away as heat radiation.

The triode-connected 6GE5s are capable of passing very heavy currents but are restrained to useful loads by an adjustable negative grid bias. The grids draw no current, so large powers can be controlled by adjusting a 2-watt pot. A second pot, in series with the first, serves as vernier and for control of grid voltage at near-zero bias.

Two cathode resistors weaken the dependence of tube currents upon grid bias by adding a little self-bias. This also serves to stabilize and equalize tube loads. By switching two additional resistors across the two permanently installed ones, somewhat higher currents may be drawn for low-voltage test work.

A simple power supply provides heater voltage for the 6GE5s and a well-filtered bias potential. The thorough filtering and neon regulator guarantee that ripple observed in the loaded supply's output originates in the supply rather than the load.

Ballpark accuracy is sufficient for most amateur purposes. Two inexpensive meters serve to indicate voltage and current. The voltmeter multiplier resistors were estimated from the roughly known meter resistance, and then chosen by bench test.

The milliammeter circuit is based upon a 10 ohm wirewound resistor. The current is measured as a small voltage appears across this resistor. Multipliers were chosen to read the voltage, a slight departure from usual practice. A 500 mA range, which may seem a little high, helps avoid pinning the meter when setting up for a test.

When the grid bias is near zero at high anode voltages, relaxation oscillations may occur. Their cause is not apparent. The cathode and anode resistors minimize the nuisance value of these oscillations, which occur only under extreme operating conditions.

Construction

A 5″ × 7″ × 2″ reinforced aluminum chassis provides just enough room for construction. The 6GE5s are placed to the rear, not too close together, for best ventilation. The 4″ × 7″ front panel was made of .050 sheet aluminum by a local sheet metal shop. It was bent square with a 1/2″ lip at the bottom and a 1 1/4″ lip at the top, bent over again at the back. This arrangement makes a very convenient handle. During construction the right angles were adjusted for a rearward tilt of about 5/8″. A pair of 1″ strips at the sides serve as panel braces.

After completing the metalwork, all parts were cleaned in strong detergent, roughened with wet sandpaper and spray-painted with Rustoleum enamel. Hand lettering with waterproof India ink was covered by a finish coat of clear enamel.

Parts arrangement is not critical, except for one circuit. The high-impedance 6GE5 grid wiring should be kept well clear of the AC lines to avoid imposing 60 Hz hum on the load box's resistance characteristic. Five lug strips bearing a total of 37 tie points were mounted under the chassis, and 23 points were used in construction.

High voltage DC is hard to turn off without a special switch. A ceramic HV switch is indicated for this application.

Using The Load Box

The load box is very useful for new work, debugging, and for servicing. When initiating a new project, the power supply can be stuck together on the bench and subjected to realistic loads and overloads. When it finally goes into the chassis it can be trusted. If you missed this simple step you may still come out ahead if you use the load box to seek out weaknesses in a supply already built. And it's good for service work too.

If necessary, disconnect the supply from its normal load while the load box is warming up. Choose meter ranges for higher voltage and current than the supply is expected to produce. Turn both adjust controls full anticlockwise and attach the load box to the power supply.

Turn on the supply and the voltmeter should indicate the supply's voltage. Turn on the load switch and advance the coarse load control until the milliammeter indicates a current is flowing. An unregulated supply's output voltage should fall as load increases. Make appropriate corrections to meter ranges and proceed with the test.

If very accurate results are required, the patch-panel input arrangement simplifies the connection of outside meters. These terminals are also handy for attaching an AC voltmeter or an oscilloscope.

The fine load control is used at very low supply voltages, and for zeroing in on preferred voltage or current values when making tests with accurate instruments.

Figure 8-9 shows the results of a power supply test. It was a good workout for the load box too, which was used far beyond its ratings.

The upper line is the voltage found at the rectifier output terminal. Regulation here is not good. The lower line shows the regulating circuit output. The regulator tube's 26 watts maximum dissipation will not be exceeded in normal operation (150 mA), and there is reserve against aging and overload. It looks like this supply should be reliable for long periods of time.

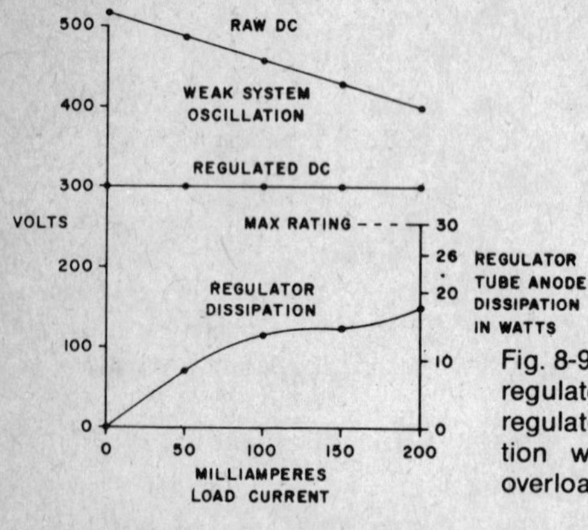

Fig. 8-9. Test results taken from a regulated power supply, showing regulator dissipation, and regulation well into the anticipated overload region.

3000V DC SUPPLY

Have you ever built a piece of gear with available parts and done an immaculate job, only to have a transformer or component go bad after a period of time? When you go to purchase a replacement you usually find it is sold out or no longer available. So you shop and shop, trying to find something that will do the job and also fit into your cramped dimensions. Perhaps you'll be lucky, or possibly end up rebuilding.

The described 3000V power supply incorporates rugged design specifications coupled with generous dimensions that provide versatility in accommodating transformers and related components found on the surplus market.

Construction

The high voltage power supply is easily constructed, as all mechanical work can be performed with a metal munching tool, pop rivet gun, good soldering gun, and ordinary hand tools. An electric drill with variable speed control will save much time.

The main aluminum chassis and front are 33.02cm × 43.18cm × 7.62cm (standard) and the back aluminum wall has a 2 cm inside lip at the top and bottom for attachment to the main chassis and cover. The front chassis and rear wall are attached to the main chassis by generous use of pop rivets. The main chassis is reinforced on the bottom with a thick steel plate with one caster at each corner and one in the middle to support the weight. The line cord is fed to the rear through a steel conduit.

The cover is manufactured by hand bending a sheet of aluminum to tightly fit the chassis assembly and is held in place by sheet metal screws. Right angle aluminum brackets were installed on the back plate along the sides to accommodate the fastening of the cover. Air is exhausted by mounting home air vent assemblies on the sides of the cover. A local lumber yard has the vents.

Before attaching aluminum to aluminum rough each contact surface with fine sandpaper to assure a good electrical connection. Also, connect each chassis and the back together electrically with copper braid.

The front of the supply contains a voltmeter, on-off switch and pilot light. The rear of the supply is designed with safety in mind. The B+ and B− connections are in a Minibox with two grommetted holes in the bottom. The large insulated feed-through was fitted on a small plexiglass sheet and the hole in the aluminum made extra large to prevent high voltage breakdown. High voltage cables should have a minimum rating of two to three times the DC output voltage.

The diode stacks are made by mounting eight diodes on four pre-punched epoxy boards. The insulated spacers for the boards are nothing more than self-tapping plastic expansion tubes, available at most hardware stores. The boards are connected to the spacers and the spacers to the chassis by nylon screws.

The filter capacitors are mounted in holes drilled in plexiglass with a hole saw. The plexiglass is held in place by self-tapping plastic expansion tubes and nylon screws. To prevent the capacitors from arcing to the chassis, the area below the capacitors has a sheet of punched epoxy paper board cemented to it—also the cutout plastic circles are cemented to the bottom of each capacitor.

To keep air circulating, a small fan is mounted in the rear of the supply. The fan is fused and the fuse is located in the front under chassis where it can be changed without removing the cover. The AC line is terminated in the front chassis at the switch and at this point the thyrector is also located across the line.

Located at the lower right of the rear chassis is a heavy duty ground connector *which should always be utilized for maximum safety*.

All lettering is accomplished by the application of white dry transfers over black wrinkle paint.

Circuit

The circuit (Fig. 8-10) utilizes a full wave bridge rectifier circuit with a capacitor filter of 50 μF. This provides approximately 5% regulation with a 3 kΩ load. Ten 500 μF, 450V capacitors provide a total voltage rating of 4500V.

The high voltage diodes, capacitors, and transformer are protected from excessive current when the power supply is first turned on by a series limiting resistor R1. The time delay for relay pickup is determined by R4, which adjusts the time required for C1 to charge and energize the relay K1 which closes its contacts and shorts out R1. Too much delay causes R1 to overheat. One second proved satisfactory.

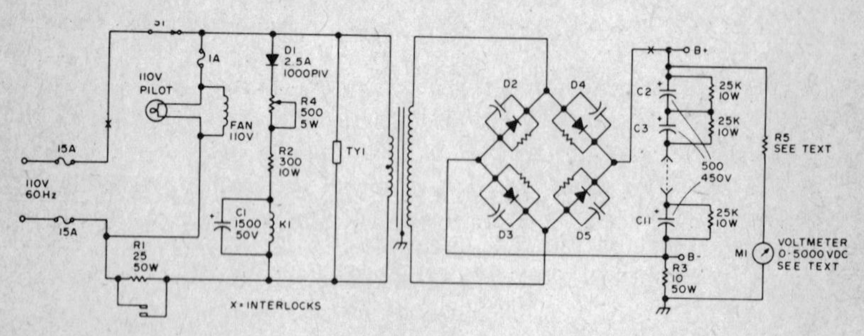

Fig. 8-10. Schematic fo the 3000V power supply. The diode stacks D2—D5 are constructed of 8 – 2.5A 1000 PIV series connected diodes each. Shunted across each diode is a 470K 1W resistor and a .01 1000V disc capacitor. C2—C11 should be 500μF with a minimum voltage rating of 450V DC. K1 is a P&B type PR3DY, 24V DC coil with 25 amp contacts. T1 has a 2200V rms secondary with a 500 mA minimum rating. The thyrector is a G.E. 6R520SP4B4.

The supply also incorporates a voltmeter that measures the output voltage. An inexpensive meter can be utilized as the supply incorporates a resistor multiplier string to increase the range of the basic meter movement, *but never use a meter with a metal zero adjusting screw in high voltage circuits!* To choose the correct value of R5 for your meter, use the following formula:

$$R5 = \frac{\text{full scale desired}}{\text{meter reading in amps}}$$

A resistor is not a high voltage device; therefore, to achieve the desired resistance of R5, many series resistors must be used to handle the voltage. 10

1MΩ, 1W resistors in series mounted on a strip of epoxy board were used, thereby distributing the voltage equally across ten resistors. The epoxy board is mounted to the front chassis on two ceramic insulators.

To protect the supply diodes from transients, a thyrector-diode assembly is installed at the line input. Also, each side of the line is fused to provide adequate protection to the supply and station line circuits.

Interlocks

All high voltage power supplies should contain an interlock or interlocks. Basically there are two types: the primary interlock and the secondary interlock.

The primary interlock is similar to the power cord assembly on a television receiver. When you remove the back of the set you open the AC line and the television cannot be energized by unauthorized personnel without a special line cord. See Fig. 8-11.

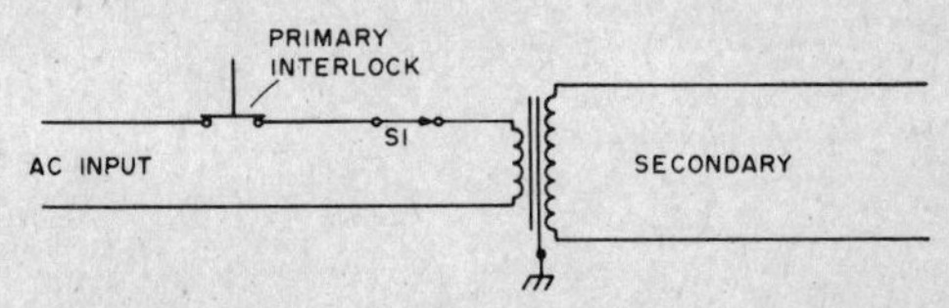

Fig. 8-11. Primary interlock. When removing cover or panel of supply the primary interlock should open thereby preventing the supply from being accidentally energized. A homebrew spring operated switch or commercial pressure switch works well. The primary interlock must have ample current carrying capability.

The secondary interlock Fig. 8-12 normally shorts the secondary out, thereby discharging any residual charge on the high voltage capacitor string

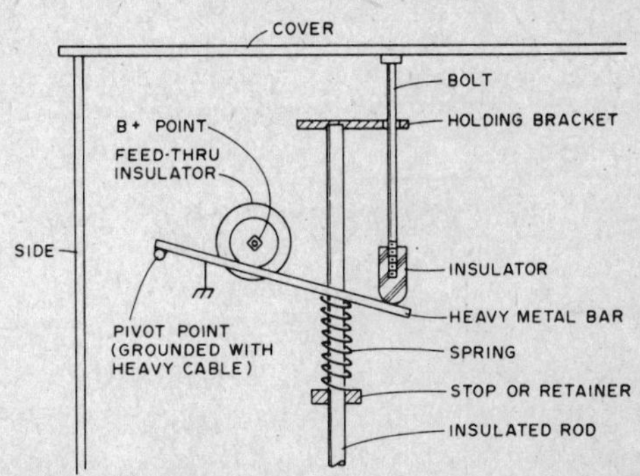

Fig. 8-12. Secondary interlock. Removing the cover permits the metal shorting bar to move up and contact the B+ point, thereby shorting any dangerous voltage to ground. As long as the cover is removed the B+ point will be grounded. This assembly must be mechanically strong and not subject to movement or bending.

and protecting the amateur from electrical shock due to an open bleeder or equalization resistor.

Neither a primary nor secondary interlock alone will give 100% protection, but utilization of both in one supply will come close.

In essence, in respect to safety, it can be said that a power supply that does not break down requires minimum service—therefore the best protection is to build high voltage supplies with generously designed safety factors.

Transformers

This supply can accept transformers with a secondary voltage of up to 2500V RMS with no design changes. A 2500V RMS secondary will give an unloaded output of 3500V DC. Even at this DC level there is an ample power supply design safety factor. For a transformer with a 220V primary, see Fig. 8-13 for wiring details.

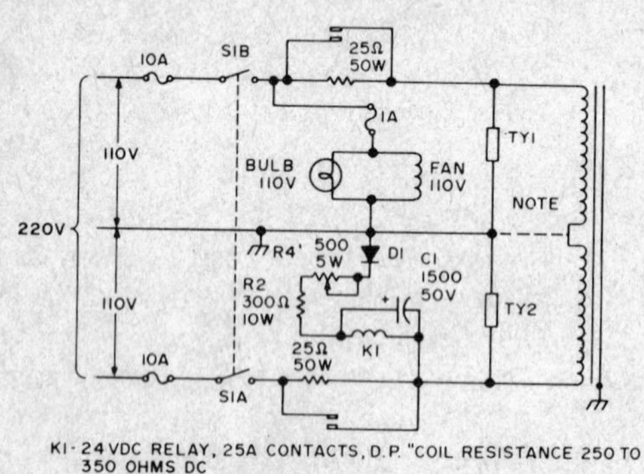

Fig. 8-13. Alternate 220V primary circuit for use with 220V transformers. The components are similar to those used in the original circuit except two thyrectors are used and the relay K1 is a double pole type.

Testing

When the circuit is completed, check the wiring and look for any possible short circuits. Between ground and any positive voltage points, look for a minimum of 3 cm separation when the insulation is solely air.

Also, inspect each electrolytic capacitor to make sure none of the exhaust ports are obstructed by construction. A defective electrolytic or an electrolytic with a plugged or blocked exhaust port can explode violently.

Before energizing the circuit, review the basic rules of safety.

1. Never bypass an interlock.
2. Fuse the circuit properly.
3. Never operate or test the supply with the cover removed or high voltage terminal exposed.

4. Make sure others in the household are aware of the location and operation of the master power cut-off switch so they can disconnect circuit from line in an emergency.
5. Label *all* high voltage points and equipment as such: DANGER—HIGH VOLTAGE.
6. Voltmeter should read zero and main AC line should be disconnected before removing cover or changing high voltage leads.
7. Make sure the family members know the basics of artificial respiration. Many shock victims die of suffocation before professional help arrives.
8. If you don't understand something, get the facts before proceeding.
9. Properly connect the power supply to a good and permanent earth ground.

Even when the circuit has been inspected and rules followed, there is the potential danger of defective new or used components. A 3000V DC supply with a 50 μF filter is a lethal device. Always assume that *all* points in a circuit of this type are dangerous and proceed with that in mind.

VOLTAGE LIMIT SENSOR

The voltage limit sensor (VLS) is a compact, self-contained go, no-go indicator which tells at a glance if the voltage in an automobile or boat electrical system is satisfactory. Many of the latest state-of-the-art electronics instruments have incorporated into them various sensors which continuously monitor test points. Whenever one of these test points deviates outside prescribed limits, a warning light or indicator of some type alerts the operator. In the go, no-go variety of indicators, similar to the oil pressure, generator and temperature lights on automobiles, only a critical condition will necessitate some action on the part of the human operator. The low cost sensor described in this article provides an alerting indicator if the voltage in the electrical system falls outside safe limits.

Theory of Operation

Operation of the sensor is very straightforward. Referring to the schematic diagram, Fig. 8-14, the voltage input provides both the sense voltage and the supply voltage needed to operate the VLS.

The undervoltage part of the VLS consists of Q1, Q2, DS1 and D1 along with three resistors R1, R2 and R3. When the input voltage is less than the breakdown voltage of D1, transistor Q1 is turned off. This in turn allows the current flowing through Q1's collector load resistor, R3, to flow into the base of Q2. This causes Q2 to go into saturation, acting like a switch to light the amber indicator DS1. As the input voltage goes through the zener breakdown point, current begins flowing into the base of Q1. When Q1 has gone into saturation, no base current is available for Q2, which turns off. Indicator, DS1 also goes

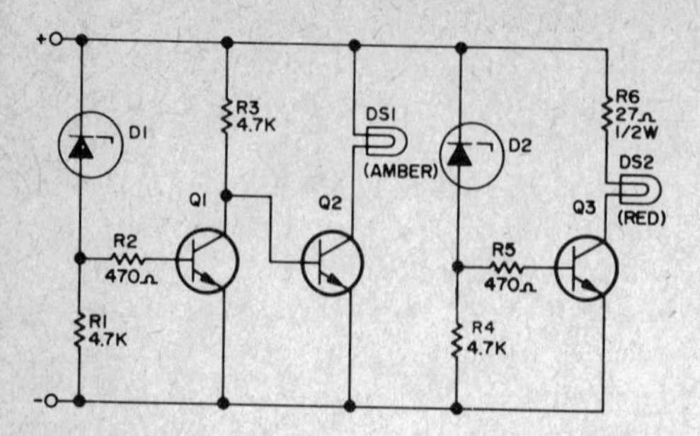

Fig. 8-14. VLS schematic diagram.

Parts List. Q1, Q2, Q3—Motorola MPS 3704; D1—Motorola IN 5243B, 13V 5 zener diode; D2—Motorola IN 6245B, 15V 5 zener diode; PL1—Dialco MS 25256 Pilot Lamp Assembly (Amber Lens) with 330 Bulb (T-1 3/4, 14V at 80 mA); PL2—Dialco MS 25256 Pilot Lamp Assembly (Red Lens) with 330 Built (T-1 3/4. 14V at 80 mA); R1, R3, R4—4.7 K , 1/4W; R2, R5—470 , 1/4W; R6—27 , 1/2W; misc. hardware—CU 2101A minibox, terminal strips, rubber grommet, etc.

out since Q2 is cutoff. Resistor R1 assures a sharp turn on of Q1 and also provides a path for collector leakage of Q1. R2 limits the base current into Q1 after D1 is conducting.

The overvoltage part of the VLS consists of Q3, DS2, D2 and resistors R4, R5, and R6. Q3 is cut off until the input voltage exceeds the zener breakdown of D2. At this point, base current flows into Q3, turning on the indicator DS2. R4 and R5 serve similar functions as R1 and R2 described above. R6 is a current limiting resistor, so DS2 does not burn out for the higher input voltages.

Design Criteria and Construction

Silicon transistors are used to assure stable operation over a wide range of temperatures. The transistors, Q1, Q2 and Q3, were chosen to have high beta of 100 to 300, a high collector current rating of at least 100 mA. Voltage breakdown can be 20-25V or more. A collector power dissipation rating of 300 mW or greater is also desirable.

The voltage at which an automobile operates its primary low voltage system is a function of temperature. For example, a typical GM product will have a range from 13.5V at 150°F to 15.2V at 0°F. The combination of zener diodes, D1 and D2, plus the small base to emitter voltage drops of Q1 and Q3 were chosen such that any voltage less than 13.5V would light the amber indicator and any voltage more than 15.2V would light the red indicator. The 5% zener diodes assure an accurate turn on and turn off without any adjustments.

Since the VLS detects a voltage falling outside these defined limits, it was felt that tracking as a function of battery temperature was not justified. An elaborate temperature sensing circuit was deemed unnecessary and beyond the basic requirements of the VLS.

The circuit is constructed in a small Bud Minibox without crowding. The two pilot lamp indicators are mounted in one end and a rubber grommet in the other end for the two wires. If the VLS is to be used on a negative ground system and the unit is to be securely fastened to the metal ground of the automobile or boat, then the negative lead can be grounded to the case internally and only one wire, the positive lead, brought out of the unit. No special wiring precautions are necessary; however, if sockets are not used for the transistor it is recommended that a heat sink be used on the leads during the soldering operation. This will prevent the possibility of damage to the transistors from excessive heat.

Checkout and Installation

Since the VLS draws a negligible amount of current during normal operation and only 80 mA during the time an indicator is on, power can be obtained from almost any point in the low voltage electrical system. However, it should be switched on and off with normal ignition and accessories, since with just 12V input, the amber indicator will be on and drawing continuous current.

During operation at an ambient temperature of about 75°F where the voltage input will be about 14.2V or so, it is possible to use a 1.5V dry cell battery placed in series with the positive lead to check the VLS. With the 1.5V battery positive terminal connected to the VLS (battery voltage adding), the red indicator should light. With the 1.5V battery negative terminal connected to the VLS (battery voltage subtracting), the amber indicator should light. This test will generally work unless the automobile low voltage is not adjusted properly or the ambient temperature is very high or very low. In these cases, a bench-type variable-voltage power supply could be used for final checkout.

The voltage limit sensor will monitor your 12V battery and charging system, alerting you only to potential unsafe conditions not indicated on the usual idiot light.

A SIMPLE REVERSE CURRENT BATTERY CHARGER

The reverse current charging technique is very effective but it is awkward to change the forward- and reverse-current resistor values whenever the battery type or the number of cells to be charged is changed. A resistor switching arrangement was next tried but lacked the flexibility of continuously variable controls. Also variable resistors of the required power ratings were found to be too bulky and expensive.

Try using manually operated current-limiting transistors for both reverse and forward current control as shown in the schematic (Fig. 8-15). The control potentiometers are now low-wattage units, since they only have to handle transistor base currents. Current adjustment is also smooth and noncritical for all forward- and reverse-current values ordinarily needed to charge all types of batteries. A current metering circuit is added to permit accurate current monitoring over the range of 2-500 mA. Note that a DPST switch is used for on-off. This is so the batteries will not accidentally discharge through the reverse-current transistor and transformer secondary winding if the charger should be turned off without first disconnecting the batteries.

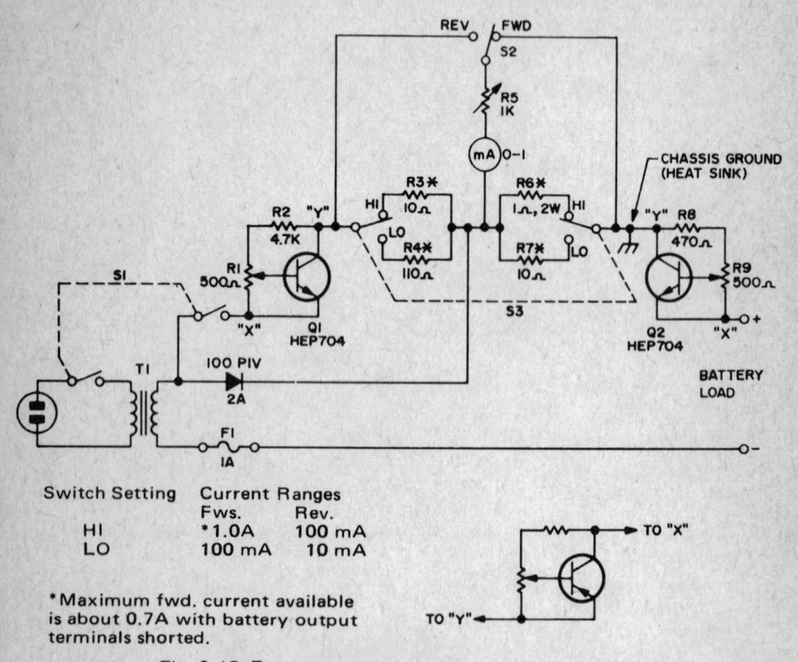

Switch Setting	Current Ranges Fws.	Rev.
HI	*1.0A	100 mA
LO	100 mA	10 mA

*Maximum fwd. current available is about 0.7A with battery output terminals shorted.

Fig. 8-15. Reverse-current battery charger schematic.

Construction of the unit presents no problem since layout and lead lengths are not critical. Panel layout of the original model is shown in Fig. 8-16. The forward-current transistor (Q2) is bolted directly to the chassis which acts as a heat sink. For this reason both charger output terminals are above chassis potential and must be insulated from the panel and chassis. The reverse-current transistor (Q1) is merely insulated from the chassis by insulated shoulder washers and requires no heat sink. It never has to dissipate more than 1W under any operating conditions, which is far below the dissipation ratings of the 2N3055.

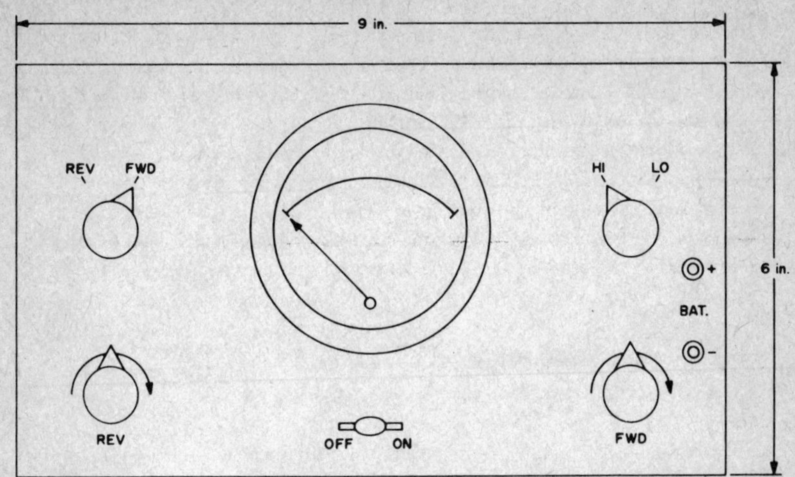

Fig. 8-16. Battery-charger panel layout.

If you happen to have equivalent PNP silicon transistors of the proper ratings, they may be used in place of the NPN types shown by just reversing the transistor and its voltage divider connections as shown in the insert on the schematic. Do not attempt to substitute germanium power transistors for the silicon types, however; germanium power transistors are quite temperature sensitive and will require constant readjustment with the simple circuit used in this charger.

The metering circuit uses a 1 mA meter in a simple current measuring circuit. If the recommended resistors shown are used, no calibration will be required other than adjusting the meter series resistor (R5) on the lowest current range. This is most readily accomplished by using another milliammeter in series with the 10 mA shunt and adjusting R5 for full scale with 10 mA flowing through the circuit.

In case you are wondering about the odd value of shunt resistor R4, remember that the meter circuit takes about 10% of the total current. This requires that the shunt resistor be increased in value to give a true meter reading. On the higher current ranges this compensation is unnecessary since the meter current is 1% or less of the total current in the circuit.

Other meter movements may be used instead of a 0–1 mA meter. For instance, if a 50 μA meter is used, change R5 from 1 kΩ to 20 kΩ and R4 to 100Ω since meter loading on the circuit would be negligible, even on the 10 mA range.

Operating the charger is simple enough: Select the proper current range, adjust the forward current potentiometer to the desired current value, throw the switch to the reverse-current position, and adjust the reverse-current potentiometer to 10% of the forward-current value. It is a good idea at this point to again check the forward current and readjust as required. It will be found that the simple transistor regulators have adequate temperature stability for this application and will require little or no readjustment after warm-up.

Results have been very satisfactory when using this charger in a variety of ways including trickle charging. All battery types have responded well when recommended charging rates and duty cycles as recommended by the battery manufacturers have been adhered to. The only variable results experienced were when attempts were made to "recharge" the lowly carbon–zinc cell. Apparently rejuvenation of these cells is quite dependent upon duty cycle and age.

A GALLON AND A HALF IN A GALLON BUCKET

Any number of articles have been written on the subject of using TV transformers in high voltage supplies. The usual method is to voltage double the entire HV winding and disregard all other windings in order to reduce the load on the transformer and thereby minimize heating of the core. This arrangement works well as evidenced by the number of these supplies in everyday use.

This section will deal with a method (Fig. 8-17) of dissipating the heat generated by the transformer, thus allowing greater current to be drawn from a given transformer without danger of catastrophic failure.

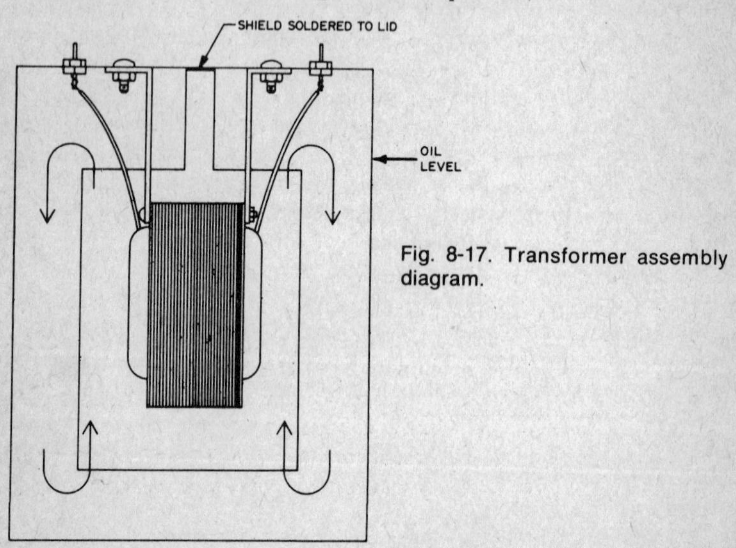

Fig. 8-17. Transformer assembly diagram.

Taking a page from the Heath Cantenna, a gallon bucket was procured as the starting point. The transformer was stripped of its end bells and unused windings to improve heat transfer. Angle brackets were then attached to each side of the core at one end for mounting purposes. A fruit juice can was modified to form a shield clearing the core by 1/4 in. or so. 1 1/2 in. tabs were cut in the top of the can and the end 1/4 in. bent to a 90° angle in order to space the main portion of the can about 1 1/4 in. from the mounting surface. Do not decrease this dimension, it provides room for the hot oil to expand.

The transformer is mounted in the center of the lid of the bucket by means of screws running through the lid and the previously mentioned brackets. The tabs of the shield are soldered to the underside of the lid after it is aligned with the transformer. The primary and secondary leads are brought out through feedthrough bushings.

Select a square of medium to heavy sheet metal the same size as the diameter of the bucket and after drilling a hole in each corner, silver solder or epoxy it to the bottom of the bucket. This will serve as a base for mounting purposes. Clean the entire assembly and paint the outside flat black. Fill with transformer oil to 1/4 in. above the top of the shield with the transformer in place. This will require a slight bit of guesswork. Transformer oil is often available from the local power company in small quantities and at a very reasonable price.

As the transformer heats up, the oil directly in contact with the windings becomes warm and rises to the surface within the shield only to be replaced by cooler oil from the bottom of the bucket. The hot oil is cooled by contact with the outside of the bucket, which acts as a heat sink. By the time it reaches the bottom again the oil is much cooler and is ready to absorb more heat.

It is a conservative estimate that in SSB service a 50% increase in output current is available without running the transformer at elevated temperatures. There is some loss in output voltage due to the resistance of the secondary winding, but this loss is small in relation to the increased current available. I have been using a 1400W PEP arrangement without much difficulty.

SOLID-STATE SSB POWER SUPPLY

At the risk of over-generalization, it can be said that all ham power supplies can be broken down into three types: 1, low power receiver supplies; 2, medium power transmitter supplies; and 3, high power transmitter supplies. The recognized trend today is to "semiconductorize" all three types, sending countless 5Y3s, 5R4s, and 866s to the junk box.

While many hams would hesitate to attempt the construction of an SSB transceiver, $50 to $100 or more can be saved by building at least the power supply, a comparatively simple job.

A typical ham's junk box (if there is such a thing) may likely contain close to all the components required, keeping the cash outlay to an absolute minimum. Every necessary component is also available on the surplus market at large savings.

Table 8-1. Power Supply Component Values.

	Choke Input Filter		Capacitor Input Filter	
	Full Wave Rect.	Full Wave Bridge	Full Wave Rect.	Full Wave Bridge
PIV	2.8 x vac	1.4 x vac	2.8 x vac	1.4 x vac
Rectifier Current rating	.5 x idc	.5 x idc	.5 x idc	.5 x idc
vac	1.13 x vdc	1.13 x vdc	.85 x vdc	.85 x vdc

The power supply described here (Fig. 8-18) was built for the Heath HW-12, 22, and 32 single band 200 watt PEP transceivers. The requirements, however, are representative of many on the market. It delivers about 800 VDC at 250 mA peak, 250 VEC at 100 mA—124 VDC for grid bias, and 12.6 VAC filament voltage. The bias features zener regulation, and the entire supply has proven very satisfactory in service.

T1, the power transformer, was acquired as "new surplus." It is rated at 800 VCT + 300 mA DC. The filament windings are rated at 6 amps. The bias transformer, T2, is simply a 6.3 volt filament transformer wired backwards. If this is not available, a 1:1 120V transformer could be used across the AC line.

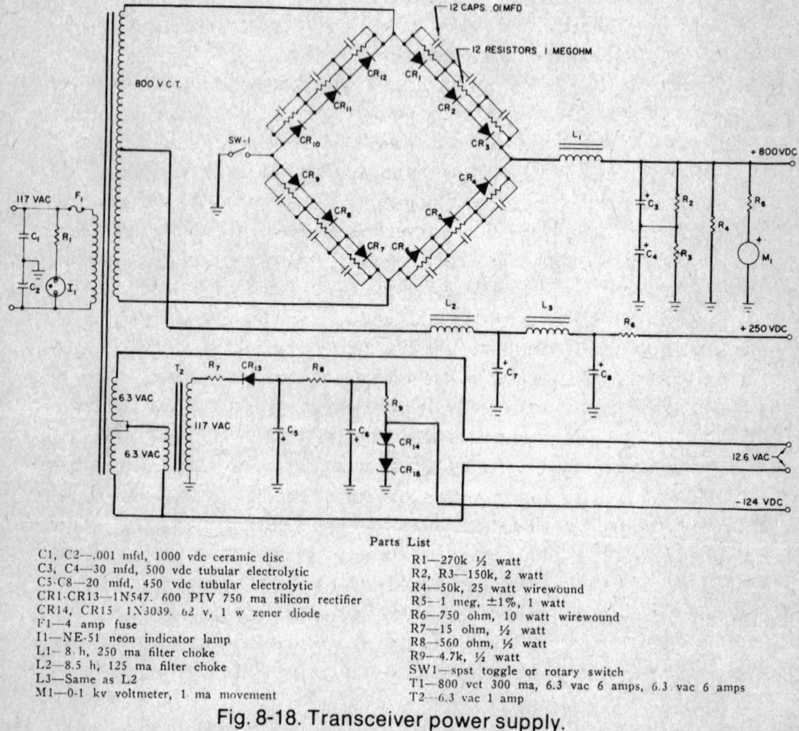

Parts List

C1, C2—.001 mfd, 1000 vdc ceramic disc
C3, C4—30 mfd, 500 vdc tubular electrolytic
C5-C8—20 mfd, 450 vdc tubular electrolytic
CR1-CR13—1N547, 600 PIV 750 ma silicon rectifier
CR14, CR15 1N3039, 62 v, 1 w zener diode
F1—4 amp fuse
I1—NE-51 neon indicator lamp
L1—8 h, 250 ma filter choke
L2—8.5 h, 125 ma filter choke
L3—Same as L2
M1—0-1 kv voltmeter, 1 ma movement
R1—270k ½ watt
R2, R3—150k, 2 watt
R4—50k, 25 watt wirewound
R5—1 meg, ±1%, 1 watt
R6—750 ohm, 10 watt wirewound
R7—15 ohm, ½ watt
R8—560 ohm, ½ watt
R9—4.7k, ½ watt
SW1—spst toggle or rotary switch
T1—800 vct 300 ma, 6.3 vac 6 amps, 6.3 vac 6 amps
T2—6.3 vac 1 amp

Fig. 8-18. Transceiver power supply.

The high voltage portion is a full wave bridge rectifier, using 12 600V PIV silicon rectifiers. It is filtered by a single "L" filter consisting of L1, C3 and C4. R2 and R3 equalize the voltage across the filter capacitors. R4 is a bleeder resistor, and has three functions: it discharges the filter capacitors for safety, it helps to regulate the output voltage, and it keeps the output voltage, when unloaded, from climbing above the capacitor voltage ratings. Because of the quiescent current of the final amplifier, very little bleeder current is required.

M1 is a 1000V voltmeter. A milliammeter in series with the load would have been more typical, but it was desired to monitor voltage regulation rather than load current. It is optional, and either, or both, can be used.

The low voltage from the transformer center tap, after being well filtered, supplies the receiver section and the low lever transmitter stages. L2 and L3 are also surplus units, and are actually one tapped inductor. R6 is a dropping resistor to obtain the correct output voltage.

The bias voltage is half-wave rectified by CR 13. R7 is a surge resistor and protects the rectifier. CR14 and CR15 are 62 volt, 1 watt zener diodes, giving constant bias voltage regardless of line variations. Very little output current is drawn from the bias supply.

When wiring the filament windings in series, a voltmeter across the output will read either zero volts or 12.6 volts AC. If zero is read, reverse one of the windings so that they add rather then cancel.

One side of the primary winding is connected through the transceiver function switch to the line, so that the power supply can be switched on with the transceiver. SW 1 is a high/low power switch. Since the power supply is solid state, there are no filaments to warm up, and the full output voltage is at the terminals immediately upon switching on. This is very hard on the transmitter tubes, since their filaments have not yet heated. With SW 1 open, there will be no voltage at the low voltage output, and only half the normal voltage at the high voltage terminals. Full bias, however, will be supplied. After allowing filaments to heat for a minute or so, the switch can be closed, applying full voltage. This should lengthen tube life appreciably. While an SPST toggle switch has worked fine in the writer's supply with no sign of arcing, it may be desired to use a ceramic wafer switch with a higher voltage rating. Of course, the same thing can be done with a time delay relay.

Much wiring can be eliminated, if desired, by purchasing a packaged bridge rectifier unit. These contain the silicon rectifier bridge with all the required voltage equalizing resistors and capacitors in a compact potted package with four terminals.

This power supply exhibits many advantages over vacuum tube types. The rectifiers run stone cold, there is much less voltage drop across them, they don't require amperes of filament current, they don't shatter when you drop them, and they certainly take up less room.

Index